Pharmacogenomics

" The Use of Genetics in Prescribing Medications "

Edited by Paul F. Kisak

Contents

Chapter 1

Pharmacogenomics

Pharmacogenomics is the study of the role of the genome in drug response. Its name (*pharmaco-* + *genomics*) reflects its combining of pharmacology and genomics. Pharmacogenomics analyzes how the genetic makeup of an individual affects his/her response to drugs.[1] It deals with the influence of acquired and inherited genetic variation on drug response in patients by correlating gene expression or single-nucleotide polymorphisms with pharmacokinetics and pharmacodynamics (drug absorption, distribution, metabolism, and elimination), as well as drug receptor target effects.[2][3][4] The term *pharmacogenomics* is often used interchangeably with *pharmacogenetics*. Although both terms relate to drug response based on genetic influences, pharmacogenetics focuses on single drug-gene interactions, while pharmacogenomics encompasses a more genome-wide association approach, incorporating genomics and epigenetics while dealing with the effects of multiple genes on drug response.[5][6][7]

Pharmacogenomics aims to develop rational means to optimize drug therapy, with respect to the patients' genotype, to ensure maximum efficacy with minimal adverse effects.[8] Through the utilization of pharmacogenomics, it is hoped that pharmaceutical drug treatments can deviate from what is dubbed as the "one-dose-fits-all" approach. Pharmacogenomics also attempts to eliminate the trial-and-error method of prescribing, allowing physicians to take into consideration their patient's genes, the functionality of these genes, and how this may affect the efficacy of the patient's current or future treatments (and where applicable, provide an explanation for the failure of past treatments).[5] Such approaches promise the advent of precision medicine and even personalized medicine, in which drugs and drug combinations are optimized for narrow subsets of patients or even for each individual's unique genetic makeup.[9][10] Whether used to explain a patient's response or lack thereof to a treatment, or act as a predictive tool, it hopes to achieve better treatment outcomes, greater efficacy, minimization of the occurrence of drug toxicities and adverse drug reactions (ADRs). For patients who have lack of therapeutic response to a treatment, alternative therapies can be prescribed that would best suit their requirements. In order to provide pharmacogenomic recommendations for a given drug, two possible types of input can be used: genotyping or exome or whole genome sequencing.[11] Sequencing provides many more data points, including detection of mutations that prematurely terminate the synthesized protein (early stop codon).[11]

1.1 History

Pharmacogenomics was first recognized by Pythagoras around 510 BC when he made a connection between the dangers of fava bean ingestion with hemolytic anemia and oxidative stress. Interestingly, this identification was later validated and attributed to deficiency of G6PD in the 1950s and called favism.[12][13] Although the first official publication dates back to 1961,[14] circa 1950s marked the unofficial beginnings of this science. Reports of prolonged paralysis and fatal reactions linked to genetic variants in patients who lacked butyryl-cholinesterase ('pseudocholinesterase') following administration of succinylcholine injection during anesthesia were first reported in 1956.[2][15] The term pharmacogenetic was first coined in 1959 by Friedrich Vogel of Heidelberg, Germany (although some papers suggest it was 1957). In the late 1960s, twin studies supported the inference of genetic involvement in drug metabolism, with identical twins sharing remarkable similarities to drug response compared to fraternal twins.[16] The term pharmacogenomics first began appearing around the 1990s.[12]

The first FDA approval of a pharmacogenetic test was in 2005[17] (for alleles in CYP2D6 and CYP2C19).

1.2 Drug-metabolizing enzymes

There are several known genes which are largely responsible for variances in drug metabolism and response. The focus of this article will remain on the genes that are more widely accepted and utilized clinically for brevity.

- Cytochrome P450s
- VKORC1
- TPMT

1.2.1 Cytochrome P450

The most prevalent drug-metabolizing enzymes (DME) are the Cytochrome P450 (CYP) enzymes. The term Cytochrome P450 was coined by Omura and Sato in 1962 to describe the membrane-bound, heme-containing protein characterized by 450 nm spectral peak when complexed with carbon monoxide.[18] The human CYP family consists of 57 genes, with 18 families and 44 subfamilies. CYP proteins are conveniently arranged into these families and subfamilies on the basis of similarities identified between the amino acid sequences. Enzymes that share 35-40% identity are assigned to the same family by an Arabic numeral, and those that share 55-70% make up a particular subfamily with a designated letter.[19] For example, CYP2D6 refers to family 2, subfamily D, and gene number 6.

From a clinical perspective, the most commonly tested CYPs include: CYP2D6, CYP2C19, CYP2C9, CYP3A4 and CYP3A5. These genes account for the metabolism of approximately 80-90% of currently available prescription drugs.[20][21] The table below provides a summary for some of the medications that take these pathways.

CYP2D6

Also known as debrisoquine hydroxylase (named after the drug that led to its discovery), CYP2D6 is the most well-known and extensively studied CYP gene.[24] It is a gene of great interest also due to its highly polymorphic nature, and involvement in a high number of medication metabolisms (both as a major and minor pathway). More than 100 CYP2D6 genetic variants have been identified.[23]

CYP2C19

Discovered in the early 1980s, CYP2C19 is the second most extensively studied and well understood gene in pharmacogenomics.[22] Over 28 genetic variants have been identified for CYP2C19,[25] of which affects the metabolism of several classes of drugs, such as antidepressants and proton pump inhibitors.[26]

CYP2C9

CYP2C9 constitutes the majority of the CYP2C subfamily, representing approximately 20% of the liver content. It is involved in the metabolism of approximately 10% of all drugs, which include medications with narrow therapeutic windows such as warfarin and tolbutamide.[26][27] There are approximately 57 genetic variants associated with CYP2C9.[25]

CYP3A4 and CYP3A5

The CYP3A family is the most abundantly found in the liver, with CYP3A4 accounting for 29% of the liver content.[22] These enzymes also cover between 40-50% of the current prescription drugs, with the CYP3A4 accounting for 40-45% of these medications.[13] CYP3A5 has over 11 genetic variants identified at the time of this publication.[25]

1.2.2 VKORC1

The vitamin K epoxide reductase complex subunit 1 (VKORC1) is responsible for the pharmacodynamics of warfarin.[28] VKORC1 along with CYP2C9 are useful for identifying the risk of bleeding during warfarin administration. Warfarin works by inhibiting VKOR, which is encoded by the VKORC1 gene. Individuals with polymorphism in this have an affected response to warfarin treatment.[29]

1.2.3 TPMT

Thiopurine methyltransferase (TPMT) catalyzes the S-methylation of thiopurines, thereby regulating the balance between cytotoxic thioguanine nucleotide and inactive metabolites in hematopoietic cells.[30] TPMT is highly involved in 6-MP metabolism and TMPT activity and TPMT genotype is known to affect the risk of toxicity. Excessive levels of 6-MP can cause myelosuppression and myelotoxicity.[31]

Codeine, clopidogrel, tamoxifen, and warfarin a few examples of medications that follow the above metabolic pathways.

1.3 Predictive prescribing

Patient genotypes are usually categorized into the following predicted phenotypes:

- Ultra-Rapid Metabolizer: Patients with substantially increased metabolic activity.
- Extensive Metabolizer: Normal metabolic activity;
- Intermediate Metabolizer: Patients with reduced metabolic activity; and
- Poor Metabolizer: Patients with little to no functional metabolic activity.

The two extremes of this spectrum are the Poor Metabolizers and Ultra-Rapid Metabolizers. Efficacy of a medication is not only based on the above metabolic statuses, but also the type of drug consumed. Drugs can be classified into two main groups: active drugs and prodrugs. Active drugs refer to drugs that are inactivated during metabolism, and prodrugs are inactive until they are metabolized.

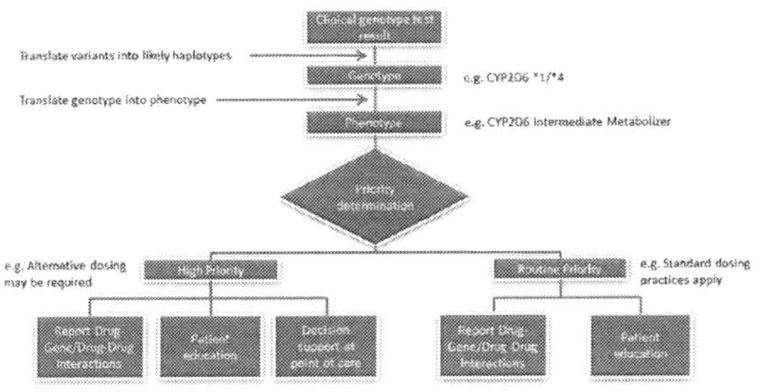

An overall process of how pharmacogenomics functions in a clinical practice. From the raw genotype results, this is then translated to the physical trait, the phenotype. Based on these observations, optimal dosing is evaluated.[30]

For example, we have two patients who are taking codeine for pain relief. Codeine is a prodrug, so it requires conversion from its inactive form to its active form. The active form of codeine is morphine, which provides the therapeutic effect of pain relief. If person A receives one *1 allele each from mother and father to code for the CYP2D6 gene, then that person is considered to have an extensive metabolizer (EM) phenotype, as allele *1 is considered to have a normal-function (this would be represented as CYP2D6 *1/*1). If person B on the other hand had received one *1 allele from the mother and a *4 allele from the father, that individual would be an Intermediate Metabolizer (IM) (the genotype would be CYP2D6 *1/*4). Although both individuals are taking the same dose of codeine, person B could potentially lack the therapeutic benefits of codeine due to the decreased conversion rate of codeine to its active counterpart morphine.

Each phenotype is based upon the allelic variation within the individual genotype. However, several genetic events can influence a same phenotypic trait, and establishing genotype-to-phenotype relationships can thus be far from consensual with many enzymatic patterns. For instance, the influence of the CYP2D6*1/*4 allelic variant on the clinical outcome in patients treated with Tamoxifen remains debated today. In oncology, genes coding for DPD, UGT1A1, TPMT, CDA involved in the pharmacokinetics of 5-FU/capecitabine, irinotecan, 6-mercaptopurine and gemcitabine/cytarabine, respectively, have all been described as being highly polymorphic. A strong body of evidence suggests that patients affected by these genetic polymorphisms will experience severe/lethal toxicities upon drug intake, and that pre-therapeutic screening does help to reduce the risk of treatment-related toxicities through adaptive dosing strategies.[32]

1.4 Applications

The list below provides a few more commonly known applications of pharmacogenomics:[33]

- Improve drug safety, and reduce ADRs;
- Tailor treatments to meet patients' unique genetic predisposition, identifying optimal dosing;
- Improve drug discovery targeted to human disease; and
- Improve proof of principle for efficacy trials.

Pharmacogenomics may be applied to several areas of medicine, including Pain Management, Cardiology, Oncology, and Psychiatry. A place may also exist in Forensic Pathology, in which pharmacogenomics can be used to determine the cause of death in drug-related deaths where no findings emerge using autopsy.[34]

In cancer treatment, pharmacogenomics tests are used to identify which patients are most likely to respond to certain cancer drugs. In behavioral health, pharmacogenomic tests provide tools for physicians and care givers to better manage medication selection and side effect amelioration. Pharmacogenomics is also known as companion diagnostics, meaning tests being bundled with drugs. Examples include KRAS test with cetuximab and EGFR test with gefitinib. Beside efficacy, germline pharmacogenetics can help to identify patients likely to undergo severe toxicities when given cytotoxics showing impaired detoxification in relation with genetic polymorphism, such as canonical 5-FU.[35]

In cardiovascular disorders, the main concern is response to drugs including warfarin, clopidogrel, beta blockers, and statins.[11]

1.5 Example case studies

Case A – Antipsychotic adverse reaction[36]

Patient A suffers from schizophrenia. Their treatment included a combination of ziprasidone, olanzapine, trazodone and benzotropine. The patient experienced dizziness and sedation, so they were tapered off ziprasidone and olanzapine, and transition to quetiapine. Trazodone was discontinued. The patient then experienced excessive sweating, tachycardia and neck pain, gained considerable weight and had hallucinations. Five months later, quetiapine was tapered and discontinued, with ziprasidone re-introduction into their treatment due to the excessive weight gain. Although the patient lost the excessive weight they gained, they then developed muscle stiffness, cogwheeling, tremor and night sweats. When benztropine was added they experienced blurry vision. After an additional five months, the patient was switched from ziprasidone to aripiprazole. Over the course of 8 months, patient A gradually experienced more weight gain, sedation, developed difficulty with their gait, stiffness, cogwheel and dyskinetic ocular movements. A pharmacogenomics test later proved the patient had a CYP2D6 *1/*41, with has a predicted phenotype of IM and CYP2C19 *1/*2 with predicted phenotype of IM as well.

Case B – Pain Management [37]

Patient B is a woman who gave birth by caesarian section. Her physician prescribed codeine for post-caesarian pain. She took the standard prescribed dose, however experienced nausea and dizziness while she was taking codeine. She also noticed that her breastfed infant was lethargic and feeding poorly. When the patient mentioned these symptoms to her physician, they recommended that she discontinue codeine use. Within a few days, both the patient and her infant's symptoms were no longer present. It is assumed that if the patient underwent a pharmacogenomic test, it would have revealed she may have had a duplication of the gene CYP2D6 placing her in the Ultra-rapid metabolizer (UM) category, explaining her ADRs to codeine use.

Case C – FDA Warning on Codeine Overdose for Infants[38]

On February 20, 2013, the FDA released a statement addressing a serious concern regarding the connection between children who are known as CYP2D6 UM and fatal reactions to codeine following tonsillectomy and/or adenoidectomy (surgery to remove the tonsils and/or adenoids). They released their strongest Boxed Warning to elucidate the dangers of CYP2D6 UMs consuming codeine. Codeine is converted to morphine by CYP2D6, and those who have UM phenotypes are at danger of producing large amounts of morphine due to the increased function of the gene. The morphine can elevate to life-threatening or fatal amounts, as became evident with the death of three children in August 2012.

1.6 Polypharmacy

A potential role pharmacogenomics may play would be to reduce the occurrence of polypharmacy. It is theorized that with tailored drug treatments, patients will not have the need to take several medications that are intended to treat the same condition. In doing so, they could potentially minimize the occurrence of ADRs, have improved treatment outcomes, and can save costs by avoiding purchasing extraneous medications. An example of this can be found in Psychiatry, where patients tend to be receiving more medications than even age-matched non-psychiatric patients. This has been associated with an increased risk of inappropriate prescribing.[39]

The need for pharmacogenomics tailored drug therapies may be most evident in a survey conducted by the Slone Epidemiology Center at Boston University from February 1998 to April 2007. The study elucidated that an average of 82% of adults in the United States are taking at least one medication (prescription or nonprescription drug, vitamin/mineral, herbal/natural supplement), and 29% are taking five or more. The study suggested that those aged 65 years or older continue to be the biggest consumers of medications, with 17-19 % in this age group taking at least ten medications in a given week. Polypharmacy has also shown to have increased since 2000 from 23% to 29%.[40]

1.7 Drug labeling

The U.S. Food and Drug Administration (FDA) appears to be very invested in the science of pharmacogenomics[41] as is demonstrated through the 120 and more FDA-approved drugs that include pharmacogenomic biomarkers in their labels.[42] On May 22, 2005, the FDA issued its first *Guidance for Industry: Pharmacogenomic Data Submissions*, which clarified the type of pharmacogenomic data required to be submitted to the FDA and when.[43] Experts recognized the importance of the FDA's acknowledgement that pharmacogenomics experiments will not bring negative regulatory consequences.[44] The FDA had released its latest guide *Clinical Pharmacogenomics (PGx): Premarket Evaluation in Early-Phase Clinical Studies and Recommendations for Labeling* in January, 2013. The guide is intended to address the use of genomic information during drug development and regulatory review processes.

1.8 Challenges

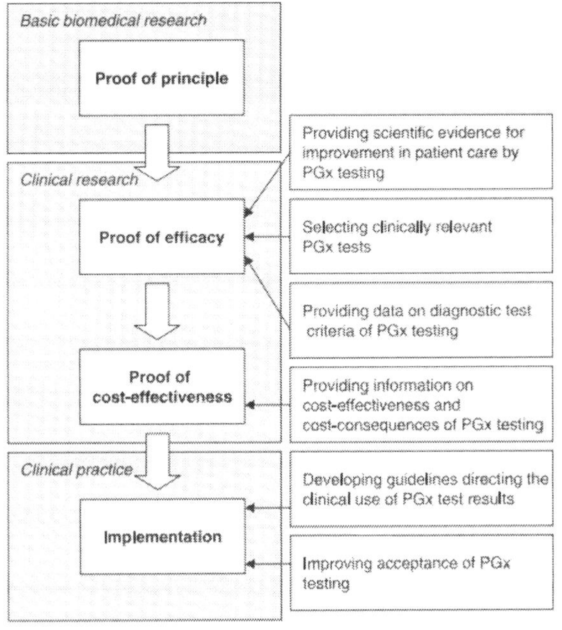

Consecutive phases and associated challenges in Pharmacogenomics.[45]

Although there appears to be a general acceptance of the basic tenet of pharmacogenomics amongst physicians and healthcare professionals,[46] several challenges exist that slow the uptake, implementation, and standardization of pharmacogenomics. Some of the concerns raised by physicians include:[7][46][47]

- Limitation on how to apply the test into clinical practices and treatment;
- A general feeling of lack of availability of the test;
- The understanding and interpretation of evidence-based research; and
- Ethical, legal and social issues.

Issues surrounding the availability of the test include:[45]

- *The lack of availability of scientific data*: Although there are considerable number of DME involved in the metabolic pathways of drugs, only a fraction have sufficient scientific data to validate their use within a clinical setting; and
- *Demonstrating the cost-effectiveness of pharmacogenomics*: Publications for the pharmacoeconomics of pharmacogenomics are scarce, therefore sufficient evidence does not at this time exist to validate the cost-effectiveness and cost-consequences of the test.

Although other factors contribute to the slow progression of pharmacogenomics (such as developing guidelines for clinical use), the above factors appear to be the most prevalent.

1.9 Controversies

Some alleles that vary in frequency between specific populations have been shown to be associated with differential responses to specific drugs. The beta blocker atenolol is an anti-hypertensive medication that is shown to more significantly lower the blood pressure of Caucasian patients than African American patients in the United States. This observation suggests that Caucasian and African American populations have different alleles governing oleic acid biochemistry, which react differentially with atenolol.[48] Similarly, hypersensitivity to the antiretroviral drug abacavir is strongly associated with a single-nucleotide polymorphism that varies in frequency between populations.[49]

The FDA approval of the drug BiDil (isosorbide dinitrate/hydralazine) with a label specifying African-Americans with congestive heart failure, produced a storm of controversy over race-based medicine and fears of genetic stereotyping,[50] even though the label for BiDil did not specify any genetic variants but was based on racial self-identification.[51][52]

1.10 Future

Computational advances in pharmacogenomics has proven to be a blessing in research. As a simple example, for nearly a decade the ability to store more information on a hard drive has enabled us to investigate a human genome sequence cheaper and in more detail with regards to the effects/risks/safety concerns of drugs and other such substances. Such computational advances are expected to continue in the future.[53] The aim is to use the genome sequence data to effectively make decisions in order to minimise the negative impacts on, say, a patient or the health industry in general. A large amount of research in the biomedical sciences regarding Pharmacogenomics as of late stems from combinatorial chemistry,[54] genomic mining, omic technologies and high throughput screening. In order for the field to grow, rich knowledge enterprises and business must work more closely together and adopt simulation strategies. Consequently, more importance must be placed on the role of computational biology with regards to safety and risk assessments. Here, we can find the growing need and importance of being able to manage large, complex data sets, being able to extract information by integrating disparate data so that developments can be made in improving human health.

1.11 Web-based resources

1.12 See also

- Genomics
  - Pharmacogenetics
  - Toxicogenomics
  - Clinomics
  - Genetic engineering
- Population groups in biomedicine
- Toxgnostics

1.13 References

[1] Ermak, Gennady (2015). *Emerging Medical Technologies*. World Scientific. ISBN 978-981-4675-80-2.

[2] Johnson JA (November 2003). "Pharmacogenetics: potential for individualized drug therapy through genetics.". *Trends Genet.* **19** (11): 660–6. PMID 14585618. doi:10.1016/j.tig.2003.09.008.

[3] "Center for Pharmacogenomics and Individualized Therapy". Retrieved 2014-06-25.

[4] "overview of pharmacogenomics". Up-to-Date. May 16, 2014. Retrieved 2014-06-25.

[5] Sheffield LJ, Phillimore HE (2009). "Clinical use of pharmacogenomic tests in 2009". *Clin Biochem Rev.* **30** (2): 55–65. PMC 2702214. PMID 19565025.

[6] Shin J, Kayser SR, Langaee TY (April 2009). "Pharmacogenetics: from discovery to patient care.". *Am J Health Syst Pharm.* **66** (7): 625–37. PMID 19299369. doi:10.2146/ajhp080170.

[7] "Center for Genetics Education".

[8] Becquemont L (June 2009). "Pharmacogenomics of adverse drug reactions: practical applications and perspectives". *Pharmacogenomics.* **10** (6): 961–9. PMID 19530963. doi:10.2217/pgs.09.37.

[9] "Guidance for Industry Pharmacogenomic Data Submissions" (PDF). U.S. Food and Drug Administration. March 2005. Retrieved 2008-08-27.

[10] Squassina A, Manchia M, Manolopoulos VG, Artac M, Lappa-Manakou C, Karkabouna S, Mitropoulos K, Del Zompo M, Patrinos GP (August 2010). "Realities and expectations of pharmacogenomics and personalized medicine: impact of translating genetic knowledge into clinical practice". *Pharmacogenomics.* **11** (8): 1149–67. PMID 20712531. doi:10.2217/pgs.10.97.

[11] Huser V, Cimino JJ (2013). "Providing pharmacogenomics clinical decision support using whole genome sequencing data as input". *AMIA Summits on Translational Science Proceedings.* **2013**: 81. PMID 24303303.

[12] Pirmohamed M (2001). "Pharmacogenetics and pharmacogenomics". *Br J Clin Pharmacol.* **52** (4): 345–7. PMC 2014592. PMID 11678777. doi:10.1046/j.0306-5251.2001.01498.x.

[13] Prasad K (2009). "Role of regulatory agencies in translating pharmacogenetics to the clinics". *Clin Cases Miner Bone Metab.* **6** (1): 29–34. PMC 2781218. PMID 22461095.

[14] Evans DA, Clarke CA (1961). "Pharmacogenetics". *Br Med Bull.* **17**: 234–40. PMID 13697554.

[15] Kalow W (2006). "Pharmacogenetics and pharmacogenomics: origin, status, and the hope for personalized medicine". *Pharmacogenomics J.* **6** (3): 162–5. PMID 16415920. doi:10.1038/sj.tpj.6500361.

[16] Motulsky AG, Qi M (2006). "Pharmacogenetics, pharmacogenomics and ecogenetics". *J Zhejiang Univ Sci B.* **7** (2): 169–70. PMC 1363768. PMID 16421980. doi:10.1631/jzus.2006.B0169.

[17] Realities and Expectations of Pharmacogenomics and Personalized Medicine: Impact of Translating Genetic Knowledge into Clinical Practice. 2010

[18] Debose-Boyd RA (February 2007). "A helping hand for cytochrome p450 enzymes". *Cell Metab.* **5** (2): 81–3. PMID 17276348. doi:10.1016/j.cmet.2007.01.007.

[19] Nebert DW, Russell DW (October 12, 2002). "Clinical importance of the cytochromes P450". *Lancet.* **360** (9340): 1155–62. PMID 12387968. doi:10.1016/s0140-6736(02)11203-7.

[20] Hart SN, Wang S, Nakamoto K, Wesselman C, Li Y, Zhong XB (January 2008). "Genetic polymorphisms in cytochrome P450 oxidoreductase influence microsomal P450-catalyzed drug metabolism". *Pharmacogenet Genomics.* **18** (1): 11–24. PMID 18216718. doi:10.1097/FPC.0b013e3282f2f121.

[21] Gomes AM, Winter S, Klein K, Turpeinen M, Schaeffeler E, Schwab M, Zanger UM (April 2009). "Pharmacogenomics of human liver cytochrome P450 oxidoreductase: multifactorial analysis and impact on microsomal drug oxidation". *Pharmacogenomics.* **10** (4): 579–99. PMID 19374516. doi:10.2217/pgs.09.7.

[22] Hasler JA (February 1999). "Pharmacogenetics of cytochromes P450". *Mol Aspects Med.* **20** (1–2): 25–137. PMID 10575648.

[23] Ingelman-Sundberg M (April 2004). "Pharmacogenetics of cytochrome P450 and its applications in drug therapy: the past, present and future". *Trends Pharmacol Sci.* **25** (4): 193–200. PMID 15063083. doi:10.1016/j.tips.2004.02.007.

[24] Badyal DK, Dadhich AP (October 2001). "Cytochrome P450 and drug interactions" (PDF). *Indian Journal of Pharmacology*. **33**: 248–259.

[25] Ingelman-Sundberg, M; Nebert, DW; Sim, SC. "The Human Cytochrome P450 (CYP) Allele Nomenclature Database". Retrieved 2014-09-03.

[26] Ingelman-Sundberg M, Sim SC, Gomez A, Rodriguez-Antona C (December 2007). "Influence of cytochrome P450 polymorphisms on drug therapies: pharmacogenetic, pharmacoepigenetic and clinical aspects". *Pharmacol Ther.* **116** (3): 496–526. PMID 18001838. doi:10.1016/j.pharmthera.2007.09.004.

[27] Sikka R, Magauran B, Ulrich A, Shannon M (December 2005). "Bench to bedside: Pharmacogenomics, adverse drug interactions, and the cytochrome P450 system". *Acad Emerg Med.* **12** (12): 1227–35. PMID 16282513. doi:10.1111/j.1553-2712.2005.tb01503.x.

[28] Teh LK, Langmia IM, Fazleen Haslinda MH, Ngow HA, Roziah MJ, Harun R, Zakaria ZA, Salleh MZ (April 2012). "Clinical relevance of VKORC1 (G-1639A and C1173T) and CYP2C9*3 among patients on warfarin". *J Clin Pharm Ther.* **37** (2): 232–6. PMID 21507031. doi:10.1111/j.1365-2710.2011.01262.x.

[29] U.S. Food and Drug Administration (FDA). "Table of Pharmacogenomic Biomarkers in Drug Labels.". Retrieved 2014-09-03.

[30] Crews KR, Hicks JK, Pui CH, Relling MV, Evans WE (October 2012). "Pharmacogenomics and individualized medicine: translating science into practice". *Clin Pharmacol Ther.* **92** (4): 467–75. PMC 3589526 ⓐ. PMID 22948889. doi:10.1038/clpt.2012.120.

[31] Sim SC, Kacevska M, Ingelman-Sundberg M (February 2013). "Pharmacogenomics of drug-metabolizing enzymes: a recent update on clinical implications and endogenous effects". *Pharmacogenomics.* **13** (1): 1–11. PMID 23089672. doi:10.1038/tpj.2012.45.

[32] Lee SY, McLeod HL (January 2011). "Pharmacogenetic tests in cancer chemotherapy: what physicians should know for clinical application". *J Pathol.* **223** (1): 15–27. PMID 20818641. doi:10.1002/path.2766.

[33] Cohen, Nadine (November 2008). *Pharmacogenomics and Personalized Medicine (Methods in Pharmacology and Toxicology)*. Totowa, NJ: Humana Press. p. 6. ISBN 978-1934115046.

[34] Pelotti, Susan; Bini, Carla (September 12, 2011). *Forensic Pharmacogenetics (Forensic Medicine - From Old Problems to New Challenges)*. INTECH Open Access Publisher. p. 268. ISBN 978-953-307-262-3. Retrieved 2014-09-03.

[35] Ciccolini J, Gross E, Dahan L, Lacarelle B, Mercier C (October 2010). "Routine dihydropyrimidine dehydrogenase testing for anticipating 5-fluorouracil-related severe toxicities: hype or hope?". *Clin Colorectal Cancer.* **9** (4): 224–8. PMID 20920994. doi:10.3816/CCC.2010.n.033.

[36] Foster A, Wang Z, Usman M, Stirewalt E, Buckley P (December 2007). "Pharmacogenetics of antipsychotic adverse effects: Case studies and a literature review for clinicians". *Neuropsychiatr Dis Treat.* **3** (6): 965–973. PMC 2656342 ⓐ. PMID 19300635. doi:10.2147/ndt.s1752.

[37] "Pharmacogenetics: increasing the safety and effectiveness of drug therapy [Brochure]" (PDF). American Medical Association. 2011.

[38] "FDA Drug Safety Communication: Safety review update of codeine use in children; new Boxed Warning and Contraindication on use after tonsillectomy and/or adenoidectomy". United States Food and Drug Administration. 2013-02-20.

[39] Ritsner, Michael (2013). *Polypharmacy in Psychiatry Practice, Volume I. Multiple Medication Strategies*. Dordrecht: Springer Science and Business Media. ISBN 978-94-007-5804-9.

[40] "Patterns of Medication Use in the United States". Boston University, Slone Epidemiology Center. 2006.

[41] "Pharmacogenetics and Pharmacogenomics: State-of-the-art and potential socio-economic impacts in the EU". European Commission, Joint Research Centre, Institute for Prospective Technological Studies. 2006-04-01.

[42] "Table of Pharmacogenomic Biomarkers in Drug Labels". United States Food and Drug Administration. 2013-06-19.

[43] Xie HG, Frueh FW (2005). "Pharmacogenomics steps toward personalized medicine." (PDF). *Personalized Medicine.* **2** (4): 325–337. doi:10.2217/17410541.2.4.325.

[44] Katsnelson A (2005). "Cautious welcome for FDA pharmacogenomics guidance". *Nat Biotechnol.* **23** (5): 510. PMID 15877053. doi:10.1038/nbt0505-510.

[45] Swen JJ, Huizinga TW, Gelderblom H, de Vries EG, Assendelft WJ, Kirchheiner J, Guchelaar HK. "Translating Pharmacogenomics: Challenges on the Road to the Clinic". *PLoS Medicine.* **4** (8): 1317–24. PMC 1945038 ⓐ. PMID 17696640. doi:10.1371/journal.pmed.0040209.

[46] Stanek EJ, Sanders CL, Taber KA, Khalid M, Patel A, Verbrugge RR, Agatep BC, Aubert RE, Epstein RS, Frueh FW (Mar 2012). "Adoption of pharmacogenomic testing by US physicians: results of a nationwide survey.". *Clin Pharmacol Ther.* **91** (3): 450–8. PMID 22278335. doi:10.1038/clpt.2011.306.

[47] Ma JD, Lee KC, Kuo GM (August 2012). "Clinical application of pharmacogenomics". *J Pharm Pract.* **25** (4): 417–27. PMID 22689709. doi:10.1177/0897190012448309.

[48] Wikoff WR, Frye RF, Zhu H, Gong Y, Boyle S, Churchill E, Cooper-Dehoff RM, Beitelshees AL, Chapman AB, Fiehn O, Johnson JA, Kaddurah-Daouk R (2013). "Pharmacometabolomics reveals racial differences in response to atenolol treatment". *PLoS ONE.* **8** (3): e57639. PMC 3594230. PMID 23536766. doi:10.1371/journal.pone.0057639.

[49] Rotimi CN, Jorde LB (2010). "Ancestry and disease in the age of genomic medicine". *N Engl J Med.* **363** (16): 1551–8. PMID 20942671. doi:10.1056/NEJMra0911564.

[50] Bloche MG (2004). "Race-based therapeutics". *N Engl J Med.* **351** (20): 2035–7. PMID 15533852. doi:10.1056/NEJMp048271.

[51] Frank R (March 30 – April 1, 2006). "Back with a Vengeance: the Reemergence of a Biological Conceptualization of Race in Research on Race/Ethnic Disparities in Health". *Annual Meeting of the Population Association of America.* Los Angeles, California. Retrieved 2008-11-20.

[52] Crawley L (2007). "The paradox of race in the Bidil debate". *J Natl Med Assoc.* **99** (7): 821–2. PMC 2574363. PMID 17668653.

[53] Kalow, Werner (2005). *Pharmacogenomics.* New York: Taylor & Francis. pp. 552–3. ISBN 1-57444-878-1.

[54] Thorpe DS (2001). "Combinatorial chemistry: starting the second decade.". *Pharmacogenomics J.* **1** (4): 229–32. PMID 11908762. doi:10.1038/sj.tpj.6500045.

[55] Barh, Debmalya; Dhawan, Dipali; Ganguly, Nirmal Kumar (2013). *Omics for Personalized Medicine.* India: Springer Media. ISBN 978-81-322-1183-9. doi:10.1007/978-81-322-1184-6.

[56] Stram, Daniel (2014). *Design, Analysis, and Interpretation of Genome-Wide Association Scans.* Los Angeles: Springer Science and Business Media. ISBN 978-1-4614-9442-3. doi:10.1007/978-1-4614-9443-0_8.

1.14 Further reading

- Katsnelson A (August 2005). "A Drug to Call One's Own: Will medicine finally get personal?". *Scientific American.*
- Karczewski KJ, Daneshjou R, Altman RB (2012). "Chapter 7: Pharmacogenomics". *PLoS Comput Biol.* **8** (12): e1002817. PMC 3531317. PMID 23300409. doi:10.1371/journal.pcbi.1002817.

1.15 External links

- "Pharmacogenomics Factsheet". National Center for Biotechnology Information (NCBI), U.S. National Library of Medicine. Retrieved 2011-07-11. a quick introduction to customised drugs
- "Pharmacogenomics Education Initiatives". U.S. Food and Drug Administration. 2010-09-24. Retrieved 2011-07-11.
- "Personalized Medicine (Pharmacogenetics)". University of Utah's Genetic Science Learning Center. Retrieved 2011-07-11.
- "Center for Pharmacogenomics and Individualized Therapy". University of North Carolina at Chapel Hill Center for Pharmacogenomics and Individualized Therapy. Retrieved 2014-06-25.

Journals:

- "Future Medicine - Pharmacogenomics". *Journal.* Future Medicine Ltd. ISSN 1462-2416.
- "Pharmacogenetics and Genomics". *Journal (previously Pharmacogenetics).* Lippincott Williams & Wilkins. ISSN 1744-6872. Retrieved 2011-07-11.
- "The Pharmacogenomics Journal". Nature Publishing Group. ISSN 1470-269X. Retrieved 2011-07-11.
- "Pharmacogenomics: Subjects : Omics Gateway". Nature Publishing Group. Retrieved 2011-07-11.

Chapter 2

Metabolic pathway

In biochemistry, a **metabolic pathway** is a linked series of chemical reactions occurring within a cell. The reactants, products, and intermediates of an enzymatic reaction are known as metabolites, which are modified by a sequence of chemical reactions catalyzed by enzymes.[1] In a metabolic pathway, the product of one enzyme acts as the substrate for the next. These enzymes often require dietary minerals, vitamins, and other cofactors to function.

Different metabolic pathways function based on the position within a eukaryotic cell and the significance of the pathway in the given compartment of the cell.[2] For instance, the citric acid cycle, electron transport chain, and oxidative phosphorylation all take place in the mitochondrial membrane. In contrast, glycolysis, pentose phosphate pathway, and fatty acid biosynthesis all occur in the cytosol of a cell.[3]

There are two types of metabolic pathways that are characterized by their ability to either synthesize molecules with the utilization of energy (anabolic pathway) or break down of complex molecules by releasing energy in the process (catabolic pathway).[4] The two pathways complement each other in that the energy released from one is used up by the other. The degradative process of a catabolic pathway provides the energy required to conduct a biosynthesis of an anabolic pathway.[4] In addition to the two distinct metabolic pathways is the amphibolic pathway, which can be either catabolic or anabolic based on the need for or the availability of energy.[5]

Pathways are required for the maintenance of homeostasis within an organism and the flux of metabolites through a pathway is regulated depending on the needs of the cell and the availability of the substrate. The end product of a pathway may be used immediately, initiate another metabolic pathway or be stored for later use. The metabolism of a cell consists of an elaborate network of interconnected pathways that enable the synthesis and breakdown of molecules (anabolism and catabolism)

2.1 Overview

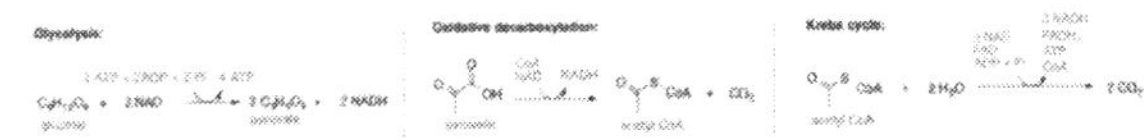

Net reactions of common metabolic pathways

Each metabolic pathway consists of a series of biochemical reactions that are connected by their intermediates: the products of one reaction are the substrates for subsequent reactions, and so on. Metabolic pathways are often considered to flow in one direction. Although all chemical reactions are technically reversible, conditions in the cell are often such that it is thermodynamically more favorable for flux to flow in one direction of a reaction. For example, one pathway may be responsible for the synthesis of a particular amino acid, but the breakdown of that amino acid may occur via a separate and distinct pathway. One example of an exception to this "rule" is the metabolism of glucose. Glycolysis results in the breakdown of glucose, but several reactions in the glycolysis pathway are reversible and participate in the re-synthesis of glucose (gluconeogenesis).

- Glycolysis was the first metabolic pathway discovered:

1. As glucose enters a cell, it is immediately phosphorylated by ATP to glucose 6-phosphate in the irreversible first step.
2. In times of excess lipid or protein energy sources, certain reactions in the glycolysis pathway may run in reverse to produce glucose 6-phosphate, which is then used for storage as glycogen or starch.

- Metabolic pathways are often regulated by feedback inhibition.
- Some metabolic pathways flow in a 'cycle' wherein each component of the cycle is a substrate for the subsequent reaction in the cycle, such as in the Krebs Cycle (see below).

- Anabolic and catabolic pathways in eukaryotes often occur independently of each other, separated either physically by compartmentalization within organelles or separated biochemically by the requirement of different enzymes and co-factors.

2.2 Major metabolic pathways

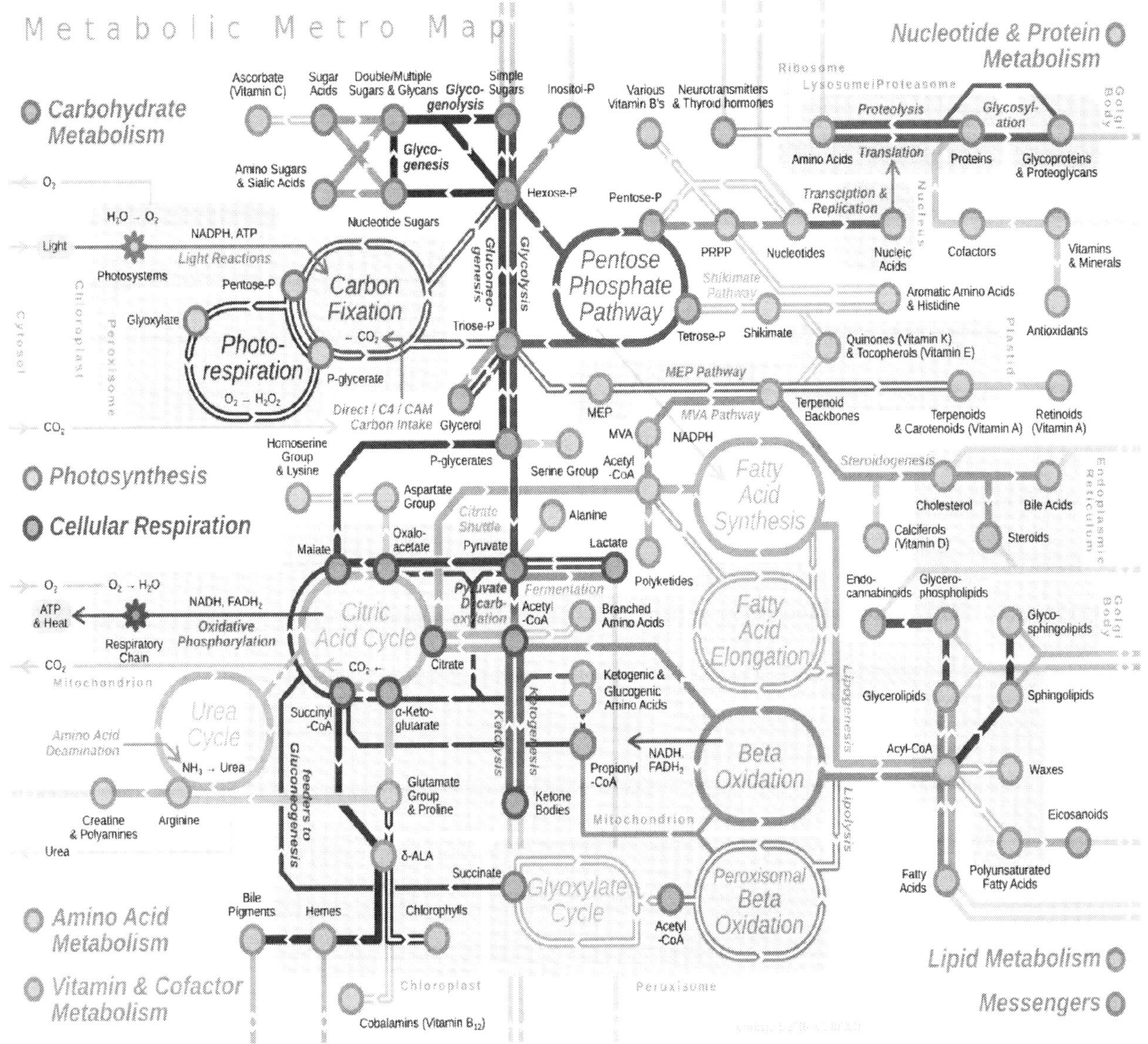

Major metabolic pathways in metro-style map.

Single lines: pathways common to most lifeforms. Double lines: pathways not in humans (occurs in e.g. plants, fungi, prokaryotes). Orange nodes: carbohydrate metabolism. Violet nodes: photosynthesis. Red nodes: cellular respiration. Pink nodes: cell signaling. Blue nodes: amino acid metabolism. Grey nodes: vitamin and cofactor metabolism. Brown nodes: nucleotide and protein metabolism. Green nodes: lipid metabolism.

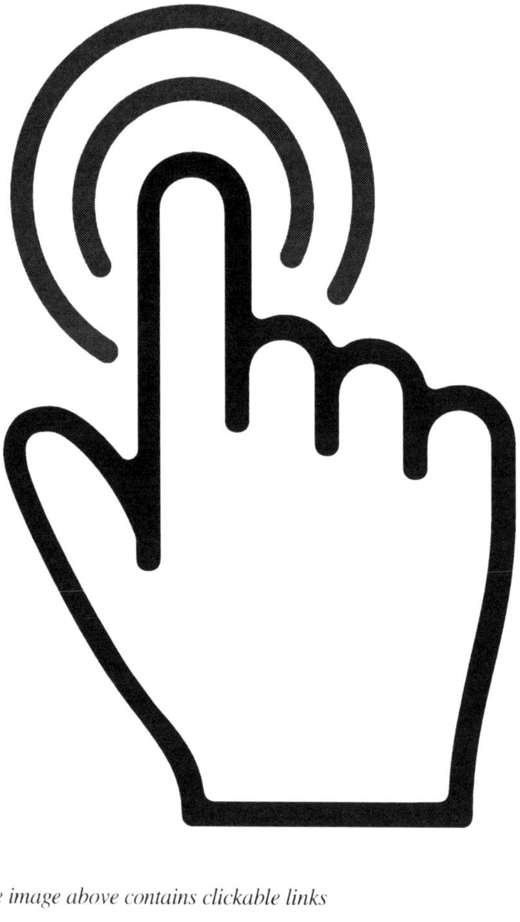

The image above contains clickable links

2.2.1 Catabolic pathway (catabolism)

A **catabolic pathway** is a series of reactions that bring about a net release of energy in the form of a high energy phosphate bond formed with the energy carriers Adenosine Diphosphate (ADP) and Guanosine Diphosphate (GDP) to produce Adenosine Triphosphate (ATP) and Guanosine Triphosphate (GTP), respectively. The net reaction is, therefore, thermodynamically favorable, for it results in a lower free energy for the final products.[6] A catabolic pathway is an exergonic system that produces chemical energy in the form of ATP, GTP, NADH, NADPH, FADH2, etc. from energy containing sources such as carbohydrates, fats, and proteins. The end products are often carbon dioxide, water, and ammonia. Coupled with an endergonic reaction of anabolism, the cell can synthesize new macromolecules using the original precursors of the anabolic pathway.[7] An example of a coupled reaction is the phosphorylation of fructose-6-phosphate to form the intermediate fructose-1,6-bisphosphate by the enzyme phsophofructokinase accompanied by the hydrolysis of ATP in the pathway of glycolysis. The resulting chemical reaction within the metabolic pathway is highly thermodynamically favorable and, as a result, irreversible in the cell.[8]

$$Fructose - 6 - Phosphate + ATP \longrightarrow Fructose - 1,6 - Bisphosphate + ADP$$

Cellular respiration

Main article: Cellular respiration

A core set of energy-producing catabolic pathways occur within all living organisms in some form. These pathways transfer the energy released by breakdown of nutrients into ATP and other small molecules used for energy (e.g. GTP, NADPH, FADH). All cells can perform anaerobic respiration by glycolysis. Additionally, most organisms can perform more efficient aerobic respiration through the citric acid cycle and oxidative phosphorylation. Additionally plants, algae and cyanobacteria are able to use sunlight to anabolically synthesize compounds from non-living matter by photosynthesis.

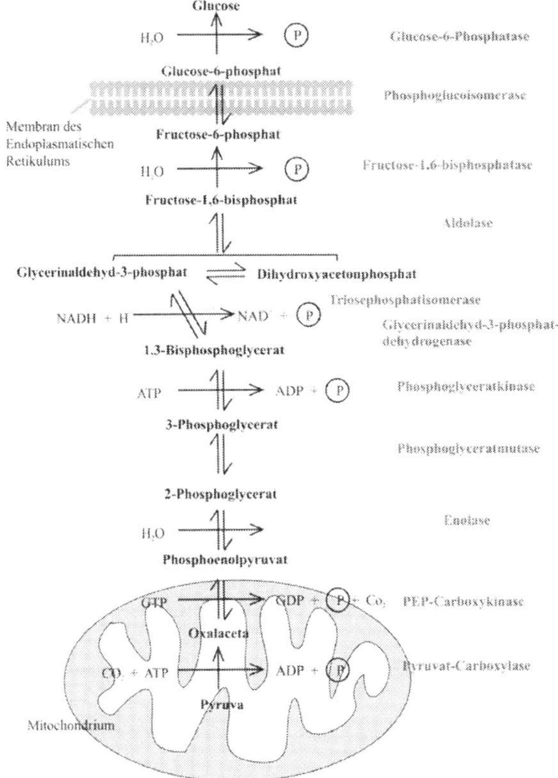

Gluconeogenesis Mechanism

2.2.2 Anabolic pathway (anabolism)

In contrast to catabolic pathways, **anabolic pathways** require an energy input to construct macromolecules such as polypeptides, nucleic acids, proteins, polysaccharides, and lipids. The isolated reaction of anabolism is unfavorable in a cell due to a positive Gibbs Free Energy ($+\Delta G$). Thus, an input of chemical energy through a coupling with an exergonic reaction is necessary.[9] The coupled reaction of the catabolic pathway affects the thermodynamics of the reaction by lowering the overall activation energy of an anabolic pathway and allowing the reaction to take place.[10] Otherwise, an endergonic reaction is non-spontaneous.

An anabolic pathway is a biosynthetic pathway, meaning that it combines smaller molecules to form larger and more complex ones.[11] An example is the reversed pathway of glycolysis, otherwise known as gluconeogenesis, which occurs in the liver and sometimes in the kidney to maintain proper glucose concentration in the blood and supply the brain and muscle tissues with adequate amount of glucose. Although gluconeogenesis is similar to the reverse pathway of glycolysis, it contains three distinct enzymes from glycolysis that allow the pathway to occur spontaneously.[12] An example of the pathway for gluconeogenesis is illustrated in the image titled "Gluconeogenesis Mechanism".

2.2.3 Amphibolic pathway

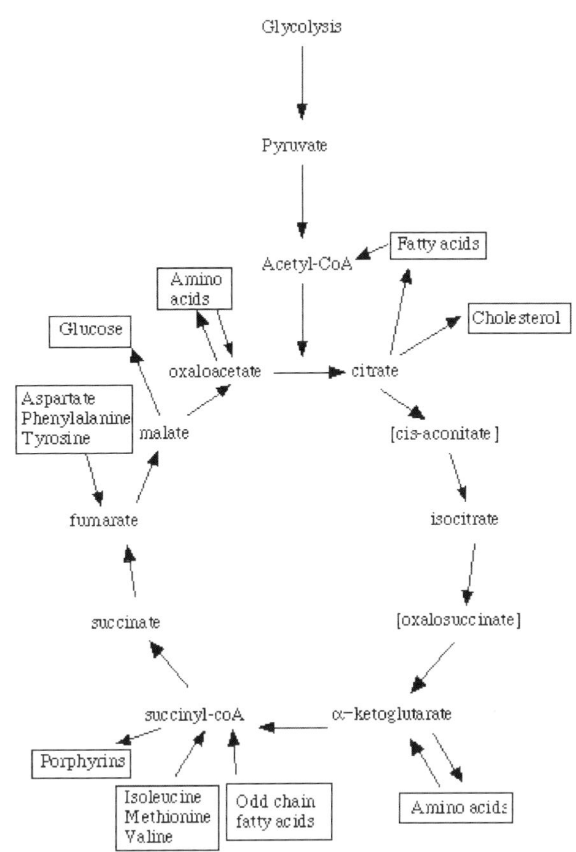

Amphibolic Properties of the Citric Acid Cycle

An **amphibolic pathway** is one that can be either catabolic or anabolic based on the availability of or the need for energy.[13] The currency of energy in a biological cell is adenosine triphosphate (ATP), which stores its energy in the phosphoanhydride bonds. The energy is utilized to conduct biosynthesis, facilitate movement, and regulate active transport inside of the cell.[14] Examples of amphibolic pathways are the citric acid cycle and the glyoxylate cycle. These sets of chemical reactions contain both energy producing and utilizing pathways.[15] To the right is an illustration of the amphibolic properties of the TCA cycle.

The glyoxylate shunt pathway is an alternative to the tricarboxylic acid (TCA) cycle, for it redirects the pathway of TCA to prevent full oxidation of carbon compounds, and to preserve high energy carbon sources as future energy sources. This pathway occurs only in plants and bacteria and transpires in the absence of glucose molecules.[16]

2.3 Regulation

The flux of the entire pathway is regulated by the rate-determining steps.[17] These are the slowest steps in a network of reactions. The rate-limiting step occurs near the beginning of the pathway and is regulated by feedback inhibition, which ultimately controls the overall rate of the pathway.[18] The metabolic pathway in the cell is regulated by covalent or non-covalent modifications. A covalent modification involves an addition or removal of a chemical bond, whereas a non-covalent modification (also known as allosteric regulation) is the binding of the regulator to the enzyme via hydrogen bonds, electrostatic interactions, and Van Der Waals forces.[19]

The rate of turnover in a metabolic pathway, otherwise known as the metabolic flux, is regulated based on the stoichiometric reaction model, the utilization rate of metabolites, and the translocation pace of molecules across the lipid bilayer.[20] The regulation methods are based on experiments involving 13C-labeling, which is then analyzed by Nuclear Magnetic Resonance (NMR) or gas chromatography-mass spectrometry (GC-MS)-derived mass compositions. The aforementioned techniques synthesize a statistical interpretation of mass distribution in proteinogenic amino acids to the catalytic activities of enzymes in a cell.[21]

2.4 See also

- Metabolism
- Metabolic network
- Metabolic network modelling
- Metabolic engineering

2.5 References

[1] Cox, David L. Nelson, Michael M. (2008). *Lehninger principles of biochemistry* (5th ed.). New York: W.H. Freeman. p. 26. ISBN 978-0-7167-7108-1.

[2] Nicholson, Donald E. (March 1971). *An Introduction to Metabolic Pathways by S. DAGLEY* (Vol. 59, No. 2 ed.). Sigma Xi, The Scientific Research Society. p. 266.

[3] Pratt, Donald Voet, Judith G. Voet, Charlotte W. (2013). *Fundamentals of Biochemistry: Life at the Molecular Level* (4th ed.). Hoboken, NJ: Wiley. pp. 441–442. ISBN 978-0470-54784-7.

[4] Reece, Jane B. (2011). *Campbell biology / Jane B. Reece ... [et al.]*. (9th ed.). Boston: Benjamin Cummings. p. 143. ISBN 978-0-321-55823-7.

[5] Berg, Jeremy M.; Tymoczko, John L.; Stryer, Lubert; Gatto, Gregory J. (2012). *Biochemistry* (7th ed.). New York: W.H. Freeman. p. 429. ISBN 1429229365.

[6] Clarke, Jeremy M. Berg; John L. Tymoczko; Lubert Stryer. Web content by Neil D. (2002). *Biochemistry* (5. ed., 4. print. ed.). New York, NY [u.a.]: W. H. Freeman. pp. 578–579. ISBN 0716730510.

[7] Eichhorn, Peter H. Raven ; Ray F. Evert ; Susan E. (2011). *Biology of plants* (8. ed.). New York, NY: Freeman. pp. 100–106. ISBN 978-1-4292-1961-7.

[8] Pratt, Donald Voet, Judith G. Voet, Charlotte W. (2013). *Fundamentals of biochemistry : life at the molecular level* (4th ed.). Hoboken, NJ: Wiley. pp. 474–478. ISBN 978-0470-54784-7.

[9] Cox, David L. Nelson, Michael M. (2008). *Lehninger principles of biochemistry* (5th ed.). New York: W.H. Freeman. pp. 25–27. ISBN 978-0-7167-7108-1.

[10] Cox, David L. Nelson, Michael M. (2008). *Lehninger principles of biochemistry* (5th ed.). New York: W.H. Freeman. p. 25. ISBN 978-0-7167-7108-1.

[11] Clarke, Jeremy M. Berg; John L. Tymoczko; Lubert Stryer. Web content by Neil D. (2002). *Biochemistry* (5. ed., 4. print. ed.). New York, NY [u.a.]: W. H. Freeman. p. 570. ISBN 0716730510.

[12] Berg, Jeremy M.; Tymoczko, John L.; Stryer, Lubert; Gatto, Gregory J. (2012). *Biochemistry* (7th ed.). New York: W.H. Freeman. pp. 480–482. ISBN 9781429229364.

[13] Clarke, Jeremy M. Berg; John L. Tymoczko; Lubert Stryer. Web content by Neil D. (2002). *Biochemistry* (5. ed., 4. print. ed.). New York, NY [u.a.]: W. H. Freeman. p. 570. ISBN 0-7167-3051-0.

[14] Clarke, Jeremy M. Berg; John L. Tymoczko; Lubert Stryer. Web content by Neil D. (2002). *Biochemistry* (5. ed., 4. print. ed.). New York, NY [u.a.]: W. H. Freeman. p. 571. ISBN 0-7167-3051-0.

[15] Pratt, Donald Voet, Judith G. Voet, Charlotte W. (2013). *Fundamentals of biochemistry : life at the molecular level* (4th ed.). Hoboken, NJ: Wiley. p. 572. ISBN 978-0470-54784-7.

[16] Choffnes, Eileen R.; Relman,, David A.; Academies, Leslie Pray, rapporteurs, Forum on Microbial Threat, Board on Global Health, Institute of Medicine of the National (2011).

The science and applications of synthetic and systems biology workshop summary. Washington, D.C.: National Academies Press. p. 135. ISBN 978-0-309-21939-6.

[17] Cox, David L. Nelson, Michael M. (2008). *Lehninger principles of biochemistry* (5th ed.). New York: W.H. Freeman. pp. 577–578. ISBN 978-0-7167-7108-1.

[18] Kruger, ed. by Nicholas J.; Hill,, Steve A.; Ratcliffe, R. George (1999). *Regulation of primary metabolic pathways in plants : [proceedings of an international conference held on 9 - 11 January 1997 at St Hugh's College, Oxford under the auspices of the Phytochemical Society of Europe].* Dordrecht [u.a.]: Kluwer. p. 258. ISBN 079235494X.

[19] White, David (1995). *The physiology and biochemistry of prokaryotes.* New York [u.a.]: Oxford Univ. Press. p. 133. ISBN 0-19-508439-X.

[20] Weckwerth, edited by Wolfram (2006). *Metabolomics methods and protocols.* Totowa, N.J.: Humana Press. p. 177. ISBN 1597452440.

[21] Weckwerth, edited by Wolfram (2006). *Metabolomics methods and protocols.* Totowa, N.J.: Humana Press. p. 178. ISBN 1597452440.

2.6 External links

- Full map of metabolic pathways
- BioCyc: Metabolic network models for thousands of sequenced organisms
- KEGG: Kyoto Encyclopedia of Genes and Genomes
- Reactome, a database of reactions, pathways and biological processes
- MetaCyc: A database of experimentally elucidated metabolic pathways (2,200+ pathways from more than 2,500 organisms).
- The Pathway Localization database (PathLocdb)
- Metabolism, Cellular Respiration and Photosynthesis - The Virtual Library of Biochemistry, Molecular Biology and Cell Biology
- DAVID: Visualize genes on pathway maps
- Wikipathways: pathways for the people
- ConsensusPathDB
- *metpath*: Integrated interactive metabolic chart

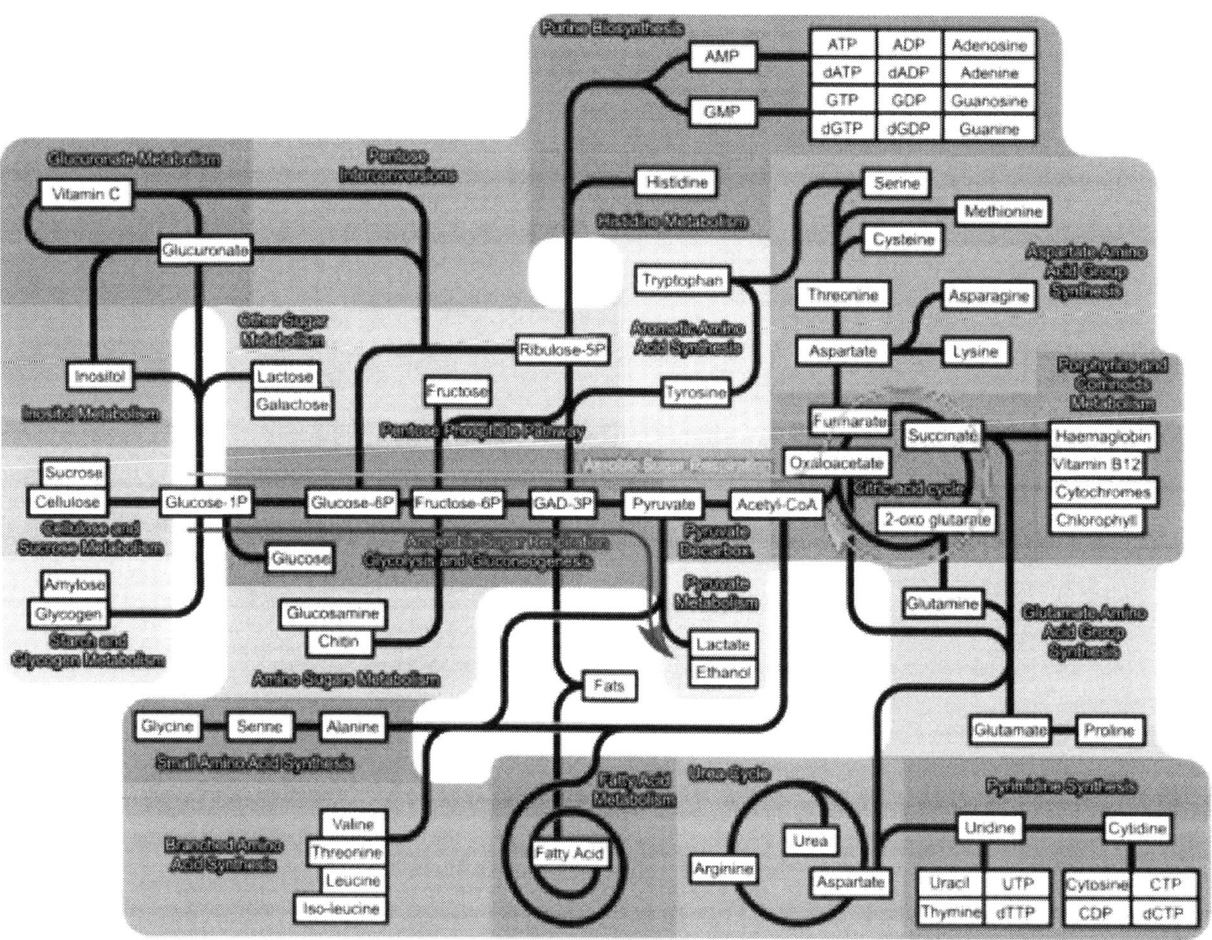
Purine Biosynthesis
AMP
GMP
ATP
ADP
Adenosine
dATP
dADP
Adenine
GTP
GDP
Guanosine
dGTP
dGDP
Guanine
Glucuronate Metabolism
Pentose Interconversions
Vitamin C
Glucuronate
Histidine
Histidine Metabolism
Serine
Methionine
Cysteine
Aspartate Amino Acid Group Synthesis
Tryptophan
Threonine
Asparagine
Other Sugar Metabolism
Aromatic Amino Acid Synthesis
Inositol
Lactose
Galactose
Ribulose-5P
Fructose
Tyrosine
Aspartate
Lysine
Porphyrins and Corrinoids Metabolism
Inositol Metabolism
Pentose Phosphate Pathway
Fumarate
Succinate
Haemaglobin
Vitamin B12
Cytochromes
Chlorophyll
Oxaloacetate
Citric acid cycle
2-oxo glutarate
Sucrose
Cellulose
Glucose-1P
Glucose-6P
Fructose-6P
GAD-3P
Pyruvate
Acetyl-CoA
Cellulose and Sucrose Metabolism
Glucose
Glycolysis and Gluconeogenesis
Pyruvate Decarbox.
Pyruvate Metabolism
Amylose
Glycogen
Starch and Glycogen Metabolism
Glucosamine
Chitin
Amino Sugars Metabolism
Lactate
Ethanol
Glutamine
Glutamate Amino Acid Group Synthesis
Fats
Glycine
Serine
Alanine
Glutamate
Proline
Small Amino Acid Synthesis
Fatty Acid Metabolism
Urea Cycle
Pyrimidine Synthesis
Branched Amino Acid Synthesis
Valine
Threonine
Leucine
Iso-leucine
Fatty Acid
Urea
Arginine
Aspartate
Uridine
Cytidine
Uracil
UTP
Thymine
dTTP
Cytosine
CTP
CDP
dCTP

Chapter 3

Pharmacokinetics

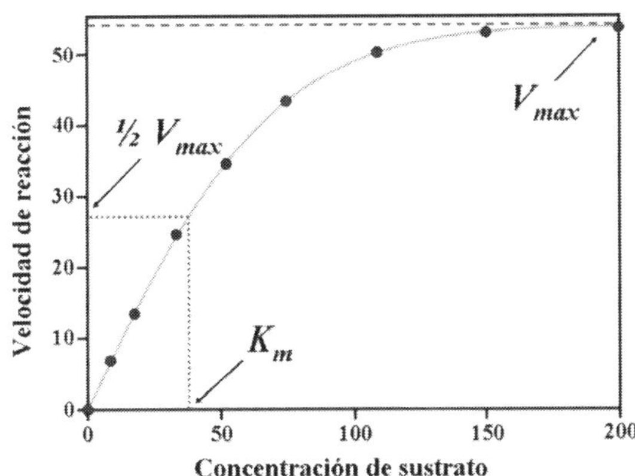

Graph that demonstrates the Michaelis-Menten kinetics model for the relationship between an enzyme and a substrate: one of the parameters studies in pharmacokinetics, where the substrate is a pharmaceutical drug.

Pharmacokinetics (from Ancient Greek *pharmakon* "drug" and *kinetikos* "moving, putting in motion"; see chemical kinetics), sometimes abbreviated as **PK**, is a branch of pharmacology dedicated to determining the fate of substances administered to a living organism. The substances of interest include any chemical xenobiotic such as: pharmaceutical drugs, pesticides, food additives, cosmetic ingredients, etc. It attempts to analyze chemical metabolism and to discover the fate of a chemical from the moment that it is administered up to the point at which it is completely eliminated from the body. Pharmacokinetics is the study of how an organism affects a drug, whereas pharmacodynamics is the study of how the drug affects the organism. Both together influence dosing, benefit, and adverse effects, as seen in PK/PD models.

3.1 Overview

Pharmacokinetics describes how the body affects a specific xenobiotic/chemical after administration through the mechanisms of absorption and distribution, as well as the metabolic changes of the substance in the body (e.g. by metabolic enzymes such as cytochrome P450 or glucuronosyltransferase enzymes), and the effects and routes of excretion of the metabolites of the drug.[1] Pharmacokinetic properties of chemicals are affected by the route of administration and the dose of administered drug. These may affect the absorption rate.[2]

Models have been developed to simplify conceptualization of the many processes that take place in the interaction between an organism and a chemical substance. One of these, the multi-compartmental model, are the most commonly used approximations to reality; however, the complexity involved in adding parameters with that modelling approach means that *monocompartmental models* and above all *two compartmental models* are the most-frequently used. The various compartments that the model is divided into are commonly referred to as the ADME scheme (also referred to as LADME if liberation is included as a separate step from absorption):

- Liberation - the process of release of a drug from the pharmaceutical formulation.[3][4] See also IVIVC.
- Absorption - the process of a substance entering the blood circulation.
- Distribution - the dispersion or dissemination of substances throughout the fluids and tissues of the body.
- Metabolism (or biotransformation, or inactivation) – the recognition by the organism that a foreign substance is present and the irreversible transformation of parent compounds into daughter metabolites.
- Excretion - the removal of the substances from the body. In rare cases, some drugs irreversibly accumulate in body tissue.

The two phases of metabolism and excretion can also be grouped together under the title elimination. The study of these distinct phases involves the use and manipulation of basic concepts in order to understand the process dynamics. For this reason in order to fully comprehend the *kinetics* of a drug it is necessary to have detailed knowledge of a number of factors such as: the properties of the substances that act as excipients, the characteristics of the appropriate biological membranes and the way that substances can cross them, or the characteristics of the enzyme reactions that inactivate the drug.

All these concepts can be represented through mathematical formulas that have a corresponding graphical representation. The use of these models allows an understanding of the characteristics of a molecule, as well as how a particular drug will behave given information regarding some of its basic characteristics such as its acid dissociation constant (pKa), bioavailability and solubility, absorption capacity and distribution in the organism.

The model outputs for a drug can be used in industry (for example, in calculating bioequivalence when designing generic drugs) or in the clinical application of pharmacokinetic concepts. Clinical pharmacokinetics provides many performance guidelines for effective and efficient use of drugs for human-health professionals and in veterinary medicine.

3.2 Metrics

The following are the most commonly measured pharmacokinetic metrics:[5]

In pharmacokinetics, *steady state* refers to the situation where the overall intake of a drug is fairly in dynamic equilibrium with its elimination. In practice, it is generally considered that steady state is reached when a time of 4 to 5 times the half-life for a drug after regular dosing is started.

The following graph depicts a typical time course of drug plasma concentration and illustrates main pharmacokinetic metrics:

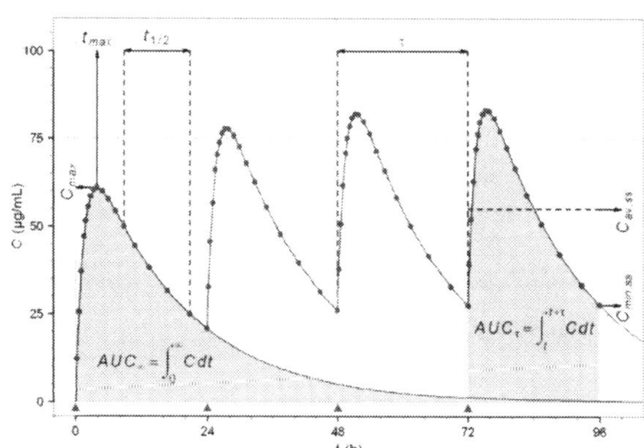

The time course of drug plasma concentrations over 96 hours following oral administrations every 24 hours. Note that the AUC in steady state equals AUC∞ after the first dose.

3.3 Pharmacokinetic models

Pharmacokinetic modelling is performed by noncompartmental or compartmental methods. Noncompartmental methods estimate the exposure to a drug by estimating the area under the curve of a concentration-time graph. Compartmental methods estimate the concentration-time graph using kinetic models. Noncompartmental methods are often more versatile in that they do not assume any specific compartmental model and produce accurate results also acceptable for bioequivalence studies. The final outcome of the transformations that a drug undergoes in an organism and the rules that determine this fate depend on a number of interrelated factors. A number of functional models have been developed in order to simplify the study of pharmacokinetics. These models are based on a consideration of an organism as a number of related compartments. The simplest idea is to think of an organism as only one homogenous compartment. This *monocompartmental model* presupposes that blood plasma concentrations of the drug are a true reflection of the drug's concentration in other fluids or tissues and that the elimination of the drug is directly proportional to the drug's concentration in the organism (first order kinetics).

However, these models do not always truly reflect the real situation within an organism. For example, not all body tissues have the same blood supply, so the distribution of the drug will be slower in these tissues than in others with a better blood supply. In addition, there are some tissues (such as the brain tissue) that present a real barrier to the distribution of drugs, that can be breached with greater or lesser ease depending on the drug's characteristics. If these relative conditions for the different tissue types are considered along with the rate of elimination, the organism can be considered to be acting like two compartments: one that we can call the *central compartment* that has a more rapid distribution, comprising organs and systems with a well-developed blood supply; and a *peripheral compartment* made up of organs with a lower blood flow. Other tissues, such as the brain, can occupy a variable position depending on a drug's ability to cross the barrier that separates the organ from the blood supply.

This *two compartment model* will vary depending on which compartment elimination occurs in. The most common situation is that elimination occurs in the central compartment as the liver and kidneys are organs with a good blood supply. However, in some situations it may be that elimina-

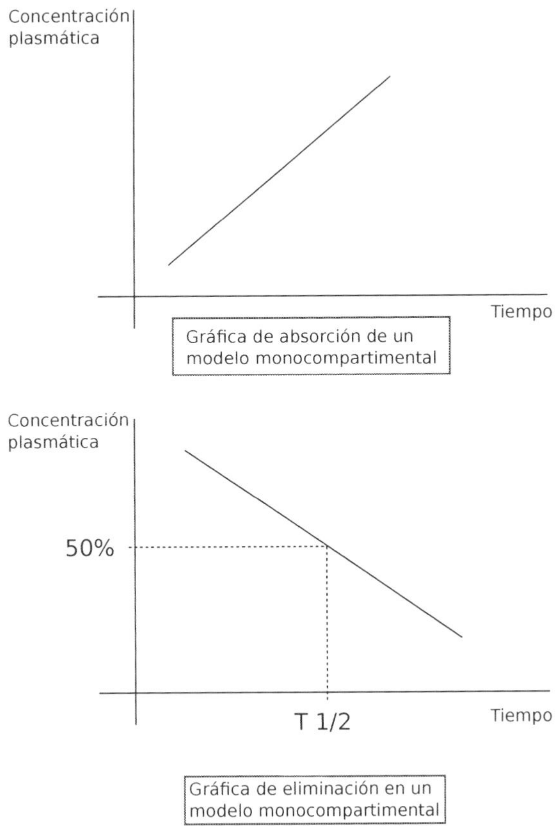

Graph representing the monocompartmental action model.

tion occurs in the peripheral compartment or even in both. This can mean that there are three possible variations in the two compartment model, which still do not cover all possibilities.[6]

This model may not be applicable in situations where some of the enzymes responsible for metabolizing the drug become saturated, or where an active elimination mechanism is present that is independent of the drug's plasma concentration. In the real world each tissue will have its own distribution characteristics and none of them will be strictly linear. If we label the drug's volume of distribution within the organism **VdF** and its volume of distribution in a tissue **VdT** the former will be described by an equation that takes into account all the tissues that act in different ways, that is:

$$Vd_F = Vd_{T1} + Vd_{T2} + Vd_{T3} + ... + Vd_{Tn}$$

This represents the *multi-compartment model* with a number of curves that express complicated equations in order to obtain an overall curve. A number of computer programs have been developed to plot these equations.[6] However complicated and precise this model may be, it still does not truly represent reality despite the effort involved in obtaining various distribution values for a drug. This is because the concept of distribution volume is a relative concept that is not a true reflection of reality. The choice of model therefore comes down to deciding which one offers the lowest margin of error for the drug involved.

3.3.1 Noncompartmental analysis

Noncompartmental PK analysis is highly dependent on estimation of total drug exposure. Total drug exposure is most often estimated by area under the curve (AUC) methods, with the trapezoidal rule (numerical integration) the most common method. Due to the dependence on the length of 'x' in the trapezoidal rule, the area estimation is highly dependent on the blood/plasma sampling schedule. That is, the closer time points are, the closer the trapezoids reflect the actual shape of the concentration-time curve.

3.3.2 Compartmental analysis

Compartmental PK analysis uses kinetic models to describe and predict the concentration-time curve. PK compartmental models are often similar to kinetic models used in other scientific disciplines such as chemical kinetics and thermodynamics. The advantage of compartmental over some noncompartmental analyses is the ability to predict the concentration at any time. The disadvantage is the difficulty in developing and validating the proper model. Compartment-free modelling based on curve stripping does not suffer this limitation. The simplest PK compartmental model is the one-compartmental PK model with IV bolus administration and first-order elimination. The most complex PK models (called PBPK models) rely on the use of physiological information to ease development and validation.

3.3.3 Single-compartment model

Linear pharmacokinetics is so-called because the graph of the relationship between the various factors involved (dose, blood plasma concentrations, elimination, etc.) gives a straight line or an approximation to one. For drugs to be effective they need to be able to move rapidly from blood plasma to other body fluids and tissues.

The change in concentration over time can be expressed as $C = C_{initial} \times e^{-k_{el} \times t}$

3.3.4 Multi-compartmental models

The graph for the non-linear relationship between the various factors is represented by a curve; the relationships be-

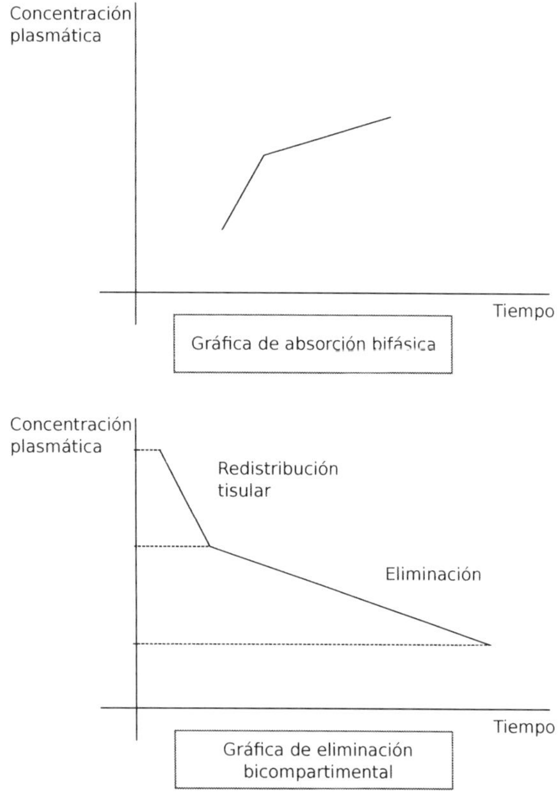

Graphs for absorption and elimination for a non-linear pharmacokinetic model.

tween the factors can then be found by calculating the dimensions of different areas under the curve. The models used in *'non linear pharmacokinetics* are largely based on Michaelis-Menten kinetics. A reaction's factors of non linearity include the following:

- Multiphasic absorption: Drugs injected intravenously are removed from the plasma through two primary mechanisms: (1) Distribution to body tissues and (2) metabolism + excretion of the drugs. The resulting decrease of the drug's plasma concentration follows a biphasic pattern (see figure).
    - **Alpha phase**: An initial phase of rapid decrease in plasma concentration. The decrease is primarily attributed to drug distribution from the central compartment (circulation) into the peripheral compartments (body tissues). This phase ends when a pseudo-equilibrium of drug concentration is established between the central and peripheral compartments.
    - **Beta phase**: A phase of gradual decrease in plasma concentration after the alpha phase. The decrease is primarily attributed to drug metabolism and excretion.[7]

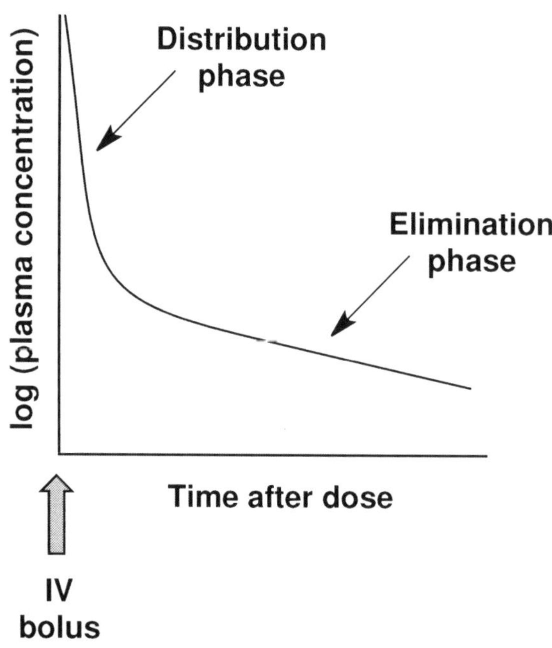

Plasma drug concentration vs time after an IV dose

    - Additional phases (gamma, delta, etc.) are sometimes seen.[8]

- A drug's characteristics make a clear distinction between tissues with high and low blood flow.

- Enzymatic saturation: When the dose of a drug whose elimination depends on biotransformation is increased above a certain threshold the enzymes responsible for its metabolism become saturated. The drug's plasma concentration will then increase disproportionately and its elimination will no longer be constant.

- Induction or enzymatic inhibition: Some drugs have the capacity to inhibit or stimulate their own metabolism, in negative or positive feedback reactions. As occurs with fluvoxamine, fluoxetine and phenytoin. As larger doses of these pharmaceuticals are administered the plasma concentrations of the unmetabolized drug increases and the elimination half-life increases. It is therefore necessary to adjust the dose or other treatment parameters when a high dosage is required.

- The kidneys can also establish active elimination mechanisms for some drugs, independent of plasma concentrations.

It can therefore be seen that non-linearity can occur because of reasons that affect the entire pharmacokinetic sequence: absorption, distribution, metabolism and elimination.

3.3.5 Variable volume in time models

Variable volume pharmacokinetic models can be drug centered models that imply a volume of drug distribution to be that volume in which the drug is distributed at that elapsed time following drug administration.[9][10] Another possibility occurs when the body volume is changing in time, which would occur, for example, during dialysis when the volume in which a drug can be distributed is itself changing in time.[11]

3.4 Bioavailability

Main article: Bioavailability

At a practical level, a drug's bioavailability can be defined as the proportion of the drug that reaches its site of action. From this perspective the intravenous administration of a drug provides the greatest possible bioavailability, and this method is considered to yield a bioavailability of 1 (or 100%). Bioavailability of other delivery methods is compared with that of intravenous injection («absolute bioavailability») or to a standard value related to other delivery methods in a particular study («relative bioavailability»).

$$B_A = \frac{[ABC]_P.D_{IV}}{[ABC]_{IV}.D_P}$$

$$B_R = \frac{[ABC]_A.dose_B}{[ABC]_B.dose_A}$$

Once a drug's bioavailability has been established it is possible to calculate the changes that need to be made to its dosage in order to reach the required blood plasma levels. Bioavailability is therefore a mathematical factor for each individual drug that influences the administered dose. It is possible to calculate the amount of a drug in the blood plasma that has a real potential to bring about its effect using the formula: $De = B.Da$; where *De* is the effective dose, *B* bioavailability and *Da* the administered dose.

Therefore, if a drug has a bioavailability of 0.8 (or 80%) and it is administered in a dose of 100 mg, the equation will demonstrate the following:

De = 0.8 x 100 mg = 80 mg

That is the 100 mg administered represents a blood plasma concentration of 80 mg that has the capacity to have a pharmaceutical effect.

This concept depends on a series of factors inherent to each drug, such as:[12]

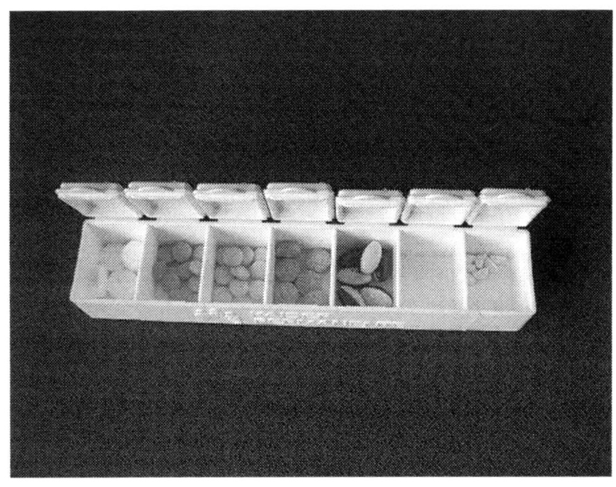

Different forms of tablets, which will have different pharmacokinetic behaviours after their administration.

- Pharmaceutical form
- Chemical form
- Route of administration
- Stability
- Metabolism

These concepts, which are discussed in detail in their respective titled articles, can be mathematically quantified and integrated to obtain an overall mathematical equation:

$$De = Q.Da.B$$

where **Q** is the drug's purity.[12]

$$Va = \frac{Da.B.Q}{\tau}$$

where Va is the drug's rate of administration and τ is the rate at which the absorbed drug reaches the circulatory system.

Finally, using the Henderson-Hasselbalch equation, and knowing the drug's pKa (pH at which there is an equilibrium between its ionized and non ionized molecules), it is possible to calculate the non ionized concentration of the drug and therefore the concentration that will be subject to absorption:

$$pH = pKa + log\frac{B}{A}$$

When two drugs have the same bioavailability, they are said to be biological equivalents or bioequivalents. This concept of bioequivalence is important because it is currently used as a yardstick in the authorization of generic drugs in many countries .

3.5 LADME

Main article: ADME

A number of phases occur once the drug enters into contact with the organism, these are described using the acronym LADME:

- Liberation of the active substance from the delivery system,
- Absorption of the active substance by the organism,
- Distribution through the blood plasma and different body tissues,
- Metabolism that is, inactivation of the xenobiotic substance, and finally
- Excretion or elimination of the substance or the products of its metabolism.

Some textbooks combine the first two phases as the drug is often administered in an active form, which means that there is no liberation phase. Others include a phase that combines distribution, metabolism and excretion into a «disposition phase». Other authors include the drug's toxicological aspect in what is known as *ADME-Tox* or *ADMET*.

Each of the phases is subject to physico-chemical interactions between a drug and an organism, which can be expressed mathematically. Pharmacokinetics is therefore based on mathematical equations that allow the prediction of a drug's behavior and which place great emphasis on the relationships between drug plasma concentrations and the time elapsed since the drug's administration.

3.6 Analysis

3.6.1 Bioanalytical methods

Bioanalytical methods are necessary to construct a concentration-time profile. Chemical techniques are employed to measure the concentration of drugs in biological matrix, most often plasma. Proper bioanalytical methods should be selective and sensitive. For example, microscale thermophoresis can be used to quantify how the biological matrix/liquid affects the affinity of a drug to its target.[13][14]

Mass spectrometry

Pharmacokinetics is often studied using mass spectrometry because of the complex nature of the matrix (often plasma or urine) and the need for high sensitivity to observe concentrations after a low dose and a long time period. The most common instrumentation used in this application is LC-MS with a triple quadrupole mass spectrometer. Tandem mass spectrometry is usually employed for added specificity. Standard curves and internal standards are used for quantitation of usually a single pharmaceutical in the samples. The samples represent different time points as a pharmaceutical is administered and then metabolized or cleared from the body. Blank samples taken before administration are important in determining background and ensuring data integrity with such complex sample matrices. Much attention is paid to the linearity of the standard curve; however it is common to use curve fitting with more complex functions such as quadratics since the response of most mass spectrometers is not linear across large concentration ranges.[15][16][17]

There is currently considerable interest in the use of very high sensitivity mass spectrometry for microdosing studies, which are seen as a promising alternative to animal experimentation.[18]

3.7 Population pharmacokinetics

Population pharmacokinetics is the study of the sources and correlates of variability in drug concentrations among individuals who are the target patient population receiving clinically relevant doses of a drug of interest.[19][20][21] Certain patient demographic, pathophysiological, and therapeutical features, such as body weight, excretory and metabolic functions, and the presence of other therapies, can regularly alter dose-concentration relationships. For example, steady-state concentrations of drugs eliminated mostly by the kidney are usually greater in patients suffering from renal failure than they are in patients with normal renal function receiving the same drug dosage. Population pharmacokinetics seeks to identify the measurable pathophysiologic factors that cause changes in the dose-concentration relationship and the extent of these changes so that, if such changes are associated with clinically significant shifts in the therapeutic index, dosage can be appropriately modified. An advantage of population pharmacokinetic modelling is its ability to analyse sparse data sets (sometimes only one concentration measurement per patient is available).

Software packages used in population pharmacokinetics modelling include NONMEM, which was developed at the UCSF and newer packages which incorporate GUIs like

Monolix as well as graphical model building tools; Phoenix NLME.

3.8 Clinical pharmacokinetics

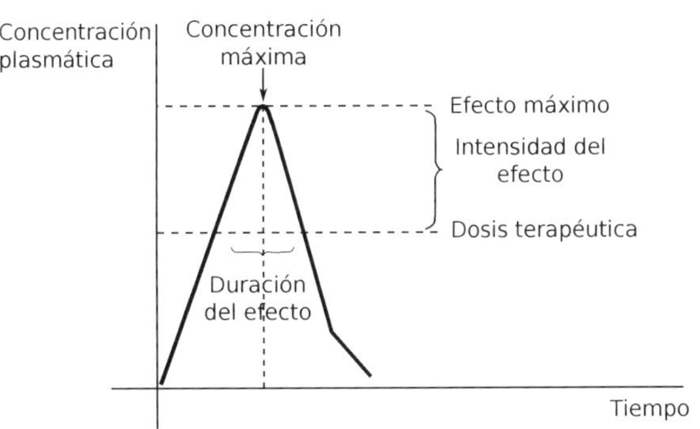

Basic graph for evaluating the therapeutic implications of pharmacokinetics.

Clinical pharmacokinetics (arising from the clinical use of population pharmacokinetics) is the direct application to a therapeutic situation of knowledge regarding a drug's pharmacokinetics and the characteristics of a population that a patient belongs to (or can be ascribed to).

An example is the relaunch of the use of ciclosporin as an immunosuppressor to facilitate organ transplant. The drug's therapeutic properties were initially demonstrated, but it was almost never used after it was found to cause nephrotoxicity in a number of patients.[22] However, it was then realized that it was possible to individualize a patient's dose of ciclosporin by analysing the patients plasmatic concentrations (pharmacokinetic monitoring). This practice has allowed this drug to be used again and has facilitated a great number of organ transplants.

Clinical monitoring is usually carried out by determination of plasma concentrations as this data is usually the easiest to obtain and the most reliable. The main reasons for determining a drug's plasma concentration include:[23]

- Narrow therapeutic range (difference between toxic and therapeutic concentrations)
- High toxicity
- High risk to life.

Drugs where pharmacokinetic monitoring is recommended include:

3.9 Ecotoxicology

Ecotoxicology is the branch of science that deals with the nature, effects, and interactions of substances that are harmful to the environment.[24][25]

3.10 See also

3.11 References

[1] Pharmacokinetics. (2006). In *Mosby's Dictionary of Medicine, Nursing & Health Professions*. Philadelphia, PA: Elsevier Health Sciences. Retrieved December 11, 2008, from http://www.credoreference.com/entry/6686418

[2] Knights K, Bryant B (2002). *Pharmacology for Health Professionals*. Amsterdam: Elsevier. ISBN 0-7295-3664-5.

[3] Koch HP, Ritschel WA (1986). "Liberation". *Synopsis der Biopharmazie und Pharmakokinetik* (in German). Landsberg, München: Ecomed. pp. 99–131. ISBN 3-609-64970-4.

[4] Ruiz-Garcia A, Bermejo M, Moss A, Casabo VG (February 2008). "Pharmacokinetics in drug discovery". *Journal of Pharmaceutical Sciences*. **97** (2): 654–90. PMID 17630642. doi:10.1002/jps.21009.

[5] AGAH working group PHARMACOKINETICS (2004-02-16). "Collection of terms, symbols, equations, and explanations of common pharmacokinetic and pharmacodynamic parameters and some statistical functions" (PDF). Arbeitsgemeinschaft für Angewandte Humanpharmakologie (AGAH) (Association for Applied Human Pharmacology). Retrieved 2011-04-04.

[6] Milo Gibaldi, Donald Perrier. *Farmacocinética*Reverté 1982 pages 1-10. ISBN 84-291-5535-X, 9788429155358

[7] Gill SC, Moon-Mcdermott L, Hunt TL, Deresinski S, Blaschke T, Sandhaus RA (Sep 1999). "Phase I Pharmacokinetics of Liposomal Amikacin (MiKasome) in Human Subjects: Dose Dependence and Urinary Clearance". *Abstr Intersci Conf Antimicrob Agents Chemother*. **39**: 33 (abstract no. 1195).

[8] Weiner D, Gabrielsson J (2000). "PK24 - Non-linear kinetics - flow II". *Pharmacokinetic/pharmacodynamic data analysis: concepts and applications*. Apotekarsocieteten. pp. 527–36. ISBN 91-86274-92-9.

[9] Niazi S (March 1976). "Volume of distribution as a function of time". *Journal of Pharmaceutical Sciences*. **65** (3): 452–4. PMID 1263103. doi:10.1002/jps.2600650339.

[10] Wesolowski CA, Wesolowski MJ, Babyn PS, Wanasundara SN (2016). "Time Varying Apparent Volume of Distribution and Drug Half-Lives Following Intravenous Bolus Injections". *PLOS ONE*. **11** (7): e0158798. PMC 4942076 . PMID 27403663. doi:10.1371/journal.pone.0158798.

[11] Kim DK, Lee JC, Lee H, Joo KW, Oh KH, Kim YS, Yoon HJ, Kim HC (April 2016). "Calculation of the clearance requirements for the development of a hemodialysis-based wearable artificial kidney". *Hemodialysis International. International Symposium on Home Hemodialysis*. **20** (2): 226–34. PMID 26245302. doi:10.1111/hdi.12343.

[12] Michael E. Winter, Mary Anne Koda-Kimple, Lloyd Y. Young, Emilio Pol Yanguas *Farmacocinética clínica básica* Ediciones Díaz de Santos, 1994 pgs. 8-14 ISBN 84-7978-147-5, 9788479781477 (in Spanish)

[13] Baaske P, Wienken CJ, Reineck P, Duhr S, Braun D (March 2010). "Optical thermophoresis for quantifying the buffer dependence of aptamer binding". *Angewandte Chemie*. **49** (12): 2238–41. PMID 20186894. doi:10.1002/anie.200903998. Lay summary – *Phsyorg.com*.

[14] Wienken CJ, Baaske P, Rothbauer U, Braun D, Duhr S (October 2010). "Protein-binding assays in biological liquids using microscale thermophoresis". *Nature Communications*. **1** (7): 100. Bibcode:2010NatCo...1E.100W. PMID 20981028. doi:10.1038/ncomms1093.

[15] Hsieh Y, Korfmacher WA (June 2006). "Increasing speed and throughput when using HPLC-MS/MS systems for drug metabolism and pharmacokinetic screening". *Current Drug Metabolism*. **7** (5): 479–89. PMID 16787157. doi:10.2174/138920006777697963.

[16] Covey TR, Lee ED, Henion JD (October 1986). "High-speed liquid chromatography/tandem mass spectrometry for the determination of drugs in biological samples". *Analytical Chemistry*. **58** (12): 2453–60. PMID 3789400. doi:10.1021/ac00125a022.

[17] Covey TR, Crowther JB, Dewey EA, Henion JD (February 1985). "Thermospray liquid chromatography/mass spectrometry determination of drugs and their metabolites in biological fluids". *Analytical Chemistry*. **57** (2): 474–81. PMID 3977076. doi:10.1021/ac50001a036.

[18] Committee for Medicinal Products for Human Use (CHMP) (December 2009). "ICH guideline M3(R2) on non-clinical safety studies for the conduct of human clinical trials and marketing authorisation for pharmaceuticals" (PDF). European Medicines Agency, Evaluation of Medicines for Human Use. EMA/CPMP/ICH/286/1995. Retrieved 4 May 2013.

[19] Sheiner LB, Rosenberg B, Marathe VV (October 1977). "Estimation of population characteristics of pharmacokinetic parameters from routine clinical data". *Journal of Pharmacokinetics and Biopharmaceutics*. **5** (5): 445–79. PMID 925881. doi:10.1007/BF01061728.

[20] Sheiner LB, Beal S, Rosenberg B, Marathe VV (September 1979). "Forecasting individual pharmacokinetics". *Clinical Pharmacology and Therapeutics*. **26** (3): 294–305. PMID 466923. doi:10.1002/cpt1979263294.

[21] Bonate PL (October 2005). "Recommended reading in population pharmacokinetic pharmacodynamics". *The AAPS Journal*. **7** (2): E363–73. PMC 2750974 . PMID 16353916. doi:10.1208/aapsj070237.

[22] R. García del Moral, M. Andújar y F. O'Valle Mecanismos de nefrotoxicidad por ciclosporina A a nivel celular (in Spanish). NEFROLOGIA. Vol. XV. Supplement 1, 1995. Consulted 23 February 2008.

[23] Joaquín Herrera Carranza Manual de farmacia clínica y Atención Farmacéutica (in Spanish). Published by Elsevier España, 2003; page 159. ISBN 84-8174-658-4

[24] Jager T, Albert C, Preuss TG, Ashauer R (April 2011). "General unified threshold model of survival--a toxicokinetic-toxicodynamic framework for ecotoxicology". *Environmental Science & Technology*. **45** (7): 2529–40. Bibcode:2011EnST...45.2529J. PMID 21366215. doi:10.1021/es103092a.

[25] Ashauer R. "Toxicokinetic-Toxicodynamic Models - Ecotoxicology and Models". Swiss Federal Institute of Aquatic Science and Technology. Retrieved 2011-12-03.

3.12 External links

3.12.1 Software

Noncompartmental

- Freeware: bear and PK for R
- Commercial: MLAB, EquivTest, Kinetica, MATLAB/SimBiology, Phoenix/WinNonlin, PK Solutions, RapidNCA.

Compartment based

- Freeware: ADAPT, Boomer (GUI), SBPKPD.org (Systems Biology Driven Pharmacokinetics and Pharmacodynamics), WinSAAM, PKfit for R, PharmaCalc and PharmaCalcCL, Java applications.
- Commercial: Imalytics, Kinetica, MATLAB/SimBiology, Phoenix/WinNonlin, PK Solutions, PottersWheel, ProcessDB, SAAM II.

Physiologically based

- Freeware: MCSim

- Commercial: acslX, Cloe PK, GastroPlus, MATLAB/SimBiology, PK-Sim, ProcessDB, Simcyp, Entelos PhysioLab Phoenix/WinNonlin, ADME Workbench.

Population PK

- Freeware: WinBUGS, ADAPT, S-ADAPT / SADAPT-TRAN, Boomer, PKBugs, Pmetrics for R.
- Commercial: Kinetica, MATLAB/SimBiology, Monolix, NONMEM, Phoenix/NLME, PopKinetics for SAAM II, USC*PACK, Navigator Workbench.

Simulation

All model based software above.

- Freeware: COPASI, Berkeley Madonna, MEGen.

3.12.2 Educational centres

Global centres with the highest profiles for providing in-depth training include the Universities of Buffalo, Florida, Gothenburg, Leiden, Otago, San Francisco, Beijing, Tokyo, Uppsala, Washington, Manchester, Monash University, and University of Sheffield.[1]

[1] Tucker GT (June 2012). "Research priorities in pharmacokinetics". *British Journal of Clinical Pharmacology*. **73** (6): 924–6. PMC 3391520 ô. PMID 22360418. doi:10.1111/j.1365-2125.2012.04238.x.

Chapter 4

Germline mutation

A **germline mutation** is any detectable and heritable variation in the lineage of germ cells. Mutations in these cells are transmitted to offspring, while, on the other hand, those in somatic cells are not. A germline mutation gives rise to a **constitutional mutation** in the offspring, that is, a mutation that is present in virtually every cell. A constitutional mutation can also occur very soon after fertilisation, or continue from a previous constitutional mutation in a parent.[1]

This distinction is most important in animals, where germ cells are distinct from somatic cells. However, in plants, the reproductive cells in a particular flower will be derived from the same meristem as the cells in that flower and on the stem leading to the flower, which is a different population of cells than those that give rise to the other flowers on the plant. Single-celled organisms have no distinction between germline and somatic tissues.

In animals, mutations are more likely to occur in sperm than in ova, because a larger number of cell divisions are involved in the production of sperm.[2]

Mutations that are not *germline* are somatic mutations, which are also called *acquired mutations*.

4.1 See also

- Designer baby
- Germinal choice technology
- Germline
- Mutation

4.2 References

[1] RB1 Genetics at Daisy's Eye Cancer Fund. Retrieved May 2011

[2] Schizophrenia Risk and the Paternal Germ Line

Chapter 5

Mutation

For other uses, see Mutation (disambiguation).

In biology, a **mutation** is the permanent alteration of the nucleotide sequence of the genome of an organism, virus, or extrachromosomal DNA or other genetic elements.

Mutations result from errors during DNA replication or other types of damage to DNA, which then may undergo error-prone repair (especially microhomology-mediated end joining[1]), or cause an error during other forms of repair,[2][3] or else may cause an error during replication (translesion synthesis). Mutations may also result from insertion or deletion of segments of DNA due to mobile genetic elements.[4][5][6] Mutations may or may not produce discernible changes in the observable characteristics (phenotype) of an organism. Mutations play a part in both normal and abnormal biological processes including: evolution, cancer, and the development of the immune system, including junctional diversity.

The genomes of RNA viruses are based on RNA rather than DNA. The RNA viral genome can be double stranded (as in DNA) or single stranded. In some of these viruses (such as the single stranded human immunodeficiency virus) replication occurs quickly and there are no mechanisms to check the genome for accuracy. This error-prone process often results in mutations.

Mutation can result in many different types of change in sequences. Mutations in genes can either have no effect, alter the product of a gene, or prevent the gene from functioning properly or completely. Mutations can also occur in nongenic regions. One study on genetic variations between different species of *Drosophila* suggests that, if a mutation changes a protein produced by a gene, the result is likely to be harmful, with an estimated 70 percent of amino acid polymorphisms that have damaging effects, and the remainder being either neutral or marginally beneficial.[7] Due to the damaging effects that mutations can have on genes, organisms have mechanisms such as DNA repair to prevent or correct mutations by reverting the mutated sequence back to its original state.[4]

5.1 Description

Mutations can involve the duplication of large sections of DNA, usually through genetic recombination.[8] These duplications are a major source of raw material for evolving new genes, with tens to hundreds of genes duplicated in animal genomes every million years.[9] Most genes belong to larger gene families of shared ancestry, known as homology.[10] Novel genes are produced by several methods, commonly through the duplication and mutation of an ancestral gene, or by recombining parts of different genes to form new combinations with new functions.[11][12]

Here, protein domains act as modules, each with a particular and independent function, that can be mixed together to produce genes encoding new proteins with novel properties.[13] For example, the human eye uses four genes to make structures that sense light: three for cone cell or color vision and one for rod cell or night vision; all four arose from a single ancestral gene.[14] Another advantage of duplicating a gene (or even an entire genome) is that this increases engineering redundancy; this allows one gene in the pair to acquire a new function while the other copy performs the original function.[15][16] Other types of mutation occasionally create new genes from previously noncoding DNA.[17][18]

Changes in chromosome number may involve even larger mutations, where segments of the DNA within chromosomes break and then rearrange. For example, in the Homininae, two chromosomes fused to produce human chromosome 2; this fusion did not occur in the lineage of the other apes, and they retain these separate chromosomes.[19] In evolution, the most important role of such chromosomal rearrangements may be to accelerate the divergence of a population into new species by making populations less likely to interbreed, thereby preserving genetic differences between these populations.[20]

Sequences of DNA that can move about the genome, such as transposons, make up a major fraction of the genetic material of plants and animals, and may have been important in the evolution of genomes.[21] For example, more than a million copies of the Alu sequence are present in the human genome, and these sequences have now been recruited to perform functions such as regulating gene expression.[22] Another effect of these mobile DNA sequences is that when they move within a genome, they can mutate or delete existing genes and thereby produce genetic diversity.[5]

Nonlethal mutations accumulate within the gene pool and increase the amount of genetic variation.[23] The abundance of some genetic changes within the gene pool can be reduced by natural selection, while other "more favorable" mutations may accumulate and result in adaptive changes.

Prodryas persephone, *a Late Eocene butterfly*

For example, a butterfly may produce offspring with new mutations. The majority of these mutations will have no effect; but one might change the color of one of the butterfly's offspring, making it harder (or easier) for predators to see. If this color change is advantageous, the chance of this butterfly's surviving and producing its own offspring are a little better, and over time the number of butterflies with this mutation may form a larger percentage of the population.

Neutral mutations are defined as mutations whose effects do not influence the fitness of an individual. These can accumulate over time due to genetic drift. It is believed that the overwhelming majority of mutations have no significant effect on an organism's fitness.[24] Also, DNA repair mechanisms are able to mend most changes before they become permanent mutations, and many organisms have mechanisms for eliminating otherwise-permanently mutated somatic cells.

Beneficial mutations can improve reproductive success.

5.2 Causes

Main article: Mutagenesis

Four classes of mutations are (1) spontaneous mutations (molecular decay), (2) mutations due to error-prone replication bypass of naturally occurring DNA damage (also called error-prone translesion synthesis), (3) errors introduced during DNA repair, and (4) induced mutations caused by mutagens. Scientists may also deliberately introduce mutant sequences through DNA manipulation for the sake of scientific experimentation.

One 2017 study claimed that 66% of cancer-causing mutations are random, 29% are due to the environment (the studied population spanned 69 countries), and 5% are inherited.[25]

5.2.1 Spontaneous mutation

Spontaneous mutations occur with non-zero probability even given a healthy, uncontaminated cell. They can be characterized by the specific change:[26]

- Tautomerism — A base is changed by the repositioning of a hydrogen atom, altering the hydrogen bonding pattern of that base, resulting in incorrect base pairing during replication.
- Depurination — Loss of a purine base (A or G) to form an apurinic site (AP site).
- Deamination — Hydrolysis changes a normal base to an atypical base containing a keto group in place of the original amine group. Examples include C → U and A → HX (hypoxanthine), which can be corrected by DNA repair mechanisms; and 5MeC (5-methylcytosine) → T, which is less likely to be detected as a mutation because thymine is a normal DNA base.
- Slipped strand mispairing — Denaturation of the new strand from the template during replication, followed by renaturation in a different spot ("slipping"). This can lead to insertions or deletions.

5.2.2 Error-prone replication bypass

There is increasing evidence that the majority of spontaneously arising mutations are due to error-prone replication (translesion synthesis) past DNA damage in the template strand. Naturally occurring oxidative DNA damages arise at least 10,000 times per cell per day in humans and 50,000 times or more per cell per day in rats.[27] In mice, the majority of mutations are caused by translesion synthesis.[28] Likewise, in yeast, Kunz et al.[29] found that more than 60% of the spontaneous single base pair substitutions and deletions were caused by translesion synthesis.

5.2.3 Errors introduced during DNA repair

See also: DNA damage (naturally occurring) and DNA repair

Although naturally occurring double-strand breaks occur at a relatively low frequency in DNA, their repair often causes mutation. Non-homologous end joining (NHEJ) is a major pathway for repairing double-strand breaks. NHEJ involves removal of a few nucleotides to allow somewhat inaccurate alignment of the two ends for rejoining followed by addition of nucleotides to fill in gaps. As a consequence, NHEJ often introduces mutations.[30]

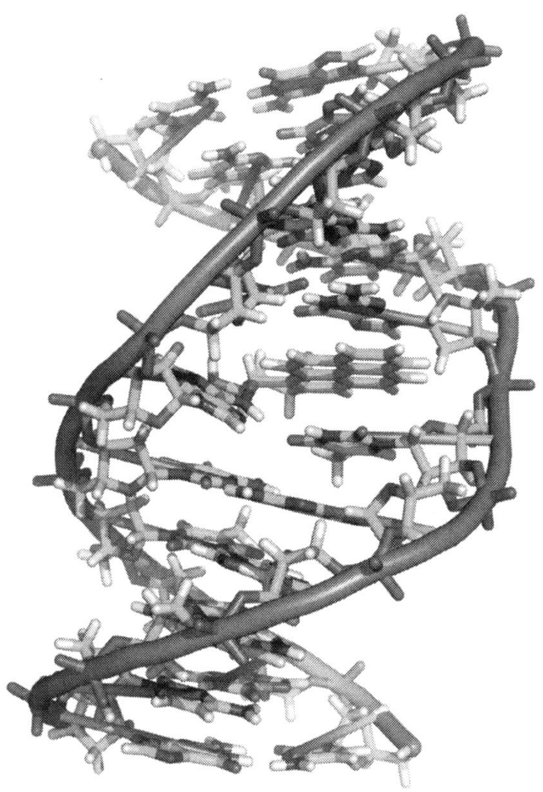

A covalent adduct between the metabolite of benzo[a]pyrene, the major mutagen in tobacco smoke, and DNA[31]

5.2.4 Induced mutation

Induced mutations are alterations in the gene after it has come in contact with mutagens and environmental causes.

Induced mutations on the molecular level can be caused by:

- Chemicals
  - Hydroxylamine
  - Base analogs (e.g., Bromodeoxyuridine (BrdU))
  - Alkylating agents (e.g., *N*-ethyl-*N*-nitrosourea (ENU)). These agents can mutate both replicating and non-replicating DNA. In contrast, a base analog can mutate the DNA only when the analog is incorporated in replicating the DNA. Each of these classes of chemical mutagens has certain effects that then lead to transitions, transversions, or deletions.
  - Agents that form DNA adducts (e.g., ochratoxin A)[32]
  - DNA intercalating agents (e.g., ethidium bromide)
  - DNA crosslinkers
  - 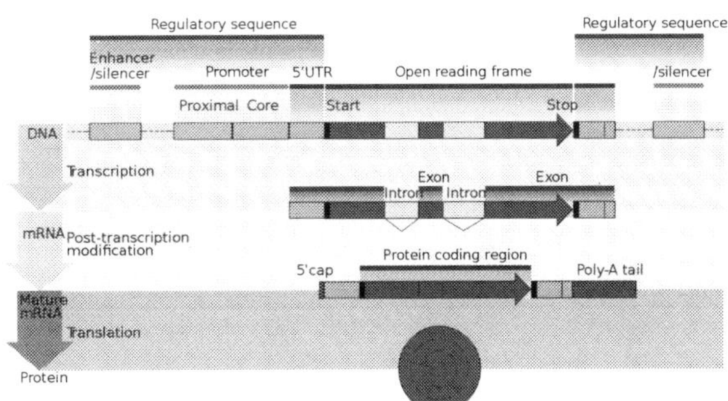

    This figure depicts the following processes of transcription, splicing, and translation of a eukaryotic gene.

    Oxidative damage
  - Nitrous acid converts amine groups on A and C to diazo groups, altering their hydrogen bonding patterns, which leads to incorrect base pairing during replication.
- Radiation
  - Ultraviolet light (UV) (non-ionizing radiation). Two nucleotide bases in DNA—cytosine and thymine—are most vulnerable to radiation that can change their properties. UV light can induce adjacent pyrimidine bases in a DNA strand to become covalently joined as a pyrimidine dimer. UV radiation, in particular longer-wave UVA, can also cause oxidative damage to DNA.[33]

- Ionizing radiation. Exposure to ionizing radiation, such as gamma radiation, can result in mutation, possibly resulting in cancer or death.

5.3 Classification of types

See also: Chromosome abnormality

5.3.1 By effect on structure

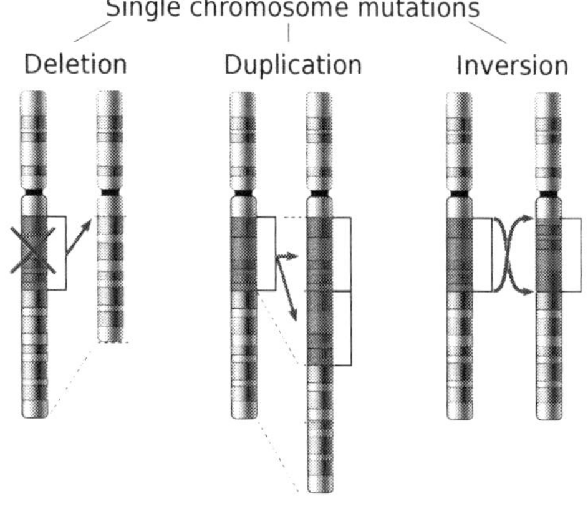

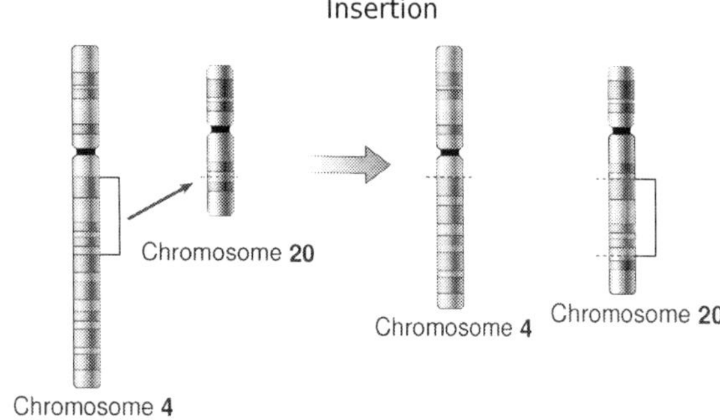

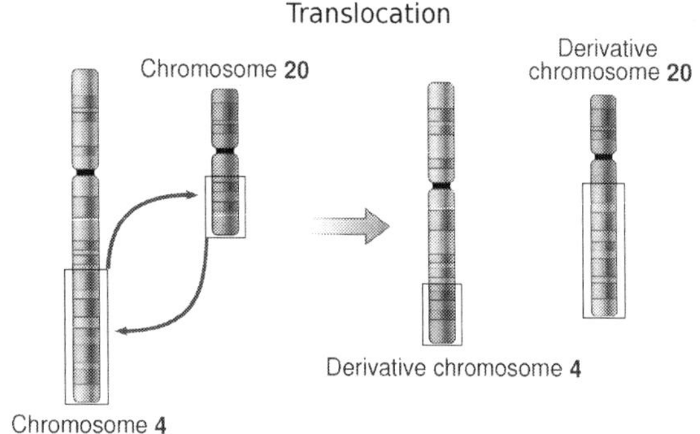

Illustrations of five types of chromosomal mutations.

The sequence of a gene can be altered in a number of ways[35]. Gene mutations have varying effects on health depending on where they occur and whether they alter the function of essential proteins. Mutations in the structure of genes can be classified as:

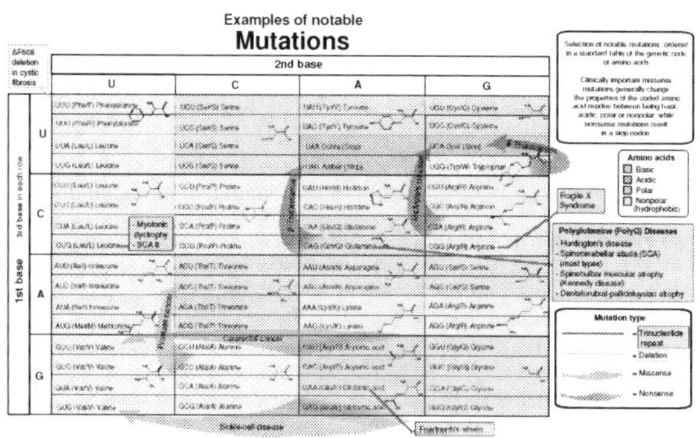

Selection of disease-causing mutations, in a standard table of the genetic code of amino acids.[34]

- Small-scale mutations, such as those affecting a small gene in one or a few nucleotides, including:
    - **Substitution mutations**, often caused by chemicals or malfunction of DNA replication, exchange a single nucleotide for another.[36] These changes are classified as transitions or transversions.[37] Most common is the transition that exchanges a purine for a purine (A ↔ G) or a pyrimidine for a pyrimidine, (C ↔ T). A transition can be caused by nitrous acid, base mispairing, or mutagenic base analogs such as BrdU. Less common is a transversion, which exchanges a purine for a pyrimidine or a pyrimidine for a purine (C/T ↔ A/G). An example of a transversion is the conversion of adenine (A) into a cytosine (C). A point mutation are modifications of single base pairs of DNA or other small base pairs within a gene. A point mutation can be reversed by another point mutation, in which the nucleotide is changed back to its original state (true reversion) or by second-site reversion (a complementary mutation elsewhere that results in regained gene functionality). Point mutations that occur within the protein coding region of a gene may be classified into three kinds, depending upon what the erroneous codon codes for:
        - Silent mutations, which code for the same (or a sufficiently similar) amino acid.
        - Missense mutations, which code for a different amino acid.
        - Nonsense mutations, which code for a stop codon and can truncate the protein.
    - **Insertions** add one or more extra nucleotides into the DNA. They are usually caused by

transposable elements, or errors during replication of repeating elements. Insertions in the coding region of a gene may alter splicing of the mRNA (splice site mutation), or cause a shift in the reading frame (frameshift), both of which can significantly alter the gene product. Insertions can be reversed by excision of the transposable element.

- **Deletions** remove one or more nucleotides from the DNA. Like insertions, these mutations can alter the reading frame of the gene. In general, they are irreversible: Though exactly the same sequence might in theory be restored by an insertion, transposable elements able to revert a very short deletion (say 1–2 bases) in *any* location either are highly unlikely to exist or do not exist at all.

- Large-scale mutations in chromosomal structure, including:
  - **Amplifications** (or gene duplications) leading to multiple copies of all chromosomal regions, increasing the dosage of the genes located within them.
  - **Deletions** of large chromosomal regions, leading to loss of the genes within those regions.
  - Mutations whose effect is to juxtapose previously separate pieces of DNA, potentially bringing together separate genes to form functionally distinct fusion genes (e.g., bcr-abl). These include:
    - **Chromosomal translocations**: interchange of genetic parts from nonhomologous chromosomes.
    - **Interstitial deletions**: an intra-chromosomal deletion that removes a segment of DNA from a single chromosome, thereby apposing previously distant genes. For example, cells isolated from a human astrocytoma, a type of brain tumor, were found to have a chromosomal deletion removing sequences between the Fused in Glioblastoma (FIG) gene and the receptor tyrosine kinase (ROS), producing a fusion protein (FIG-ROS). The abnormal FIG-ROS fusion protein has constitutively active kinase activity that causes oncogenic transformation (a transformation from normal cells to cancer cells).
    - **Chromosomal inversions**: reversing the orientation of a chromosomal segment.
  - **Loss of heterozygosity**: loss of one allele, either by a deletion or a genetic recombination event, in an organism that previously had two different alleles.

5.3.2 By effect on function

See also: Behavior mutation

- **Loss-of-function mutations**, also called **inactivating mutations**, result in the gene product having less or no function (being partially or wholly inactivated). When the allele has a complete loss of function (null allele), it is often called an amorph or amorphic mutation in the Muller's morphs schema. Phenotypes associated with such mutations are most often recessive. Exceptions are when the organism is haploid, or when the reduced dosage of a normal gene product is not enough for a normal phenotype (this is called haploinsufficiency).

- **Gain-of-function mutations**, also called **activating mutations**, change the gene product such that its effect gets stronger (enhanced activation) or even is superseded by a different and abnormal function. When the new allele is created, a heterozygote containing the newly created allele as well as the original will express the new allele; genetically this defines the mutations as dominant phenotypes. Several of Muller's morphs correspond to gain of function, including hypermorph and neomorph.

- **Dominant negative mutations** (also called **antimorphic mutations**) have an altered gene product that acts antagonistically to the wild-type allele. These mutations usually result in an altered molecular function (often inactive) and are characterized by a dominant or semi-dominant phenotype. In humans, dominant negative mutations have been implicated in cancer (e.g., mutations in genes p53,[38] ATM,[39] CEBPA[40] and PPARgamma[41]). Marfan syndrome is caused by mutations in the *FBN1* gene, located on chromosome 15, which encodes fibrillin-1, a glycoprotein component of the extracellular matrix.[42] Marfan syndrome is also an example of dominant negative mutation and haploinsufficiency.[43][44]

- **Hypomorph**, after Mullarian classification is altered gene product that acts with decreased gene expression compared to wild type allele.

- **Neomorph** is characterized by the control of new protein product synthesis.

- **Lethal mutations** are mutations that lead to the death of the organisms that carry the mutations.
- A **back mutation** or **reversion** is a point mutation that restores the original sequence and hence the original phenotype.[45]

5.3.3 By effect on fitness

See also: Fitness (biology)

In applied genetics, it is usual to speak of mutations as either harmful or beneficial.

- A **harmful**, or **deleterious**, mutation decreases the fitness of the organism.
- A **beneficial**, or **advantageous** mutation increases the fitness of the organism. Mutations that promotes traits that are desirable, are also called beneficial. In theoretical population genetics, it is more usual to speak of mutations as deleterious or advantageous than harmful or beneficial.
- A **neutral mutation** has no harmful or beneficial effect on the organism. Such mutations occur at a steady rate, forming the basis for the molecular clock. In the neutral theory of molecular evolution, neutral mutations provide genetic drift as the basis for most variation at the molecular level.
- A **nearly neutral mutation** is a mutation that may be slightly deleterious or advantageous, although most nearly neutral mutations are slightly deleterious.

Distribution of fitness effects

Attempts have been made to infer the distribution of fitness effects (DFE) using mutagenesis experiments and theoretical models applied to molecular sequence data. DFE, as used to determine the relative abundance of different types of mutations (i.e., strongly deleterious, nearly neutral or advantageous), is relevant to many evolutionary questions, such as the maintenance of genetic variation,[46] the rate of genomic decay,[47] the maintenance of outcrossing sexual reproduction as opposed to inbreeding[48] and the evolution of sex and genetic recombination.[49] In summary, the DFE plays an important role in predicting evolutionary dynamics.[50][51] A variety of approaches have been used to study the DFE, including theoretical, experimental and analytical methods.

- **Mutagenesis experiment**: The direct method to investigate the DFE is to induce mutations and then measure the mutational fitness effects, which has already been done in viruses, bacteria, yeast, and *Drosophila*. For example, most studies of the DFE in viruses used site-directed mutagenesis to create point mutations and measure relative fitness of each mutant.[52][53][54][55] In *Escherichia coli*, one study used transposon mutagenesis to directly measure the fitness of a random insertion of a derivative of Tn10.[56] In yeast, a combined mutagenesis and deep sequencing approach has been developed to generate high-quality systematic mutant libraries and measure fitness in high throughput.[57] However, given that many mutations have effects too small to be detected[58] and that mutagenesis experiments can detect only mutations of moderately large effect; DNA sequence data analysis can provide valuable information about these mutations.

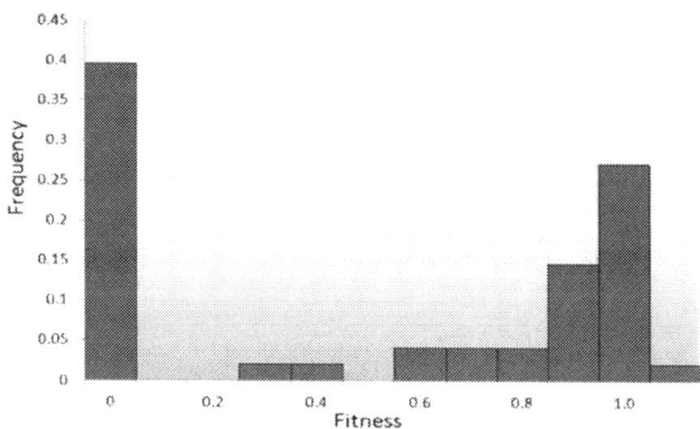

The distribution of fitness effects (DFE) of mutations in vesicular stomatitis virus. In this experiment, random mutations were introduced into the virus by site-directed mutagenesis, and the fitness of each mutant was compared with the ancestral type. A fitness of zero, less than one, one, more than one, respectively, indicates that mutations are lethal, deleterious, neutral, and advantageous.[52]

- **Molecular sequence analysis**: With rapid development of DNA sequencing technology, an enormous amount of DNA sequence data is available and even more is forthcoming in the future. Various methods have been developed to infer the DFE from DNA sequence data.[59][60][61][62] By examining DNA sequence differences within and between species, we are able to infer various characteristics of the DFE for neutral, deleterious and advantageous mutations.[23] To be specific, the DNA sequence analysis approach allows us to estimate the effects of mutations with very small effects, which are hardly detectable through mutagenesis experiments.

One of the earliest theoretical studies of the distribution of fitness effects was done by Motoo Kimura, an influential theoretical population geneticist. His neutral theory of molecular evolution proposes that most novel mutations will be highly deleterious, with a small fraction being neutral.[63][64] Hiroshi Akashi more recently proposed a bimodal model for the DFE, with modes centered around highly deleterious and neutral mutations.[65] Both theories agree that the vast majority of novel mutations are neutral or deleterious and that advantageous mutations are rare, which has been supported by experimental results. One example is a study done on the DFE of random mutations in vesicular stomatitis virus.[52] Out of all mutations, 39.6% were lethal, 31.2% were non-lethal deleterious, and 27.1% were neutral. Another example comes from a high throughput mutagenesis experiment with yeast.[57] In this experiment it was shown that the overall DFE is bimodal, with a cluster of neutral mutations, and a broad distribution of deleterious mutations.

Though relatively few mutations are advantageous, those that are play an important role in evolutionary changes.[66] Like neutral mutations, weakly selected advantageous mutations can be lost due to random genetic drift, but strongly selected advantageous mutations are more likely to be fixed. Knowing the DFE of advantageous mutations may lead to increased ability to predict the evolutionary dynamics. Theoretical work on the DFE for advantageous mutations has been done by John H. Gillespie[67] and H. Allen Orr.[68] They proposed that the distribution for advantageous mutations should be exponential under a wide range of conditions, which, in general, has been supported by experimental studies, at least for strongly selected advantageous mutations.[69][70][71]

In general, it is accepted that the majority of mutations are neutral or deleterious, with rare mutations being advantageous; however, the proportion of types of mutations varies between species. This indicates two important points: first, the proportion of effectively neutral mutations is likely to vary between species, resulting from dependence on effective population size; second, the average effect of deleterious mutations varies dramatically between species.[23] In addition, the DFE also differs between coding regions and noncoding regions, with the DFE of noncoding DNA containing more weakly selected mutations.[23]

5.3.4 By impact on protein sequence

- A **frameshift mutation** is a mutation caused by insertion or deletion of a number of nucleotides that is not evenly divisible by three from a DNA sequence. Due to the triplet nature of gene expression by codons, the insertion or deletion can disrupt the reading frame, or the grouping of the codons, resulting in a completely different translation from the original.[72] The earlier in the sequence the deletion or insertion occurs, the more altered the protein produced is.
    - For example, the code CCU GAC UAC CUA codes for the amino acids Proline, Aspartic, Tyrosine, and Leucine. If the U in CCU was deleted, the resulting sequence would be CCG ACU ACC UA, which would instead code for Proline, Threonine, and Threonine.

  In contrast, any insertion or deletion that is evenly divisible by three is termed an *in-frame mutation*

- A **nonsense mutation** is a point mutation in a sequence of DNA that results in a premature stop codon, or a *nonsense codon* in the transcribed mRNA, and possibly a truncated, and often nonfunctional protein product. This sort of mutation has been linked to different mutations, such as congenital adrenal hyperplasia. (See Stop codon.)
- **Missense mutations** or *nonsynonymous mutations* are types of point mutations where a single nucleotide is changed to cause substitution of a different amino acid. This in turn can render the resulting protein nonfunctional. Such mutations are responsible for diseases such as Epidermolysis bullosa, sickle-cell disease, and SOD1-mediated ALS.[73]
- A **neutral mutation** is a mutation that occurs in an amino acid codon that results in the use of a different, but chemically similar, amino acid. The similarity between the two is enough that little or no change is often rendered in the protein. For example, a change from AAA to AGA will encode arginine, a chemically similar molecule to the intended lysine.
- **Silent mutations** are mutations that do not result in a change to the amino acid sequence of a protein but do change the nucleotide sequence, unless the changed amino acid is sufficiently similar to the original. They may occur in a region that does not code for a protein, or they may occur within a codon in a manner that does not alter the final amino acid sequence. Silent mutations are also called *silent substitutions*, since they are not palpable changes as the changes in phenotype. The phrase *silent mutation* is often used interchangeably with the phrase *synonymous mutation*; however, synonymous mutations are a subcategory of the former, occurring only within exons (and necessarily exactly preserving the amino acid sequence of the protein). Synonymous mutations occur due to the degenerate nature of the genetic code. There can also be silent mutations in nucleotides outside of the coding regions, such as the introns, because the exact nucleotide sequence is not as crucial as it is in the coding regions.

5.3.5 By inheritance

A mutation has caused this garden moss rose to produce flowers of different colors. This is a somatic mutation that may also be passed on in the germline.

In multicellular organisms with dedicated reproductive cells, mutations can be subdivided into germline mutations, which can be passed on to descendants through their reproductive cells, and somatic mutations (also called acquired mutations),[74] which involve cells outside the dedicated reproductive group and which are not usually transmitted to descendants.

A germline mutation gives rise to a *constitutional mutation* in the offspring, that is, a mutation that is present in every cell. A constitutional mutation can also occur very soon after fertilisation, or continue from a previous constitutional mutation in a parent.[75]

The distinction between germline and somatic mutations is important in animals that have a dedicated germline to produce reproductive cells. However, it is of little value in understanding the effects of mutations in plants, which lack dedicated germline. The distinction is also blurred in those animals that reproduce asexually through mechanisms such as budding, because the cells that give rise to the daughter organisms also give rise to that organism's germline. A new germline mutation that was not inherited from either parent is called a *de novo* mutation.

Diploid organisms (e.g., humans) contain two copies of each gene—a paternal and a maternal allele. Based on the occurrence of mutation on each chromosome, we may classify mutations into three types.

- A **heterozygous mutation** is a mutation of only one allele.
- A **homozygous mutation** is an identical mutation of both the paternal and maternal alleles.
- **Compound heterozygous** mutations or a **genetic compound** consists of two different mutations in the paternal and maternal alleles.[76]

A **wild type** or **homozygous non-mutated** organism is one in which neither allele is mutated.

5.3.6 Special classes

- **Conditional mutation** is a mutation that has wild-type (or less severe) phenotype under certain "permissive" environmental conditions and a mutant phenotype under certain "restrictive" conditions. For example, a temperature-sensitive mutation can cause cell death at high temperature (restrictive condition), but might have no deleterious consequences at a lower temperature (permissive condition).[77] These mutations are non-autonomous, as their manifestation depends upon presence of certain conditions, as opposed to other mutations which appear autonomously.[78] The permissive conditions may be temperature,[79] certain chemicals,[80] light[80] or mutations in other parts of the genome.[78] *In vivo* mechanisms like transcriptional switches can create conditional mutations. For instance, association of Steroid Binding Domain can create a transcriptional switch that can change the expression of a gene based on the presence of a steroid ligand.[81] Conditional mutations have applications in research as they allow control over gene expression. This is especially useful studying diseases in adults by allowing expression after a certain period of growth, thus eliminating the deleterious effect of gene expression seen during stages of development in model organisms.[80] DNA Recombinase systems like Cre-Lox Recombination used in association with promoters that are activated under certain conditions can generate conditional mutations. Dual Recombinase technology can be used to induce multiple conditional mutations to study the diseases which manifest as a result of simultaneous mutations in multiple genes.[80] Certain inteins have been identified which splice only at certain permissive temperatures, leading to improper protein synthesis and thus, loss of function mutations at other temperatures.[82] Conditional mutations may also be used in genetic studies associated with ageing, as the expression can be changed after a certain time period in the organism's lifespan.[79]

- **Replication timing quantitative trait loci affects DNA replication.**

5.3.7 Nomenclature

In order to categorize a mutation as such, the "normal" sequence must be obtained from the DNA of a "normal" or "healthy" organism (as opposed to a "mutant" or "sick" one), it should be identified and reported; ideally, it should be made publicly available for a straightforward nucleotide-by-nucleotide comparison, and agreed upon by the scientific community or by a group of expert geneticists and biologists, who have the responsibility of establishing the *standard* or so-called "consensus" sequence. This step requires a tremendous scientific effort. Once the consensus sequence is known, the mutations in a genome can be pinpointed, described, and classified. The committee of the Human Genome Variation Society (HGVS) has developed the standard human sequence variant nomenclature,[83] which should be used by researchers and DNA diagnostic centers to generate unambiguous mutation descriptions. In principle, this nomenclature can also be used to describe mutations in other organisms. The nomenclature specifies the type of mutation and base or amino acid changes.

- Nucleotide substitution (e.g., 76A>T) — The number is the position of the nucleotide from the 5' end; the first letter represents the wild-type nucleotide, and the second letter represents the nucleotide that replaced the wild type. In the given example, the adenine at the 76th position was replaced by a thymine.
    - If it becomes necessary to differentiate between mutations in genomic DNA, mitochondrial DNA, and RNA, a simple convention is used. For example, if the 100th base of a nucleotide sequence mutated from G to C, then it would be written as g.100G>C if the mutation occurred in genomic DNA, m.100G>C if the mutation occurred in mitochondrial DNA, or r.100g>c if the mutation occurred in RNA. Note that, for mutations in RNA, the nucleotide code is written in lower case.
- Amino acid substitution (e.g., D111E) — The first letter is the one letter code of the wild-type amino acid, the number is the position of the amino acid from the N-terminus, and the second letter is the one letter code of the amino acid present in the mutation. Nonsense mutations are represented with an X for the second amino acid (e.g. D111X).
- Amino acid deletion (e.g., ΔF508) — The Greek letter Δ (delta) indicates a deletion. The letter refers to the amino acid present in the wild type and the number is the position from the N terminus of the amino acid were it to be present as in the wild type.

5.4 Mutation rates

Further information: Mutation rate

Mutation rates vary substantially across species, and the evolutionary forces that generally determine mutation are the subject of ongoing investigation.

1.

5.5 Harmful mutations

Changes in DNA caused by mutation can cause errors in protein sequence, creating partially or completely non-functional proteins. Each cell, in order to function correctly, depends on thousands of proteins to function in the right places at the right times. When a mutation alters a protein that plays a critical role in the body, a medical condition can result. Some mutations alter a gene's DNA base sequence but do not change the function of the protein made by the gene. One study on the comparison of genes between different species of *Drosophila* suggests that if a mutation does change a protein, this will probably be harmful, with an estimated 70 percent of amino acid polymorphisms having damaging effects, and the remainder being either neutral or weakly beneficial.[7] Studies have shown that only 7% of point mutations in noncoding DNA of yeast are deleterious and 12% in coding DNA are deleterious. The rest of the mutations are either neutral or slightly beneficial.[84]

If a mutation is present in a germ cell, it can give rise to offspring that carries the mutation in all of its cells. This is the case in hereditary diseases. In particular, if there is a mutation in a DNA repair gene within a germ cell, humans carrying such germline mutations may have an increased risk of cancer. A list of 34 such germline mutations is given in the article DNA repair-deficiency disorder. An example of one is albinism, a mutation that occurs in the OCA1 or OCA2 gene. Individuals with this disorder are more prone to many types of cancers, other disorders and have impaired vision. On the other hand, a mutation may occur in a somatic cell of an organism. Such mutations will be present in all descendants of this cell within the same organism, and certain mutations can cause the cell to become malignant, and, thus, cause cancer.[85]

A DNA damage can cause an error when the DNA is replicated, and this error of replication can cause a gene mutation that, in turn, could cause a genetic disorder. DNA damages are repaired by the DNA repair system of the cell. Each cell has a number of pathways through which enzymes recognize and repair damages in DNA. Because DNA can be damaged in many ways, the process of DNA repair is an important way in which the body protects itself from disease. Once DNA damage has given rise to a mutation, the mutation cannot be repaired. DNA repair pathways can only recognize and act on "abnormal" structures in the DNA. Once a mutation occurs in a gene sequence it then has normal DNA structure and cannot be repaired.

5.6 Beneficial mutations

Although mutations that cause changes in protein sequences can be harmful to an organism, on occasions the effect may be positive in a given environment. In this case, the mutation may enable the mutant organism to withstand particular environmental stresses better than wild-type organisms, or reproduce more quickly. In these cases a mutation will tend to become more common in a population through natural selection.

For example, a specific 32 base pair deletion in human CCR5 (CCR5-Δ32) confers HIV resistance to homozygotes and delays AIDS onset in heterozygotes.[86] One possible explanation of the etiology of the relatively high frequency of CCR5-Δ32 in the European population is that it conferred resistance to the bubonic plague in mid-14th century Europe. People with this mutation were more likely to survive infection; thus its frequency in the population increased.[87] This theory could explain why this mutation is not found in Southern Africa, which remained untouched by bubonic plague. A newer theory suggests that the selective pressure on the CCR5 Delta 32 mutation was caused by smallpox instead of the bubonic plague.[88]

An example of a harmful mutation is sickle-cell disease, a blood disorder in which the body produces an abnormal type of the oxygen-carrying substance hemoglobin in the red blood cells. One-third of all indigenous inhabitants of Sub-Saharan Africa carry the gene, because, in areas where malaria is common, there is a survival value in carrying only a single sickle-cell gene (sickle cell trait).[89] Those with only one of the two alleles of the sickle-cell disease are more resistant to malaria, since the infestation of the malaria *Plasmodium* is halted by the sickling of the cells that it infests.

5.7 Prion mutations

Prions are proteins and do not contain genetic material. However, prion replication has been shown to be subject to mutation and natural selection just like other forms of replication.[90] The human gene PRNP codes for the major prion protein, PrP, and is subject to mutations that can give rise to disease-causing prions.

5.8 Somatic mutations

Main article: Loss of heterozygosity
See also: Carcinogenesis

A change in the genetic structure that is not inherited from a parent, and also not passed to offspring, is called a *somatic cell genetic mutation* or *acquired mutation*.[74] Somatic mutations are not inherited because this type of mutation does not affect the reproductive cells (sperm and egg) but it can affect any other type of cell. These types of mutations are usually prompted by environmental causes, such as ultraviolet radiation or any exposure to certain harmful chemicals. This type of mutation can cause various diseases; a common one is cancer.[91]

With plants it is another case since some of the somatic mutations can be bred without the need of the seed production, for example, grafting and stem cuttings. These type of mutations in plants have brought new specimen of fruits, such as the "Delicious" apple and the "Washington" navel orange.[92]

5.9 Amorphic mutations

An Amorph (gene), a term utilized by Muller in 1932, is a mutated allele, which has lost the ability of the parent (whether wild type or any other type) allele to encode any functional protein. An amorphic mutation may be caused by the replacement of an amino acid that deactivates an enzyme or by the deletion of part of a gene that produces the enzyme.

Cells with heterozygous mutations (one good copy of gene and one mutated copy) may function normally with the unmutated copy until the good copy has been spontaneously somatically mutated. This kind of mutation happens all the time in living organisms, but it is difficult to measure the rate. Measuring this rate is important in predicting the rate at which people may develop cancer.[93]

Point mutations may arise from spontaneous mutations that occur during DNA replication. The rate of mutation may be increased by mutagens. Mutagens can be physical, such as radiation from UV rays, X-rays or extreme heat, or chemical (molecules that misplace base pairs or disrupt the helical shape of DNA). Mutagens associated with cancers are often studied to learn about cancer and its prevention.

5.10 Hypomorphic and hypermorphic mutations

A hypomorphic mutation is a replacement of amino acids that would hinder enzyme activity, which would reduce the enzyme level but not to the point of complete loss. Usually, hypomorphic mutations are recessive, but haploinsufficiency causes some alleles to be dominant.

A hypermorphic mutation changes the regulation of the gene and causes it to overproduce the gene produce causing a greater than normal enzyme levels. These type of alleles are dominant gain of function type of alleles.[92]

5.11 See also

- Aneuploidy
- Antioxidant
- Budgerigar colour genetics
- Carcinogenesis
- Ecogenetics
- Embryology
- Frameshift mutation
- Homeobox
- Muller's morphs
- Mutagenesis
- Mutant
- Mutation rate
- Polyploidy
- Robertsonian translocation
- Saltation (biology)
- Signature-tagged mutagenesis
- Site-directed mutagenesis
- TILLING (molecular biology)
- Trinucleotide repeat expansion

5.12 References

[1] Sharma S, Javadekar SM, Pandey M, Srivastava M, Kumari R, Raghavan SC (March 2015). "Homology and enzymatic requirements of microhomology-dependent alternative end joining". *Cell Death & Disease*. **6** (3): e1697. PMC 4385936. PMID 25789972. doi:10.1038/cddis.2015.58.

[2] Chen J, Miller BF, Furano AV (April 2014). "Repair of naturally occurring mismatches can induce mutations in flanking DNA". *eLife*. **3**: e02001. PMC 3999860. PMID 24843013. doi:10.7554/elife.02001.

[3] Rodgers K, McVey M (January 2016). "Error-Prone Repair of DNA Double-Strand Breaks". *Journal of Cellular Physiology*. **231** (1): 15–24. PMC 4586358. PMID 26033759. doi:10.1002/jcp.25053.

[4] Bertram JS (December 2000). "The molecular biology of cancer". *Molecular Aspects of Medicine*. Elsevier. **21** (6): 167–223. PMID 11173079. doi:10.1016/S0098-2997(00)00007-8.

[5] Aminetzach YT, Macpherson JM, Petrov DA (July 2005). "Pesticide resistance via transposition-mediated adaptive gene truncation in Drosophila". *Science*. American Association for the Advancement of Science. **309** (5735): 764–7. Bibcode:2005Sci...309..764A. PMID 16051794. doi:10.1126/science.1112699.

[6] Burrus V, Waldor MK (June 2004). "Shaping bacterial genomes with integrative and conjugative elements". *Research in Microbiology*. Elsevier. **155** (5): 376–86. PMID 15207870. doi:10.1016/j.resmic.2004.01.012.

[7] Sawyer SA, Parsch J, Zhang Z, Hartl DL (April 2007). "Prevalence of positive selection among nearly neutral amino acid replacements in Drosophila". *Proceedings of the National Academy of Sciences of the United States of America*. National Academy of Sciences. **104** (16): 6504–10. Bibcode:2007PNAS..104.6504S. PMC 1871816. PMID 17409186. doi:10.1073/pnas.0701572104.

[8] Hastings PJ, Lupski JR, Rosenberg SM, Ira G (August 2009). "Mechanisms of change in gene copy number". *Nature Reviews Genetics*. Nature Publishing Group. **10** (8): 551–64. PMC 2864001. PMID 19597530. doi:10.1038/nrg2593.

[9] Carroll, Grenier & Weatherbee 2005

[10] Harrison PM, Gerstein M (May 2002). "Studying genomes through the aeons: protein families, pseudogenes and proteome evolution". *Journal of Molecular Biology*. Elsevier. **318** (5): 1155–74. PMID 12083509. doi:10.1016/S0022-2836(02)00109-2.

[11] Orengo CA, Thornton JM (July 2005). "Protein families and their evolution-a structural perspective". *Annual Review of Biochemistry*. Annual Reviews. **74**: 867–900. PMID 15954844. doi:10.1146/annurev.biochem.74.082803.133029.

[12] Long M, Betrán E, Thornton K, Wang W (November 2003). "The origin of new genes: glimpses from the young and old". *Nature Reviews Genetics*. Nature Publishing Group. **4** (11): 865–75. PMID 14634634. doi:10.1038/nrg1204.

[13] Wang M, Caetano-Anollés G (January 2009). "The evolutionary mechanics of domain organization in proteomes and the rise of modularity in the protein world". *Structure*. Cell Press. **17** (1): 66–78. PMID 19141283. doi:10.1016/j.str.2008.11.008.

[14] Bowmaker JK (May 1998). "Evolution of colour vision in vertebrates". *Eye*. Nature Publishing Group. **12** (Pt 3b): 541–7. PMID 9775215. doi:10.1038/eye.1998.143.

[15] Gregory TR, Hebert PD (April 1999). "The modulation of DNA content: proximate causes and ultimate consequences". *Genome Research*. Cold Spring Harbor Laboratory Press. **9** (4): 317–24. PMID 10207154. doi:10.1101/gr.9.4.317 (inactive 2017-01-15).

[16] Hurles M (July 2004). "Gene duplication: the genomic trade in spare parts". *PLoS Biology*. Public Library of Science. **2** (7): E206. PMC 449868 ©. PMID 15252449. doi:10.1371/journal.pbio.0020206.

[17] Liu N, Okamura K, Tyler DM, Phillips MD, Chung WJ, Lai EC (October 2008). "The evolution and functional diversification of animal microRNA genes". *Cell Research*. Nature Publishing Group on behalf of the Shanghai Institutes for Biological Sciences. **18** (10): 985–96. PMC 2712117 ©. PMID 18711447. doi:10.1038/cr.2008.278.

[18] Siepel A (October 2009). "Darwinian alchemy: Human genes from noncoding DNA". *Genome Research*. Cold Spring Harbor Laboratory Press. **19** (10): 1693–5. PMC 2765273 ©. PMID 19797681. doi:10.1101/gr.098376.109.

[19] Zhang J, Wang X, Podlaha O (May 2004). "Testing the chromosomal speciation hypothesis for humans and chimpanzees". *Genome Research*. Cold Spring Harbor Laboratory Press. **14** (5): 845–51. PMC 479111 ©. PMID 15123584. doi:10.1101/gr.1891104.

[20] Ayala FJ, Coluzzi M (May 2005). "Chromosome speciation: humans, Drosophila, and mosquitoes". *Proceedings of the National Academy of Sciences of the United States of America*. National Academy of Sciences. 102 Suppl 1 (Suppl 1): 6535–42. Bibcode:2005PNAS..102.6535A. PMC 1131864 ©. PMID 15851677. doi:10.1073/pnas.0501847102.

[21] Hurst GD, Werren JH (August 2001). "The role of selfish genetic elements in eukaryotic evolution". *Nature Reviews Genetics*. Nature Publishing Group. **2** (8): 597–606. PMID 11483984. doi:10.1038/35084545.

[22] Häsler J, Strub K (November 2006). "Alu elements as regulators of gene expression". *Nucleic Acids Research*. Oxford University Press. **34** (19): 5491–7. PMC 1636486 ©. PMID 17020921. doi:10.1093/nar/gkl706.

[23] Eyre-Walker A, Keightley PD (August 2007). "The distribution of fitness effects of new mutations" (PDF). *Nature Reviews Genetics*. Nature Publishing Group. **8** (8): 610–8. PMID 17637733. doi:10.1038/nrg2146.

[24] Bohidar HB (January 2015). *Fundamentals of Polymer Physics and Molecular Biophysics*. Cambridge University Press. ISBN 978-1-316-09302-3.

[25] http://www.npr.org/sections/health-shots/2017/03/23/521219318/cancer-is-partly-caused-by-bad-luck-study-finds

[26] Montelone BA (1998). "Mutation, Mutagens, and DNA Repair". *www-personal.ksu.edu*. Retrieved 2015-10-02.

[27] Bernstein C, Prasad AR, Nfonsam V, Bernstein H. (2013). DNA Damage, DNA Repair and Cancer, New Research Directions in DNA Repair, Prof. Clark Chen (Ed.), ISBN 978-953-51-1114-6, InTech, http://www.intechopen.com/books/new-research-directions-in-dna-repair/dna-damage-dna-repair-and-cancer

[28] Stuart GR, Oda Y, de Boer JG, Glickman BW (March 2000). "Mutation frequency and specificity with age in liver, bladder and brain of lacI transgenic mice". *Genetics*. Genetics Society of America. **154** (3): 1291–300. PMC 1460990 ©. PMID 10757770.

[29] Kunz BA, Ramachandran K, Vonarx EJ (April 1998). "DNA sequence analysis of spontaneous mutagenesis in Saccharomyces cerevisiae". *Genetics*. Genetics Society of America. **148** (4): 1491–505. PMC 1460101 ©. PMID 9560369.

[30] Lieber MR (July 2010). "The mechanism of double-strand DNA break repair by the nonhomologous DNA end-joining pathway". *Annual Review of Biochemistry*. Annual Reviews. **79**: 181–211. PMC 3079308 ©. PMID 20192759. doi:10.1146/annurev.biochem.052308.093131.

[31] Created from PDB 1JDG

[32] Pfohl-Leszkowicz A, Manderville RA (January 2007). "Ochratoxin A: An overview on toxicity and carcinogenicity in animals and humans". *Molecular Nutrition & Food Research*. Wiley-Blackwell. **51** (1): 61–99. PMID 17195275. doi:10.1002/mnfr.200600137.

[33] Kozmin S, Slezak G, Reynaud-Angelin A, Elie C, de Rycke Y, Boiteux S, Sage E (September 2005). "UVA radiation is highly mutagenic in cells that are unable to repair 7,8-dihydro-8-oxoguanine in Saccharomyces cerevisiae". *Proceedings of the National Academy of Sciences of the United States of America*. National Academy of Sciences. **102** (38): 13538–43. Bibcode:2005PNAS..10213538K. PMC 1224634 ©. PMID 16157879. doi:10.1073/pnas.0504497102.

[34] References for the image are found in Wikimedia Commons page at: Commons:File:Notable mutations.svg#References.

[35] Rahman, Nazneen. "The clinical impact of DNA sequence changes". *Transforming Genetic Medicine Initiative*. Retrieved June 27, 2017.

[36] Freese E (April 1959). "THE DIFFERENCE BETWEEN SPONTANEOUS AND BASE-ANALOGUE INDUCED MUTATIONS OF PHAGE T4". *Proceedings of the National Academy of Sciences of the United States of America*. National Academy of Sciences. **45** (4): 622–33. PMC 222607 ⓐ. PMID 16590424. doi:10.1073/pnas.45.4.622.

[37] Freese E (June 1959). "The specific mutagenic effect of base analogues on Phage T4". *Journal of Molecular Biology*. Amsterdam, the Netherlands: Elsevier. **1** (2): 87–105. ISSN 0022-2836. doi:10.1016/S0022-2836(59)80038-3.

[38] Goh AM, Coffill CR, Lane DP (January 2011). "The role of mutant p53 in human cancer". *The Journal of Pathology*. John Wiley & Sons. **223** (2): 116–26. PMID 21125670. doi:10.1002/path.2784.

[39] Chenevix-Trench G, Spurdle AB, Gatei M, Kelly H, Marsh A, Chen X, Donn K, Cummings M, Nyholt D, Jenkins MA, Scott C, Pupo GM, Dörk T, Bendix R, Kirk J, Tucker K, McCredie MR, Hopper JL, Sambrook J, Mann GJ, Khanna KK (February 2002). "Dominant negative ATM mutations in breast cancer families". *Journal of the National Cancer Institute*. Oxford University Press. **94** (3): 205–15. PMID 11830610. doi:10.1093/jnci/94.3.205.

[40] Paz-Priel I, Friedman A (2011). "C/EBPα dysregulation in AML and ALL". *Critical Reviews in Oncogenesis*. Begell House. **16** (1–2): 93–102. PMC 3243939 ⓐ. PMID 22150310. doi:10.1615/critrevoncog.v16.i1-2.90. (Subscription required (help)).

[41] Capaccio D, Ciccodicola A, Sabatino L, Casamassimi A, Pancione M, Fucci A, Febbraro A, Merlino A, Graziano G, Colantuoni V (June 2010). "A novel germline mutation in peroxisome proliferator-activated receptor gamma gene associated with large intestine polyp formation and dyslipidemia". *Biochimica et Biophysica Acta*. Elsevier. **1802** (6): 572–81. PMID 20123124. doi:10.1016/j.bbadis.2010.01.012.

[42] McKusick VA (July 1991). "The defect in Marfan syndrome". *Nature*. Nature Publishing Group. **352** (6333): 279–81. Bibcode:1991Natur.352..279M. PMID 1852198. doi:10.1038/352279a0.

[43] Judge DP, Biery NJ, Keene DR, Geubtner J, Myers L, Huso DL, Sakai LY, Dietz HC (July 2004). "Evidence for a critical contribution of haploinsufficiency in the complex pathogenesis of Marfan syndrome". *The Journal of Clinical Investigation*. American Society for Clinical Investigation. **114** (2): 172–81. PMC 449744 ⓐ. PMID 15254584. doi:10.1172/JCI20641.

[44] Judge DP, Dietz HC (December 2005). "Marfan's syndrome". *Lancet*. Elsevier. **366** (9501): 1965–76. PMC 1513064 ⓐ. PMID 16325700. doi:10.1016/S0140-6736(05)67789-6.

[45] Ellis NA, Ciocci S, German J (February 2001). "Back mutation can produce phenotype reversion in Bloom syndrome somatic cells". *Human Genetics*. Springer-Verlag. **108** (2): 167–73. PMID 11281456. doi:10.1007/s004390000447.

[46] Charlesworth D, Charlesworth B, Morgan MT (December 1995). "The pattern of neutral molecular variation under the background selection model". *Genetics*. Genetics Society of America. **141** (4): 1619–32. PMC 1206892 ⓐ. PMID 8601499.

[47] Loewe L (April 2006). "Quantifying the genomic decay paradox due to Muller's ratchet in human mitochondrial DNA". *Genetical Research*. Cambridge University Press. **87** (2): 133–59. PMID 16709275. doi:10.1017/S0016672306008123.

[48] Bernstein H, Hopf FA, Michod RE (1987). "The molecular basis of the evolution of sex". *Advances in Genetics*. **24**: 323–70. PMID 3324702. doi:10.1016/s0065-2660(08)60012-7.

[49] Peck JR, Barreau G, Heath SC (April 1997). "Imperfect genes, Fisherian mutation and the evolution of sex". *Genetics*. Genetics Society of America. **145** (4): 1171–99. PMC 1207886 ⓐ. PMID 9093868.

[50] Keightley PD, Lynch M (March 2003). "Toward a realistic model of mutations affecting fitness". *Evolution; International Journal of Organic Evolution*. John Wiley & Sons for the Society for the Study of Evolution. **57** (3): 683–5; discussion 686–9. JSTOR 3094781. PMID 12703958. doi:10.1554/0014-3820(2003)057[0683:tarmom]2.0.co;2.

[51] Barton NH, Keightley PD (January 2002). "Understanding quantitative genetic variation". *Nature Reviews Genetics*. Nature Publishing Group. **3** (1): 11–21. PMID 11823787. doi:10.1038/nrg700.

[52] Sanjuán R, Moya A, Elena SF (June 2004). "The distribution of fitness effects caused by single-nucleotide substitutions in an RNA virus". *Proceedings of the National Academy of Sciences of the United States of America*. National Academy of Sciences. **101** (22): 8396–401. PMC 420405 ⓐ. PMID 15159545. doi:10.1073/pnas.0400146101.

[53] Carrasco P, de la Iglesia F, Elena SF (December 2007). "Distribution of fitness and virulence effects caused by single-nucleotide substitutions in Tobacco Etch virus". *Journal of Virology*. American Society for Microbiology. **81** (23): 12979–84. PMC 2169111 ⓐ. PMID 17898073. doi:10.1128/JVI.00524-07.

[54] Sanjuán R (June 2010). "Mutational fitness effects in RNA and single-stranded DNA viruses: common patterns revealed by site-directed mutagenesis studies". *Philosophical Transactions of the Royal Society of London. Series B, Biological Sciences*. Royal Society. **365** (1548): 1975–82. PMC 2880115 ⓐ. PMID 20478892. doi:10.1098/rstb.2010.0063.

[55] Peris JB, Davis P, Cuevas JM, Nebot MR, Sanjuán R (June 2010). "Distribution of fitness effects caused by single-nucleotide substitutions in bacteriophage f1". *Genetics*. Genetics Society of America. **185** (2): 603–9. PMC 2881140 . PMID 20382832. doi:10.1534/genetics.110.115162.

[56] Elena SF, Ekunwe L, Hajela N, Oden SA, Lenski RE (March 1998). "Distribution of fitness effects caused by random insertion mutations in Escherichia coli". *Genetica*. Kluwer Academic Publishers. 102–103 (1–6): 349–58. PMID 9720287. doi:10.1023/A:1017031008316.

[57] Hietpas RT, Jensen JD, Bolon DN (May 2011). "Experimental illumination of a fitness landscape". *Proceedings of the National Academy of Sciences of the United States of America*. National Academy of Sciences. **108** (19): 7896–901. PMC 3093508 . PMID 21464309. doi:10.1073/pnas.1016024108.

[58] Davies EK, Peters AD, Keightley PD (September 1999). "High frequency of cryptic deleterious mutations in Caenorhabditis elegans". *Science*. American Association for the Advancement of Science. **285** (5434): 1748–51. PMID 10481013. doi:10.1126/science.285.5434.1748.

[59] Loewe L, Charlesworth B (September 2006). "Inferring the distribution of mutational effects on fitness in Drosophila". *Biology Letters*. Royal Society. **2** (3): 426–30. PMC 1686194 . PMID 17148422. doi:10.1098/rsbl.2006.0481.

[60] Eyre-Walker A, Woolfit M, Phelps T (June 2006). "The distribution of fitness effects of new deleterious amino acid mutations in humans". *Genetics*. Genetics Society of America. **173** (2): 891–900. PMC 1526495 . PMID 16547091. doi:10.1534/genetics.106.057570.

[61] Sawyer SA, Kulathinal RJ, Bustamante CD, Hartl DL (August 2003). "Bayesian analysis suggests that most amino acid replacements in Drosophila are driven by positive selection". *Journal of Molecular Evolution*. Springer-Verlag. 57 Suppl 1 (1): S154–64. PMID 15008412. doi:10.1007/s00239-003-0022-3.

[62] Piganeau G, Eyre-Walker A (September 2003). "Estimating the distribution of fitness effects from DNA sequence data: implications for the molecular clock". *Proceedings of the National Academy of Sciences of the United States of America*. National Academy of Sciences. **100** (18): 10335–40. PMC 193562 . PMID 12925735. doi:10.1073/pnas.1833064100.

[63] Kimura M (February 1968). "Evolutionary rate at the molecular level". *Nature*. Nature Publishing Group. **217** (5129): 624–6. PMID 5637732. doi:10.1038/217624a0.

[64] Kimura 1983

[65] Akashi H (September 1999). "Within- and between-species DNA sequence variation and the 'footprint' of natural selection". *Gene*. Elsevier. **238** (1): 39–51. PMID 10570982. doi:10.1016/S0378-1119(99)00294-2.

[66] Eyre-Walker A (October 2006). "The genomic rate of adaptive evolution". *Trends in Ecology & Evolution*. Cell Press. **21** (10): 569–75. PMID 16820244. doi:10.1016/j.tree.2006.06.015.

[67] Gillespie JH (September 1984). "Molecular Evolution Over the Mutational Landscape". *Evolution*. Hoboken, NJ: John Wiley & Sons for the Society for the Study of Evolution. **38** (5): 1116–1129. ISSN 0014-3820. JSTOR 2408444. doi:10.2307/2408444.

[68] Orr HA (April 2003). "The distribution of fitness effects among beneficial mutations". *Genetics*. Genetics Society of America. **163** (4): 1519–26. PMC 1462510 . PMID 12702694.

[69] Kassen R, Bataillon T (April 2006). "Distribution of fitness effects among beneficial mutations before selection in experimental populations of bacteria". *Nature Genetics*. Nature Publishing Group. **38** (4): 484–8. PMID 16550173. doi:10.1038/ng1751.

[70] Rokyta DR, Joyce P, Caudle SB, Wichman HA (April 2005). "An empirical test of the mutational landscape model of adaptation using a single-stranded DNA virus". *Nature Genetics*. Nature Publishing Group. **37** (4): 441–4. PMID 15778707. doi:10.1038/ng1535.

[71] Imhof M, Schlotterer C (January 2001). "Fitness effects of advantageous mutations in evolving Escherichia coli populations". *Proceedings of the National Academy of Sciences of the United States of America*. National Academy of Sciences. **98** (3): 1113–7. PMC 14717 . PMID 11158603. doi:10.1073/pnas.98.3.1113.

[72] Hogan, C. Michael (October 12, 2010). "Mutation". In Monosson, Emily. *Encyclopedia of Earth*. Washington, D.C.: Environmental Information Coalition, National Council for Science and the Environment. OCLC 72808636. Retrieved 2015-10-08.

[73] Boillée S, Vande Velde C, Cleveland DW (October 2006). "ALS: a disease of motor neurons and their nonneuronal neighbors". *Neuron*. Cell Press. **52** (1): 39–59. PMID 17015226. doi:10.1016/j.neuron.2006.09.018.

[74] "Somatic cell genetic mutation". *Genome Dictionary*. Athens, Greece: Information Technology Associates. June 30, 2007. Retrieved 2010-06-06.

[75] "*RB1* Genetics". *Daisy's Eye Cancer Fund*. Oxford, UK. Archived from the original on 2011-11-26. Retrieved 2015-10-09.

[76] "Compound heterozygote". *MedTerms*. New York: WebMD. June 14, 2012. Retrieved 2015-10-09.

[77] Alberts (2014). *Molecular Biology of the Cell* (6 ed.). Garland Science. p. 487. ISBN 9780815344322.

[78] Chadov BF, Fedorova NB, Chadova EV (2015-07-01). "Conditional mutations in Drosophila melanogaster: On the occasion of the 150th anniversary of G. Mendel's report in Brünn". *Mutation Research. Reviews in Mutation Research.* **765**: 40–55. PMID 26281767. doi:10.1016/j.mrrev.2015.06.001.

[79] Landis G, Bhole D, Lu L, Tower J (July 2001). "High-frequency generation of conditional mutations affecting Drosophila melanogaster development and life span". *Genetics.* **158** (3): 1167–76. PMC 1461716. PMID 11454765.

[80] Gierut JJ, Jacks TE, Haigis KM (April 2014). "Strategies to achieve conditional gene mutation in mice". *Cold Spring Harbor Protocols.* **2014** (4): 339–49. PMC 4142476. PMID 24692485. doi:10.1101/pdb.top069807.

[81] Spencer DM (May 1996). "Creating conditional mutations in mammals". *Trends in Genetics.* **12** (5): 181–7. PMID 8984733. doi:10.1016/0168-9525(96)10013-5.

[82] Tan G, Chen M, Foote C, Tan C (September 2009). "Temperature-sensitive mutations made easy: generating conditional mutations by using temperature-sensitive inteins that function within different temperature ranges". *Genetics.* **183** (1): 13–22. PMC 2746138. PMID 19596904. doi:10.1534/genetics.109.104794.

[83] den Dunnen JT, Antonarakis SE (January 2000). "Mutation nomenclature extensions and suggestions to describe complex mutations: a discussion". *Human Mutation.* Wiley-Liss, Inc. **15** (1): 7–12. PMID 10612815. doi:10.1002/(SICI)1098-1004(200001)15:1<7::AID-HUMU4>3.0.CO;2-N.

[84] Doniger SW, Kim HS, Swain D, Corcuera D, Williams M, Yang SP, Fay JC (August 2008). Pritchard JK, ed. "A catalog of neutral and deleterious polymorphism in yeast". *PLoS Genetics.* Public Library of Science. **4** (8): e1000183. PMC 2515631. PMID 18769710. doi:10.1371/journal.pgen.1000183.

[85] Ionov Y, Peinado MA, Malkhosyan S, Shibata D, Perucho M (June 1993). "Ubiquitous somatic mutations in simple repeated sequences reveal a new mechanism for colonic carcinogenesis". *Nature.* Nature Publishing Group. **363** (6429): 558–61. Bibcode:1993Natur.363..558I. PMID 8505985. doi:10.1038/363558a0.

[86] Sullivan AD, Wigginton J, Kirschner D (August 2001). "The coreceptor mutation CCR5Delta32 influences the dynamics of HIV epidemics and is selected for by HIV". *Proceedings of the National Academy of Sciences of the United States of America.* National Academy of Sciences. **98** (18): 10214–9. Bibcode:2001PNAS...9810214S. PMC 56941. PMID 11517319. doi:10.1073/pnas.181325198.

[87] "Mystery of the Black Death". *Secrets of the Dead.* Season 3. Episode 2. October 30, 2002. PBS. Retrieved 2015-10-10. Episode background.

[88] Galvani AP, Slatkin M (December 2003). "Evaluating plague and smallpox as historical selective pressures for the CCR5-Delta 32 HIV-resistance allele". *Proceedings of the National Academy of Sciences of the United States of America.* National Academy of Sciences. **100** (25): 15276–9. Bibcode:2003PNAS..10015276G. PMC 299980. PMID 14645720. doi:10.1073/pnas.2435085100.

[89] Konotey-Ahulu, Felix. "Frequently Asked Questions [FAQ's]". *sicklecell.md.*

[90] "'Lifeless' prion proteins are 'capable of evolution'". Health. *BBC News Online.* London. January 1, 2010. Retrieved 2015-10-10.

[91] "somatic mutation | genetics". *Encyclopedia Britannica.* Retrieved 2017-03-31.

[92] Hartl, Jones, Daniel L.,Elizabeth W. (1998). *Genetics Principles and Analysis.* Sudbury, Massachusetts: Jones and Bartlett Publishers. p. 556. ISBN 0-7637-0489-X.

[93] Araten DJ, Golde DW, Zhang RH, Thaler HT, Gargiulo L, Notaro R, Luzzatto L (September 2005). "A quantitative measurement of the human somatic mutation rate". *Cancer Research.* American Association for Cancer Research. **65** (18): 8111–7. PMID 16166284. doi:10.1158/0008-5472.CAN-04-1198.

5.13 Bibliography

- Bernstein C, Prasad AR, Nfonsam V, Bernstein H (2013). "DNA Damage, DNA Repair and Cancer". In Chen C. *New Research Directions in DNA Repair.* Rijeka, Croatia: InTech. ISBN 978-953-51-1114-6. doi:10.5772/53919.

- Bernstein H, Hopf FA, Michod RE (1987). "The Molecular Basis of the Evolution of Sex". In Scandalios JG. *Molecular Genetics of Development.* Advances in Genetics. **24**. San Diego, CA: Academic Press. ISBN 0-12-017624-6. ISSN 0065-2660. LCCN 47030313. OCLC 18561279. PMID 3324702. doi:10.1016/S0065-2660(08)60012-7.

- Carroll SB, Grenier JK, Weatherbee SD (2005). *From DNA to Diversity: Molecular Genetics and the Evolution of Animal Design* (2nd ed.). Malden, MA: Blackwell Publishing. ISBN 1-4051-1950-0. LCCN 2003027991. OCLC 53972564.

- Kimura M (1983). *The Neutral Theory of Molecular Evolution.* Cambridge, UK; New York: Cambridge University Press. ISBN 0-521-23109-4. LCCN 82022225. OCLC 9081989.

5.14 External links

- den Dunnen, Johan T. "Nomenclature for the description of sequence variants". Melbourne, Australia: Human Genome Variation Society. Retrieved 2015-10-18.
- Jones, Steve; Woolfson, Adrian; Partridge, Linda (December 6, 2007). "Genetic Mutation". *In Our Time*. BBC Radio 4. Retrieved 2015-10-18.
- Liou, Stephanie (February 5, 2011). "All About Mutations". *HOPES*. Standford, CA: Huntington's Disease Outreach Project for Education at Stanford. Retrieved 2015-10-18.
- "Locus Specific Mutation Databases". Leiden, the Netherlands: Leiden University Medical Center. Retrieved 2015-10-18.
- "Welcome to the Mutalyzer website". Leiden, the Netherlands: Leiden University Medical Center. Retrieved 2015-10-18. — The Mutalyzer website.

Chapter 6

Pharmacology

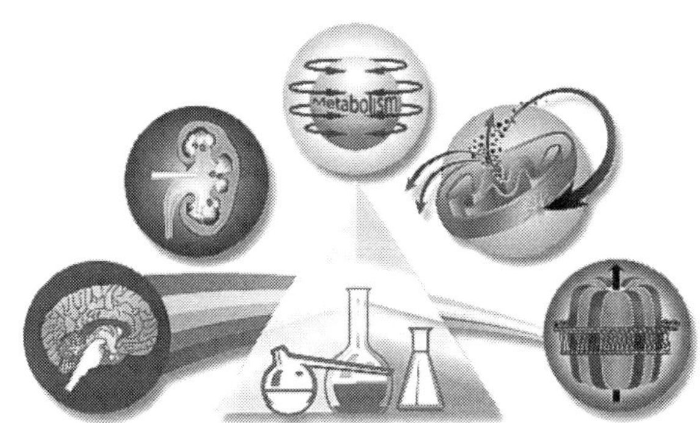

A variety of topics involved with pharmacology, including neuropharmacology, renal pharmacology, human metabolism, intracellular metabolism, and intracellular regulation

Pharmacology is the branch of biology concerned with the study of drug action,[1] where a drug can be broadly defined as any man-made, natural, or endogenous (from within body) molecule which exerts a biochemical or physiological effect on the cell, tissue, organ, or organism (sometimes the word pharmacon is used as a term to encompass these endogenous and exogenous bioactive species). More specifically, it is the study of the interactions that occur between a living organism and chemicals that affect normal or abnormal biochemical function. If substances have medicinal properties, they are considered pharmaceuticals.

The field encompasses drug composition and properties, synthesis and drug design, molecular and cellular mechanisms, organ/systems mechanisms, signal transduction/cellular communication, molecular diagnostics, interactions, toxicology, chemical biology, therapy, and medical applications and antipathogenic capabilities. The two main areas of pharmacology are pharmacodynamics and pharmacokinetics. Pharmacodynamics studies the effects of a drug on biological systems, and Pharmacokinetics studies the effects of biological systems on a drug. In broad terms, pharmacodynamics discusses the chemicals with biological receptors, and pharmacokinetics discusses the absorption, distribution, metabolism, and excretion (ADME) of chemicals from the biological systems. Pharmacology is not synonymous with pharmacy and the two terms are frequently confused. Pharmacology, a biomedical science, deals with the research, discovery, and characterization of chemicals which show biological effects and the elucidation of cellular and organismal function in relation to these chemicals. In contrast, pharmacy, a health services profession, is concerned with application of the principles learned from pharmacology in its clinical settings; whether it be in a dispensing or clinical care role. In either field, the primary contrast between the two are their distinctions between direct-patient care, for pharmacy practice, and the science-oriented research field, driven by pharmacology.

The origins of clinical pharmacology date back to the Middle Ages in Avicenna's *The Canon of Medicine*, Peter of Spain's *Commentary on Isaac*, and John of St Amand's *Commentary on the Antedotary of Nicholas*.[2] Clinical pharmacology owes much of its foundation to the work of William Withering.[3] Pharmacology as a scientific discipline did not further advance until the mid-19th century amid the great biomedical resurgence of that period.[4] Before the second half of the nineteenth century, the remarkable potency and specificity of the actions of drugs such as morphine, quinine and digitalis were explained vaguely and with reference to extraordinary chemical powers and affinities to certain organs or tissues.[5] The first pharmacology department was set up by Rudolf Buchheim in 1847, in recognition of the need to understand how therapeutic drugs and poisons produced their effects.[4]

Early pharmacologists focused on natural substances, mainly plant extracts. Pharmacology developed in the 19th century as a biomedical science that applied the principles of scientific experimentation to therapeutic contexts.[6] Today pharmacologists use genetics, molecular biology, chemistry, and other advanced tools to transform information about molecular mechanisms and targets into therapies directed against disease, defects or pathogens, and create methods for preventative care, diagnostics, and ultimately personalized medicine.

6.1 Divisions

The discipline of pharmacology can be divided into many sub disciplines each with a specific focus.

6.1.1 Clinical pharmacology

Clinical pharmacology is the basic science of pharmacology with an added focus on the application of pharmacological principles and methods in the medical clinic and towards patient care and outcomes.

6.1.2 Neuropharmacology

Neuropharmacology is the study of the effects of medication on central and peripheral nervous system functioning.

6.1.3 Psychopharmacology

Psychopharmacology, also known as behavioral pharmacology, is the study of the effects of medication on the psyche (psychology), observing changed behaviors of the body and mind, and how molecular events are manifest in a measurable behavioral form. This is similar to the closely related ethnopharmacology. Psychopharmacology is an interdisciplinary field which studies behavioral effects of psychoactive drugs. It incorporates approaches and techniques from neuropharmacology, animal behavior and behavioral neuroscience, and is interested in the behavioral and neurobiological mechanisms of action of psychoactive drugs. Another goal of behavioral pharmacology is to develop animal behavioral models to screen chemical compounds with therapeutic potentials. People in this field (called behavioral pharmacologists) typically use small animals (e.g. rodents) to study psychotherapeutic drugs such as antipsychotics, antidepressants and anxiolytics, and drugs of abuse such as nicotine, cocaine, methamphetamine, etc. study of drugs which affect behavior. *Ethopharmacology* (not to be confused with ethnopharmacology) is a term which has been in use since the 1960s[7] and derives from the Greek word ἦθος *ethos* meaning character and "pharmacology" the study of drug actions and mechanism.

6.1.4 Cardiovascular pharmacology

Cardiovascular pharmacology is the study of the effects of drugs on the entire cardiovascular system, including the heart and blood vessels.

6.1.5 Pharmacogenetics

Pharmacogenetics is clinical testing of genetic variation that gives rise to differing response to drugs.

6.1.6 Pharmacogenomics

Pharmacogenomics is the application of genomic technologies to drug discovery and further characterization of older drugs.

6.1.7 Pharmacoepidemiology

Pharmacoepidemiology is the study of the effects of drugs in large numbers of people.

6.1.8 Systems pharmacology

Systems pharmacology is the application of systems biology principles to the field of pharmacology.

6.1.9 Toxicology

Toxicology is the study of the adverse effects, molecular targets, and characterization of drugs or any chemical substance in excess (including those beneficial in lower doses).

6.1.10 Theoretical pharmacology

Theoretical pharmacology is a relatively new and rapidly expanding field of research activity in which many of the techniques of computational chemistry, in particular computational quantum chemistry and the method of molecular mechanics, are proving to be of great value. Theoretical pharmacologists aim at rationalizing the relation between the activity of a particular drug, as observed experimentally, and its structural features as derived from computer experiments. They aim to find structure—activity relations. Furthermore, on the basis of the structure of a given organic molecule, the theoretical pharmacologist aims at predicting the biological activity of new drugs that are of the same general type as existing drugs. More ambitiously, it aims to predict entirely new classes of drugs, tailor-made for specific purposes.

6.1.11 Posology

Posology is the study of how medicines are dosed. This depends upon various factors including age, climate, weight, sex, elimination rate of drug, genetic polymorphism and time of administration. It is derived from the Greek words πόσος *posos* meaning "how much?" and -λογία *-logia* "study of".[8]

6.1.12 Pharmacognosy

Pharmacognosy is a branch of pharmacology dealing especially with the composition, use, and development of medicinal substances of biological origin and especially medicinal substances obtained from plants.

6.1.13 Environmental pharmacology

Environmental pharmacology is a new discipline.[9] Focus is being given to understand gene–environment interaction, drug-environment interaction and toxin-environment interaction. There is a close collaboration between environmental science and medicine in addressing these issues, as healthcare itself can be a cause of environmental damage or remediation. Human health and ecology are intimately related. Demand for more pharmaceutical products may place the public at risk through the destruction of species. The entry of chemicals and drugs into the aquatic ecosystem is a more serious concern today. In addition, the production of some illegal drugs pollutes drinking water supply by releasing carcinogens.[10] This field is intimately linked with Public Health fields.

6.1.14 Dental pharmacology

Dental pharmacology relates to the study of drugs commonly used in the treatment of dental disease.[11]

6.2 Scientific background

The study of chemicals requires intimate knowledge of the biological system affected. With the knowledge of cell biology and biochemistry increasing, the field of pharmacology has also changed substantially. It has become possible, through molecular analysis of receptors, to design chemicals that act on specific cellular signaling or metabolic pathways by affecting sites directly on cell-surface receptors (which modulate and mediate cellular signaling pathways controlling cellular function).

A chemical has, from the pharmacological point-of-view, various properties. Pharmacokinetics describes the effect of the body on the chemical (e.g. half-life and volume of distribution), and pharmacodynamics describes the chemical's effect on the body (desired or toxic).

When describing the pharmacokinetic properties of the chemical that is the active ingredient or active pharmaceutical ingredient (API), pharmacologists are often interested in *L-ADME*:

- Liberation – How is the API disintegrated (for solid oral forms (breaking down into smaller particles)), dispersed, or dissolved from the medication?
- Absorption – How is the API absorbed (through the skin, the intestine, the oral mucosa)?
- Distribution – How does the API spread through the organism?
- Metabolism – Is the API converted chemically inside the body, and into which substances. Are these active (as well)? Could they be toxic?
- Excretion – How is the API excreted (through the bile, urine, breath, skin)?

Medication is said to have a narrow or wide *therapeutic index* or *therapeutic window*. This describes the ratio of desired effect to toxic effect. A compound with a narrow therapeutic index (close to one) exerts its desired effect at a dose close to its toxic dose. A compound with a wide therapeutic index (greater than five) exerts its desired effect at a dose substantially below its toxic dose. Those with a narrow margin are more difficult to dose and administer, and may require therapeutic drug monitoring (examples are warfarin, some antiepileptics, aminoglycoside antibiotics). Most anticancer drugs have a narrow therapeutic margin: toxic side-effects are almost always encountered at doses used to kill tumors.

6.3 Medicine development and safety testing

Development of medication is a vital concern to medicine, but also has strong economical and political implications. To protect the consumer and prevent abuse, many governments regulate the manufacture, sale, and administration of medication. In the United States, the main body that regulates pharmaceuticals is the Food and Drug Administration and they enforce standards set by the United States Pharmacopoeia. In the European Union, the main body that regulates pharmaceuticals is the EMA and they enforce standards set by the European Pharmacopoeia.

The metabolic stability and the reactivity of a library of candidate drug compounds have to be assessed for drug metabolism and toxicological studies. Many methods have been proposed for quantitative predictions in drug metabolism; one example of a recent computational method is SPORCalc.[12] If the chemical structure of a medicinal compound is altered slightly, this could slightly or dramatically alter the medicinal properties of the compound depending on the level of alteration as it relates to the structural composition of the substrate or receptor site on which it exerts its medicinal effect, a concept referred to as the structural activity relationship (SAR). This means that when a useful activity has been identified, chemists will make many similar compounds called analogues, in an attempt to maximize the desired medicinal effect(s) of the compound. This development phase can take anywhere from a few years to a decade or more and is very expensive.[13]

These new analogues need to be developed. It needs to be determined how safe the medicine is for human consumption, its stability in the human body and the best form for delivery to the desired organ system, like tablet or aerosol. After extensive testing, which can take up to 6 years, the new medicine is ready for marketing and selling.[13]

As a result of the long time required to develop analogues and test a new medicine and the fact that of every 5000 potential new medicines typically only one will ever reach the open market, this is an expensive way of doing things, often costing over 1 billion dollars. To recoup this outlay pharmaceutical companies may do a number of things:[13]

- Carefully research the demand for their potential new product before spending an outlay of company funds.[13]
- Obtain a patent on the new medicine preventing other companies from producing that medicine for a certain allocation of time.[13]

6.4 Drug legislation and safety

In the United States, the Food and Drug Administration (FDA) is responsible for creating guidelines for the approval and use of drugs. The FDA requires that all approved drugs fulfill two requirements:

1. The drug must be found to be effective against the disease for which it is seeking approval (where 'effective' means only that the drug performed better than placebo or competitors in at least two trials).
2. The drug must meet safety criteria by being subject to animal and controlled human testing.

Gaining FDA approval usually takes several years to attain. Testing done on animals must be extensive and must include several species to help in the evaluation of both the effectiveness and toxicity of the drug. The dosage of any drug approved for use is intended to fall within a range in which the drug produces a therapeutic effect or desired outcome.[14]

The safety and effectiveness of prescription drugs in the U.S. is regulated by the federal Prescription Drug Marketing Act of 1987.

The Medicines and Healthcare products Regulatory Agency (MHRA) has a similar role in the UK.

6.5 Education

Students of pharmacology are trained as biomedical scientists, studying the effects of drugs on living organisms. This can lead to new drug discoveries, as well as a better understanding of the way in which the human body works.

Students of pharmacology must have detailed working knowledge of aspects in physiology, pathology and chemistry. During a typical degree they will cover areas such as (but not limited to) biochemistry, cell biology, basic physiology, genetics and the Central Dogma, medical microbiology, neuroscience, and depending on the department's interests, bio-organic chemistry, or chemical biology.

Modern Pharmacology is highly interdisciplinary. Graduate programs accept students from most biological and chemical backgrounds. With the increasing drive towards biophysical and computational research to describe systems, pharmacologists may even consider themselves mainly physical scientists. In many instances, Analytical Chemistry is closely related to the studies and needs of pharmacological research. Therefore, many institutions will include pharmacology under a Chemistry or Biochemistry Department, especially if a separate Pharmacology Dept. does not exist. What makes an institutional department independent of another, or exist in the first place, is usually an artifact of historical times.

Whereas a pharmacy student will eventually work in a pharmacy dispensing medications, a pharmacologist will typically work within a laboratory setting. Careers for a pharmacologist include academic positions (medical and non-medical), governmental positions, private industrial positions, science writing, scientific patents and law, consultation, biotech and pharmaceutical employment, the alcohol industry, food industry, forensics/law enforcement, public health, and environmental/ecological sciences.

6.6 Etymology

The word "pharmacology" is derived from Greek φάρμακον, *pharmakon*, "drug, poison, spell" and -λογία, *-logia* "study of", "knowledge of"[15][16] (cf. the etymology of *pharmacy*).

6.7 See also

- Certain safety factor
- Cosmeceuticals
- Crude drugs
- Nicholas Culpeper – 17th century English Physician who translated and used 'pharmacological texts'.
- Drug design
- Drug Discovery Hit to Lead
- Drug metabolism
- Enzyme inhibitors
- Herbalism
- History of pharmacy
- International Union of Basic and Clinical Pharmacology
- Inverse benefit law
- List of abbreviations used in medical prescriptions
- List of pharmaceutical companies
- List of withdrawn drugs
- Loewe additivity
- Medical School
- Medicare Part D – the new prescription drug plan in the U.S.
- Medication
- Medicinal chemistry
- Neuropharmacology – The Molecular and Behavior study of Disease and Drugs in the Nervous System
- Neuropsychopharmacology – The detailed comprehensive study of mind, brain and drugs.
- Pharmaceutical company
- Pharmaceutical formulation
- Pharmaceuticals and personal care products in the environment
- Pharmacognosy
- Pharmacopoeia
- Pharmacotherapy
- Pharmakos
- Placebo (origins of technical term)
- Prescription drug
- Prescription Drug Marketing Act (PDMA)
- Psychopharmacology – medication for mental conditions
- Traditional Chinese Medicine

6.8 Notes and references

[1] Vallance P, Smart TG (January 2006). "The future of pharmacology". *British Journal of Pharmacology*. 147 Suppl 1 (S1): S304–7. PMC 1760753 . PMID 16402118. doi:10.1038/sj.bjp.0706454.

[2] Brater DC, Daly WJ (May 2000). "Clinical pharmacology in the Middle Ages: principles that presage the 21st century". *Clin. Pharmacol. Ther.* **67** (5): 447–50. PMID 10824622. doi:10.1067/mcp.2000.106465.

[3] Mannfred A. Hollinger (2003)."*Introduction to pharmacology*". CRC Press. p.4. ISBN 0-415-28033-8

[4] Rang HP (January 2006). "The receptor concept: pharmacology's big idea". *Br. J. Pharmacol.* 147 Suppl 1 (S1): S9–16. PMC 1760743 . PMID 16402126. doi:10.1038/sj.bjp.0706457.

[5] Maehle AH, Prüll CR, Halliwell RF (August 2002). "The emergence of the drug receptor theory". *Nat Rev Drug Discov.* **1** (8): 637–41. PMID 12402503. doi:10.1038/nrd875.

[6] Rang, H.P.; M.M. Dale; J.M. Ritter; R.J. Flower (2007). *Pharmacology*. China: Elsevier. ISBN 0-443-06911-5.

[7] Krsiak, M (1991). "Ethopharmacology: A historical perspective". *Neuroscience and biobehavioral reviews.* **15** (4): 439–45. PMID 1792005. doi:10.1016/s0149-7634(05)80124-1.

[8] "posology". *Random House Webster's Unabridged Dictionary*.

[9] Rahman, SZ; Khan, RA (Dec 2006). "Environmental pharmacology: A new discipline". *Indian J Pharmacol.* **38** (4): 229–30. doi:10.4103/0253-7613.27017.

[10] Sue Ruhoy Ilene; Daughton Christian G (2008). "Beyond the medicine cabinet: An analysis of where and why medications accumulate". *Environment International*. **34** (8): 1157–1169. doi:10.1016/j.envint.2008.05.002.

[11] "Dental Pharmacology". *nba.uth.tmc.edu*. Texas Medical Center. Retrieved 22 May 2015.

[12] James Smith; Viktor Stein (2009). "SPORCalc: A development of a database analysis that provides putative metabolic enzyme reactions for ligand-based drug design". *Computational Biology and Chemistry*. **33** (2): 149–159. PMID 19157988. doi:10.1016/j.compbiolchem.2008.11.002.

[13] Newton, David; Alasdair Thorpe; Chris Otter (2004). *Revise A2 Chemistry*. Heinemann Educational Publishers. p. 1. ISBN 0-435-58347-6.

[14] Nagle, Hinter; Barbara Nagle (2005). *Pharmacology: An Introduction*. Boston: McGraw Hill. ISBN 0-07-312275-0.

[15] Pharmacy (n.) - Online Etymology Dictionary

[16] Pharmacology - Online Etymology Dictionary

6.9 External links

- American Society for Pharmacology and Experimental Therapeutics
- British Pharmacological Society
- Pharmaceutical company profiles at NNDB
- International Conference on Harmonisation
- US Pharmacopeia
- International Union of Basic and Clinical Pharmacology
- IUPHAR Committee on Receptor Nomenclature and Drug Classification

Chapter 7

Pharmaceutical drug

"Medication" redirects here. For other uses, see Medication (disambiguation).
"Medicines" redirects here. For other uses, see Medicine (disambiguation).

A **pharmaceutical drug** (also referred to as **medicine**, **medication**, or simply as **drug**) is a drug used to diagnose, cure, treat, or prevent disease.[1][2][3] Drug therapy (pharmacotherapy) is an important part of the medical field and relies on the science of pharmacology for continual advancement and on pharmacy for appropriate management.

Drugs are classified in various ways. One of the key divisions is by level of control, which distinguishes prescription drugs (those that a pharmacist dispenses only on the order of a physician, physician assistant, or qualified nurse) from over-the-counter drugs (those that consumers can order for themselves). Another key distinction is between traditional small-molecule drugs, usually derived from chemical synthesis, and biopharmaceuticals, which include recombinant proteins, vaccines, blood products used therapeutically (such as IVIG), gene therapy, monoclonal antibodies and cell therapy (for instance, stem-cell therapies). Other ways to classify medicines are by mode of action, route of administration, biological system affected, or therapeutic effects. An elaborate and widely used classification system is the Anatomical Therapeutic Chemical Classification System (ATC system). The World Health Organization keeps a list of essential medicines.

Drug discovery and drug development are complex and expensive endeavors undertaken by pharmaceutical companies, academic scientists, and governments. Governments generally regulate what drugs can be marketed, how drugs are marketed, and in some jurisdictions, drug pricing. Controversies have arisen over drug pricing and disposal of used drugs.

7.1 Definition

In Europe, the term is "medicinal product", and it is defined by EU law as: "(a) Any substance or combination of substances presented as having properties for treating or preventing disease in human beings; or

(b) Any substance or combination of substances which may be used in or administered to human beings either with a view to restoring, correcting or modifying physiological functions by exerting a pharmacological, immunological or metabolic action, or to making a medical diagnosis."[4]:36

In the US, a "drug" is:

- A substance recognized by an official pharmacopoeia or formulary.
- A substance intended for use in the diagnosis, cure, mitigation, treatment, or prevention of disease.
- A substance (other than food) intended to affect the structure or any function of the body.
- A substance intended for use as a component of a medicine but not a device or a component, part or accessory of a device.
- Biological products are included within this definition and are generally covered by the same laws and regulations, but differences exist regarding their manufacturing processes (chemical process versus biological process.)[5]

7.2 Usage

Drug use among elderly Americans has been studied; in a group of 2377 people with average age of 71 surveyed between 2005 and 2006, 84% took at least one prescription drug, 44% took at least one over-the-counter (OTC) drug, and 52% took at least one dietary supplement; in a group of

2245 elderly Americans (average age of 71) surveyed over the period 2010 - 2011, those percentages were 88%, 38%, and 64%.[6]

7.3 Classification

Main article: Drug class

Pharmaceutical or a drug is classified on the basis of their origin.

1. Drug from natural origin: Herbal or plant or mineral origin, some drug substances are of marine origin.
2. Drug from chemical as well as natural origin: Derived from partial herbal and partial chemical synthesis Chemical, example steroidal drugs
3. Drug derived from chemical synthesis.
4. Drug derived from animal origin: For example, hormones, and enzymes.
5. Drug derived from microbial origin: Antibiotics
6. Drug derived by biotechnology genetic-engineering, hybridoma technique for example
7. Drug derived from radioactive substances.

One of the key classifications is between traditional small molecule drugs, usually derived from chemical synthesis, and biologic medical products, which include recombinant proteins, vaccines, blood products used therapeutically (such as IVIG), gene therapy, and cell therapy (for instance, stem cell therapies).

Pharmaceutical or drug or medicines are classified in various other groups besides their origin on the basis of pharmacological properties like mode of action and their pharmacological action or activity,[7] such as by chemical properties, mode or route of administration, biological system affected, or therapeutic effects. An elaborate and widely used classification system is the Anatomical Therapeutic Chemical Classification System (ATC system). The World Health Organization keeps a list of essential medicines.

A sampling of classes of medicine includes:

1. Antipyretics: reducing fever (pyrexia/pyresis)
2. Analgesics: reducing pain (painkillers)
3. Antimalarial drugs: treating malaria
4. Antibiotics: inhibiting germ growth
5. Antiseptics: prevention of germ growth near burns, cuts and wounds
6. Mood stabilizers: lithium and valpromide
7. Hormone replacements: Premarin
8. Oral contraceptives: Enovid, "biphasic" pill, and "triphasic" pill
9. Stimulants: methylphenidate, amphetamine
10. Tranquilizers: meprobamate, chlorpromazine, reserpine, chlordiazepoxide, diazepam, and alprazolam
11. Statins: lovastatin, pravastatin, and simvastatin

Pharmaceuticals may also be described as "specialty", independent of other classifications, which is an ill-defined class of drugs that might be difficult to administer, require special handling during administration, require patient monitoring during and immediately after administration, have particular regulatory requirements restricting their use, and are generally expensive relative to other drugs.[8]

7.4 Types of medicines

7.4.1 For the digestive system

- Upper digestive tract: antacids, reflux suppressants, antiflatulents, antidopaminergics, proton pump inhibitors (PPIs), H_2-receptor antagonists, cytoprotectants, prostaglandin analogues
- Lower digestive tract: laxatives, antispasmodics, antidiarrhoeals, bile acid sequestrants, opioid

7.4.2 For the cardiovascular system

- General: β-receptor blockers ("beta blockers"), calcium channel blockers, diuretics, cardiac glycosides, antiarrhythmics, nitrate, antianginals, vasoconstrictors, vasodilators.
- Affecting blood pressure/(antihypertensive drugs): ACE inhibitors, angiotensin receptor blockers, beta-blockers, α blockers, calcium channel blockers, thiazide diuretics, loop diuretics, aldosterone inhibitors
- Coagulation: anticoagulants, heparin, antiplatelet drugs, fibrinolytics, anti-hemophilic factors, haemostatic drugs
- HMG-CoA reductase inhibitors (statins) for lowering LDL cholesterol inhibitors: hypolipidaemic agents.

7.4.3 For the central nervous system

See also: Psychiatric medication and Psychoactive drug

Drugs affecting the central nervous system include: Psychedelics, hypnotics, anaesthetics, antipsychotics, eugeroics, antidepressants (including tricyclic antidepressants, monoamine oxidase inhibitors, lithium salts, and selective serotonin reuptake inhibitors (SSRIs)), antiemetics, Anticonvulsants/antiepileptics, anxiolytics, barbiturates, movement disorder (e.g., Parkinson's disease) drugs, stimulants (including amphetamines), benzodiazepines, cyclopyrrolones, dopamine antagonists, antihistamines, cholinergics, anticholinergics, emetics, cannabinoids, and 5-HT (serotonin) antagonists.

7.4.4 For pain

See also: Analgesic

The main classes of painkillers are NSAIDs, opioids and Local anesthetics.

For consciousness (anesthetic drugs)

See also: Anesthetic

Some anesthetics include Benzodiazepines and Barbiturates.

7.4.5 For musculo-skeletal disorders

The main categories of drugs for musculoskeletal disorders are: NSAIDs (including COX-2 selective inhibitors), muscle relaxants, neuromuscular drugs, and anticholinesterases.

7.4.6 For the eye

- General: adrenergic neurone blocker, astringent, ocular lubricant
- Diagnostic: topical anesthetics, sympathomimetics, parasympatholytics, mydriatics, cycloplegics
- Antibacterial: antibiotics, topical antibiotics, sulfa drugs, aminoglycosides, fluoroquinolones
- Antiviral drug
- Anti-fungal: imidazoles, polyenes
- Anti-inflammatory: NSAIDs, corticosteroids
- Anti-allergy: mast cell inhibitors
- Anti-glaucoma: adrenergic agonists, beta-blockers, carbonic anhydrase inhibitors/hyperosmotics, cholinergics, miotics, parasympathomimetics, prostaglandin agonists/prostaglandin inhibitors. nitroglycerin

7.4.7 For the ear, nose and oropharynx

Antibiotics, sympathomimetics, antihistamines, anticholinergics, NSAIDs, corticosteroids, antiseptics, local anesthetics, antifungals, cerumenolytic

7.4.8 For the respiratory system

bronchodilators, antitussives, mucolytics, decongestants inhaled and systemic corticosteroids, Beta2-adrenergic agonists, anticholinergics, Mast cell stabilizers. Leukotriene antagonists

7.4.9 For endocrine problems

androgens, antiandrogens, estrogens, gonadotropin, corticosteroids, human growth hormone, insulin, antidiabetics (sulfonylureas, biguanides/metformin, thiazolidinediones, insulin), thyroid hormones, antithyroid drugs, calcitonin, diphosponate, vasopressin analogues

7.4.10 For the reproductive system or urinary system

antifungal, alkalinizing agents, quinolones, antibiotics, cholinergics, anticholinergics, antispasmodics, 5-alpha reductase inhibitor, selective alpha-1 blockers, sildenafils, fertility medications

7.4.11 For contraception

- Hormonal contraception
- Ormeloxifene
- Spermicide

7.4.12 For obstetrics and gynecology

NSAIDs, anticholinergics, haemostatic drugs, antifibrinolytics, Hormone Replacement Therapy (HRT), bone regulators, beta-receptor agonists, follicle stimulating

hormone, luteinising hormone, LHRH
gamolenic acid, gonadotropin release inhibitor, progestogen, dopamine agonists, oestrogen, prostaglandins, gonadorelin, clomiphene, tamoxifen, Diethylstilbestrol

7.4.13 For the skin

emollients, anti-pruritics, antifungals, disinfectants, scabicides, pediculicides, tar products, vitamin A derivatives, vitamin D analogues, keratolytics, abrasives, systemic antibiotics, topical antibiotics, hormones, desloughing agents, exudate absorbents, fibrinolytics, proteolytics, sunscreens, antiperspirants, corticosteroids, immune modulators

7.4.14 For infections and infestations

antibiotics, antifungals, antileprotics, antituberculous drugs, antimalarials, anthelmintics, amoebicides, antivirals, antiprotozoals, probiotics, prebiotics, antitoxins and antivenoms.

7.4.15 For the immune system

vaccines, immunoglobulins, immunosuppressants, interferons, monoclonal antibodies

7.4.16 For allergic disorders

anti-allergics, antihistamines, NSAIDs, Corticosteroids

7.4.17 For nutrition

Tonics, electrolytes and mineral preparations (including iron preparations and magnesium preparations), parenteral nutritions, vitamins, anti-obesity drugs, anabolic drugs, haematopoietic drugs, food product drugs

7.4.18 For neoplastic disorders

cytotoxic drugs, therapeutic antibodies, sex hormones, aromatase inhibitors, somatostatin inhibitors, recombinant interleukins, G-CSF, erythropoietin

7.4.19 For diagnostics

contrast media

7.4.20 For euthanasia

See also: Barbiturate § Other non-therapeutical uses

A euthanaticum is used for euthanasia and physician-assisted suicide. Euthanasia is not permitted by law in many countries, and consequently medicines will not be licensed for this use in those countries.

7.5 Administration

Administration is the process by which a patient takes a medicine. There are three major categories of drug administration; enteral (by mouth), parenteral (into the blood stream), and other (which includes giving a drug through intranasal, topical, inhalation, and rectal means).[9]

It can be performed in various dosage forms such as pills, tablets, or capsules.

There are many variations in the routes of administration, including intravenous (into the blood through a vein) and oral administration (through the mouth).

They can be administered all at once as a bolus, at frequent intervals or continuously. Frequencies are often abbreviated from Latin, such as *every 8 hours* reading Q8H from *Quaque VIII Hora*.

7.6 Drug discovery

Main article: Drug discovery

In the fields of medicine, biotechnology and pharmacology, drug discovery is the process by which new candidate drugs are discovered.

Historically, drugs were discovered through identifying the active ingredient from traditional remedies or by serendipitous discovery. Later chemical libraries of synthetic small molecules, natural products or extracts were screened in intact cells or whole organisms to identify substances that have a desirable therapeutic effect in a process known as classical pharmacology. Since sequencing of the human genome which allowed rapid cloning and synthesis of large quantities of purified proteins, it has become common practice to use high throughput screening of large compounds libraries against isolated biological targets which are hypothesized to be disease modifying in a process known as reverse pharmacology. Hits from these screens are then tested in cells and then in animals for efficacy. Even more recently, scientists have been able to understand the shape of biological molecules at the atomic level, and to use that knowledge to design (see drug design) drug candidates.

Modern drug discovery involves the identification of screening hits, medicinal chemistry and optimization of those hits to increase the affinity, selectivity (to reduce the potential of side effects), efficacy/potency, metabolic stability (to increase the half-life), and oral bioavailability. Once a compound that fulfills all of these requirements has been identified, it will begin the process of drug development prior to clinical trials. One or more of these steps may, but not necessarily, involve computer-aided drug design.

Despite advances in technology and understanding of biological systems, drug discovery is still a lengthy, "expensive, difficult, and inefficient process" with low rate of new therapeutic discovery.[10] In 2010, the research and development cost of each new molecular entity (NME) was approximately US$1.8 billion.[11] Drug discovery is done by pharmaceutical companies, with research assistance from universities. The "final product" of drug discovery is a patent on the potential drug. The drug requires very expensive Phase I, II and III clinical trials, and most of them fail. Small companies have a critical role, often then selling the rights to larger companies that have the resources to run the clinical trials.

7.7 Development

Main article: Drug development

Drug development is the process of bringing a new drug to the market once a lead compound has been identified through the process of drug discovery. It includes preclinical research (microorganisms/animals) and clinical trials (on humans) and may include the step of obtaining regulatory approval to market the drug.[12][13]

7.8 Regulation

Main article: Regulation of therapeutic goods

The regulation of drugs varies by jurisdiction. In some countries, such as the United States, they are regulated at the national level by a single agency. In other jurisdictions they are regulated at the state level, or at both state and national levels by various bodies, as is the case in Australia. The role of therapeutic goods regulation is designed mainly to protect the health and safety of the population. Regulation is aimed at ensuring the safety, quality, and efficacy of the therapeutic goods which are covered under the scope of the regulation. In most jurisdictions, therapeutic goods must be registered before they are allowed to be marketed. There is usually some degree of restriction of the availability of certain therapeutic goods depending on their risk to consumers.

Depending upon the jurisdiction, drugs may be divided into over-the-counter drugs (OTC) which may be available without special restrictions, and prescription drugs, which must be prescribed by a licensed medical practitioner. The precise distinction between OTC and prescription depends on the legal jurisdiction. A third category, "behind-the-counter" drugs, is implemented in some jurisdictions. These do not require a prescription, but must be kept in the dispensary, not visible to the public, and only be sold by a pharmacist or pharmacy technician. Doctors may also prescribe prescription drugs for off-label use - purposes which the drugs were not originally approved for by the regulatory agency. The Classification of Pharmaco-Therapeutic Referrals helps guide the referral process between pharmacists and doctors.

The International Narcotics Control Board of the United Nations imposes a world law of prohibition of certain drugs. They publish a lengthy list of chemicals and plants whose trade and consumption (where applicable) is forbidden. OTC drugs are sold without restriction as they are considered safe enough that most people will not hurt themselves accidentally by taking it as instructed.[14] Many countries, such as the United Kingdom have a third category of "pharmacy medicines", which can only be sold in registered pharmacies by or under the supervision of a pharmacist.

7.9 Drug pricing

Main article: Prescription costs

In many jurisdictions drug prices are regulated.

7.9.1 United Kingdom

In the UK the Pharmaceutical Price Regulation Scheme is intended to ensure that the National Health Service is able to purchase drugs at reasonable prices. The prices are negotiated between the Department of Health, acting with the authority of Northern Ireland and the UK Government, and the representatives of the Pharmaceutical industry brands, the Association of the British Pharmaceutical Industry (ABPI). For 2017 this payment percentage set by the PPRS will be 4,75%.[15]

7.9.2 Canada

In Canada, the Patented Medicine Prices Review Board examines drug pricing and determines if a price is excessive or not. In these circumstances, drug manufacturers must submit a proposed price to the appropriate regulatory agency. Furthermore, "the International Therapeutic Class Comparison Test is responsible for comparing the National Average Transaction Price of the patented drug product under review"[16] different countries that the prices are being compared to are the following: France, Germany, Italy, Sweden, Switzerland, the United Kingdom, and the United States[16]

7.9.3 Brazil

In Brazil, the prices are regulated through a legislation under the name of *Medicamento Genérico* (generic drugs) since 1999.

7.9.4 India

In India, drug prices are regulated by the National Pharmaceutical Pricing Authority.

7.9.5 United States

Main article: Prescription drug prices in the United States

In the United States, drug costs are unregulated, but instead are the result of negotiations between drug companies and insurance companies. High prices have been attributed to monopolies given to manufacturers by the government and a lack of ability for organizations to negotiate prices.[17]

7.10 Blockbuster drug

Main article: List of largest selling pharmaceutical products

A blockbuster drug is a drug generating more than $1 billion of revenue for the pharmaceutical company that sells it each year.[18] Cimetidine was the first drug ever to reach more than $1 billion a year in sales, thus making it the first blockbuster drug.[19]

> "In the pharmaceutical industry, a blockbuster drug is one that achieves acceptance by prescribing physicians as a therapeutic standard for, most commonly, a highly prevalent chronic (rather than acute) condition. Patients often take the medicines for long periods."[20]

7.11 History

Main article: History of pharmacy

7.11.1 Prescription drug history

Antibiotics first arrived on the medical scene in 1932 thanks to Gerhard Domagk;[21] and coined the "wonder drugs". The introduction of the sulfa drugs led to a decline in the U.S. mortality rate from pneumonia to drop from 0.2% each year to 0.05% by 1939.[22] Antibiotics inhibit the growth or the metabolic activities of bacteria and other microorganisms by a chemical substance of microbial origin. Penicillin, introduced a few years later, provided a broader spectrum of activity compared to sulfa drugs and reduced side effects. Streptomycin, found in 1942, proved to be the first drug effective against the cause of tuberculosis and also came to be the best known of a long series of important antibiotics. A second generation of antibiotics was introduced in the 1940s: aureomycin and chloramphenicol. Aureomycin was the best known of the second generation.

Lithium was discovered in the 19th century for nervous disorders and its possible mood-stabilizing or prophylactic effect; it was cheap and easily produced. As lithium fell out of favor in France, valpromide came into play. This antibiotic was the origin of the drug that eventually created the mood stabilizer category. Valpromide had distinct psychotrophic effects that were of benefit in both the treatment of acute manic states and in the maintenance treatment of manic depression illness. Psychotropics can either be sedative or stimulant; sedatives aim at damping down the extremes of behavior. Stimulants aim at restoring normality by increasing tone. Soon arose the notion of a tranquilizer which was quite different from any sedative or stimulant. The term tranquilizer took over the notions of sedatives and became the dominant term in the West through the 1980s. In Japan, during this time, the term tranquilizer produced the notion of a psyche-stabilizer and the term mood stabilizer vanished.[23]

Premarin (conjugated estrogens, introduced in 1942) and Prempro (a combination estrogen-progestin pill, introduced in 1995) dominated the hormone replacement therapy (HRT) during the 1990s. HRT is not a life-saving drug, nor does it cure any disease. HRT has been prescribed to improve one's quality of life. Doctors prescribe estrogen for their older female patients both to treat short-term menopausal symptoms and to prevent long-term diseases. In the 1960s and early 1970s more and more physicians began to prescribe estrogen for their female patients. between 1991 and 1999, Premarin was listed as the most popular prescription and best-selling drug in America.[23]

The first oral contraceptive, Enovid, was approved by FDA in 1960. Oral contraceptives inhibit ovulation and so prevent conception. Enovid was known to be much more effective than alternatives including the condom and the diaphragm. As early as 1960, oral contraceptives were available in several different strengths by every manufacturer. In the 1980s and 1990s an increasing number of options arose including, most recently, a new delivery system for the oral contraceptive via a transdermal patch. In 1982, a new version of the Pill was introduced, known as the "biphasic" pill. By 1985, a new triphasic pill was approved. Physicians began to think of the Pill as an excellent means of birth control for young women.[23]

Stimulants such as Ritalin (methylphenidate) came to be pervasive tools for behavior management and modification in young children. Ritalin was first marketed in 1955 for narcolepsy; its potential users were middle-aged and the elderly. It wasn't until some time in the 1980s along with hyperactivity in children that Ritalin came onto the market. Medical use of methlyphanidate is predominately for symptoms of attention deficit/hyperactivity disorder (ADHD). Consumption of methylphenidate in the U.S. out-paced all other countries between 1991 and 1999. Significant growth in consumption was also evident in Canada, New Zealand, Australia, and Norway. Currently, 85% of the world's methylphanidate is consumed in America.[23]

The first minor tranquilizer was Meprobamate. Only fourteen months after it was made available, meprobamate had become the country's largest-selling prescription drug. By 1957, meprobamate had become the fastest-growing drug in history. The popularity of meprobamate paved the way for Librium and Valium, two minor tranquilizers that belonged to a new chemical class of drugs called the benzodiazepines. These were drugs that worked chiefly as antianxiety agents and muscle relaxants. The first benzodiazepine was Librium. Three months after it was approved, Librium had become the most prescribed tranquilizer in the nation. Three years later, Valium hit the shelves and was ten times more effective as a muscle relaxant and anticonvulsant. Valium was the most versatile of the minor tranquilizers. Later came the widespread adoption of major tranquilizers such as chlorpromazine and the drug reserpine. In 1970 sales began to decline for Valium and Librium, but sales of new and improved tranquilizers, such as Xanax, introduced in 1981 for the newly created diagnosis of panic disorder, soared.[23]

Mevacor (lovastatin) is the first and most influential statin in the American market. The 1991 launch of Pravachol (pravastatin), the second available in the United States, and the release of Zocor (simvastatin) made Mevacor no longer the only statin on the market. In 1998, Viagra was released as a treatment for erectile dysfunction.[23]

7.11.2 Ancient pharmacology

Using plants and plant substances to treat all kinds of diseases and medical conditions is believed to date back to prehistoric medicine.

The Kahun Gynaecological Papyrus, the oldest known medical text of any kind, dates to about 1800 BC and represents the first documented use of any kind of drug.[24][25] It and other medical papyri describe Ancient Egyptian medical practices, such as using honey to treat infections and the legs of bee-eaters to treat neck pains.

Ancient Babylonian medicine demonstrate the use of prescriptions in the first half of the 2nd millennium BC. Medicinal creams and pills were employed as treatments.[26]

On the Indian subcontinent, the Atharvaveda, a sacred text of Hinduism whose core dates from the 2nd millennium BC, although the hymns recorded in it are believed to be older, is the first Indic text dealing with medicine. It describes plant-based drugs to counter diseases.[27] The earliest foundations of ayurveda were built on a synthesis of selected ancient herbal practices, together with a massive addition of theoretical conceptualizations, new nosologies and new therapies dating from about 400 BC onwards.[28] The student of Āyurveda was expected to know ten arts that were indispensable in the preparation and application of his medicines: distillation, operative skills, cooking, horticulture, metallurgy, sugar manufacture, pharmacy, analysis and separation of minerals, compounding of metals, and preparation of alkalis.

The Hippocratic Oath for physicians, attributed to 5th century BC Greece, refers to the existence of "deadly drugs", and ancient Greek physicians imported drugs from Egypt and elsewhere.[29]

7.11.3 Medieval pharmacology

Al-Kindi's 9th century AD book, *De Gradibus* and Ibn Sina (Avicenna)'s *The Canon of Medicine* cover a range of drugs known to Medicine in the medieval Islamic world.

Medieval medicine saw advances in surgery, but few truly effective drugs existed, beyond opium (found in such extremely popular drugs as the "Great Rest" of the Antidotarium Nicolai at the time)[30] and quinine. Folklore cures and potentially poisonous metal-based compounds were popular treatments. Theodoric Borgognoni, (1205–1296), one of the most significant surgeons of the medieval period, responsible for introducing and promoting important surgical advances including basic antiseptic practice and the use of anaesthetics. Garcia de Orta described some herbal treatments that were used.

7.11.4 Modern pharmacology

For most of the 19th century, drugs were not highly effective, leading Oliver Wendell Holmes, Sr. to famously comment in 1842 that "if all medicines in the world were thrown into the sea, it would be all the better for mankind and all the worse for the fishes".[31]:21

During the First World War, Alexis Carrel and Henry Dakin developed the Carrel-Dakin method of treating wounds with an irrigation, Dakin's solution, a germicide which helped prevent gangrene.

In the inter-war period, the first anti-bacterial agents such as the sulpha antibiotics were developed. The Second World War saw the introduction of widespread and effective antimicrobial therapy with the development and mass production of penicillin antibiotics, made possible by the pressures of the war and the collaboration of British scientists with the American pharmaceutical industry.

Medicines commonly used by the late 1920s included aspirin, codeine, and morphine for pain; digitalis, nitroglycerin, and quinine for heart disorders, and insulin for diabetes. Other drugs included antitoxins, a few biological vaccines, and a few synthetic drugs. In the 1930s antibiotics emerged: first sulfa drugs, then penicillin and other antibiotics. Drugs increasingly became "the center of medical practice".[31]:22 In the 1950s other drugs emerged including corticosteroids for inflammation, rauwolfia alkaloids as tranqulizers and antihypertensives, antihistamines for nasal allergies, xanthines for asthma, and typical antipsychotics for psychosis.[31]:23–24 As of 2007, thousands of approved drugs have been developed. Increasingly, biotechnology is used to discover biopharmaceuticals.[31] Recently, multidisciplinary approaches have yielded a wealth of new data on the development of novel antibiotics and antibacterials and on the use of biological agents for antibacterial therapy.[32]

In the 1950s new psychiatric drugs, notably the antipsychotic chlorpromazine, were designed in laboratories and slowly came into preferred use. Although often accepted as an advance in some ways, there was some opposition, due to serious adverse effects such as tardive dyskinesia. Patients often opposed psychiatry and refused or stopped taking the drugs when not subject to psychiatric control.

Governments have been heavily involved in the regulation of drug development and drug sales. In the U.S., the Elixir Sulfanilamide disaster led to the establishment of the Food and Drug Administration, and the 1938 Federal Food, Drug, and Cosmetic Act required manufacturers to file new drugs with the FDA. The 1951 Humphrey-Durham Amendment required certain drugs to be sold by prescription. In 1962 a subsequent amendment required new drugs to be tested for efficacy and safety in clinical trials.[31]:24–26

Until the 1970s, drug prices were not a major concern for doctors and patients. As more drugs became prescribed for chronic illnesses, however, costs became burdensome, and by the 1970s nearly every U.S. state required or encouraged the substitution of generic drugs for higher-priced brand names. This also led to the 2006 U.S. law, Medicare Part D, which offers Medicare coverage for drugs.[31]:28–29

As of 2008, the United States is the leader in medical research, including pharmaceutical development. U.S. drug prices are among the highest in the world, and drug innovation is correspondingly high. In 2000 U.S. based firms developed 29 of the 75 top-selling drugs; firms from the second-largest market, Japan, developed eight, and the United Kingdom contributed 10. France, which imposes price controls, developed three. Throughout the 1990s outcomes were similar.[31]:30–31

7.12 Controversies

Controversies concerning pharmaceutical drugs include patient access to drugs under development and not yet approved, pricing, and environmental issues.

7.12.1 Access to unapproved drugs

Main articles: Named patient programs and Expanded access

Governments worldwide have created provisions for granting access to drugs prior to approval for patients who have exhausted all alternative treatment options and do not match clinical trial entry criteria. Often grouped under the labels of compassionate use, expanded access, or named patient supply, these programs are governed by rules which vary by country defining access criteria, data collection, promotion, and control of drug distribution.[33]

Within the United States, pre-approval demand is generally met through treatment IND (investigational new drug) applications (INDs), or single-patient INDs. These mechanisms, which fall under the label of expanded access programs, provide access to drugs for groups of patients or individuals residing in the US. Outside the US, Named Patient Programs provide controlled, pre-approval access to drugs in response to requests by physicians on behalf of specific, or "named", patients before those medicines are licensed in the patient's home country. Through these programs, patients are able to access drugs in late-stage clinical trials or approved in other countries for a genuine, unmet medical need, before those drugs have been licensed in the patient's home country.

Patients who have not been able to get access to drugs in development have organized and advocated for greater access. In the United States, ACT UP formed in the 1980s, and eventually formed its Treatment Action Group in part to pressure the US government to put more resources into discovering treatments for AIDS and then to speed release of drugs that were under development.[34]

The Abigail Alliance was established in November 2001 by Frank Burroughs in memory of his daughter, Abigail.[35] The Alliance seeks broader availability of investigational drugs on behalf of terminally ill patients.

In 2013, BioMarin Pharmaceutical was at the center of a high-profile debate regarding expanded access of cancer patients to experimental drugs.[36][37]

7.12.2 Access to medicines and drug pricing

Main articles: Essential medicines and Societal views on patents

Essential medicines as defined by the World Health Organization (WHO) are "those drugs that satisfy the health care needs of the majority of the population; they should therefore be available at all times in adequate amounts and in appropriate dosage forms, at a price the community can afford."[38] Recent studies have found that most of the medicines on the WHO essential medicines list, outside of the field of HIV drugs, are not patented in the developing world, and that lack of widespread access to these medicines arise from issues fundamental to economic development - lack of infrastructure and poverty.[39] Médecins Sans Frontières also runs a Campaign for Access to Essential Medicines campaign, which includes advocacy for greater resources to be devoted to currently untreatable diseases that primarily occur in the developing world. The Access to Medicine Index tracks how well pharmaceutical companies make their products available in the developing world.

World Trade Organization negotiations in the 1990s, including the TRIPS Agreement and the Doha Declaration, have centered on issues at the intersection of international trade in pharmaceuticals and intellectual property rights, with developed world nations seeking strong intellectual property rights to protect investments made to develop new drugs, and developing world nations seeking to promote their generic pharmaceuticals industries and their ability to make medicine available to their people via compulsory licenses.

Some have raised ethical objections specifically with respect to pharmaceutical patents and the high prices for drugs that they enable their proprietors to charge, which poor people in the developed world, and developing world, cannot afford.[40][41] Critics also question the rationale that exclusive patent rights and the resulting high prices are required for pharmaceutical companies to recoup the large investments needed for research and development.[40] One study concluded that marketing expenditures for new drugs often doubled the amount that was allocated for research and development.[42] Other critics claim that patent settlements would be costly for consumers, the health care system, and state and federal governments because it would result in delaying access to lower cost generic medicines.[43]

Novartis fought a protracted battle with the government of India over the patenting of its drug, Gleevec, in India, which ended up in India's Supreme Court in a case known as Novartis v. Union of India & Others. The Supreme Court ruled narrowly against Novartis, but opponents of patenting drugs claimed it as a major victory.[44]

7.12.3 Environmental issues

Main article: Environmental impact of pharmaceuticals and personal care products

The environmental impact of pharmaceuticals and personal care products is controversial. PPCPs are substances used by individuals for personal health or cosmetic reasons and the products used by agribusiness to boost growth or health of livestock. PPCPs comprise a diverse collection of thousands of chemical substances, including prescription and over-the-counter therapeutic drugs, veterinary drugs, fragrances, and cosmetics. PPCPs have been detected in water bodies throughout the world and ones that persist in the environment are called Environmental Persistent Pharmaceutical Pollutants. The effects of these chemicals on humans and the environment are not yet known, but to date there is no scientific evidence that they affect human health.[45]

7.13 See also

- Compliance
- Deprescribing
- Drug nomenclature

- Identification of medicinal products
- List of drugs
- List of pharmaceutical companies
- Orphan drug
- Overmedication
- Pharmaceutical code

7.14 References

[1] Definition and classification of Drug or Pharmaceutical Regulatory aspects of drug approval Accessed 30 December 2013.

[2] US Federal Food, Drug, and Cosmetic Act, SEC. 210., (g)(1)(B). Accessed 17 August 2008.

[3] Directive 2004/27/EC of the European Parliament and of the Council of 31 March 2004 amending Directive 2001/83/EC on the Community code relating to [[Medicine|medicinal products for human use. Article 1.] Published 31 March 2004. Accessed 17 August 2008.

[4] Directive 2004/27/EC Official Journal of the European Union. 30 April 2004 L136

[5] FDA Glossary

[6] Qato DM; Wilder J; Schumm L; Gillet V; Alexander G (2016-04-01). "Changes in prescription and over-the-counter medication and dietary supplement use among older adults in the united states, 2005 vs 2011". *JAMA Internal Medicine*. **176** (4): 473–482. PMID 26998708. doi:10.1001/jamainternmed.2015.8581.

[7] http://www.epgonline.org database of prescription pharmaceutical products including drug classifications

[8] Spatz I, McGee N (25 November 2013). "Specialty Pharmaceuticals". Health Policy Briefs. *Health Affairs*. Bethesda, Maryland. What's The Background?. Retrieved 28 August 2015.

[9] Finkel, Richard; Cubeddu, Luigi; Clark, Michelle (2009). *Lippencott's Illustrated Reviews: Pharmacology 4th Edition*. Lippencott Williams & Wilkins. pp. 1–4. ISBN 978-0-7817-7155-9.

[10] Anson, Blake D.; Ma, Junyi; He, Jia-Qiang (1 May 2009). "Identifying Cardiotoxic Compounds". *Genetic Engineering & Biotechnology News*. TechNote. **29** (9). Mary Ann Liebert. pp. 34–35. ISSN 1935-472X. OCLC 77706455. Archived from the original on 25 July 2009. Retrieved 25 July 2009.

[11] Steven M. Paul; Daniel S. Mytelka; Christopher T. Dunwiddie; Charles C. Persinger; Bernard H. Munos; Stacy R. Lindborg; Aaron L. Schacht (2010). "How to improve R&D productivity: the pharmaceutical industry's grand challenge". *Nature Reviews Drug Discovery*. **9** (3): 203–214. PMID 20168317. doi:10.1038/nrd3078.

[12] Strovel, Jeffrey; Sittampalam, Sitta; Coussens, Nathan P.; Hughes, Michael; Inglese, James; Kurtz, Andrew; Andalibi, Ali; Patton, Lavonne; Austin, Chris; Baltezor, Michael; Beckloff, Michael; Weingarten, Michael; Weir, Scott (July 1, 2016). "Early Drug Discovery and Development Guidelines: For Academic Researchers, Collaborators, and Start-up Companies". *Assay Guidance Manual*. Eli Lilly & Company and the National Center for Advancing Translational Sciences.

[13] Taylor, David (2015). "The Pharmaceutical Industry and the Future of Drug Development". *Issues in Environmental Science and Technology*. Royal Society of Chemistry: 1–33. doi:10.1039/9781782622345-00001.

[14] "Medication Safety". *House Findings*. Retrieved 21 July 2016.

[15] "PPRS: payment percentage 2017 - Details". 23 December 2016.

[16] "Schedule 7 - International Therapeutic Class Comparison Test".

[17] Kesselheim, Aaron S.; Avorn, Jerry; Sarpatwari, Ameet (23 August 2016). "The High Cost of Prescription Drugs in the United States". *JAMA*. **316** (8): 858. doi:10.1001/jama.2016.11237.

[18] "*"Blockbuster medicine" is defined as being one which achieves annual revenues of over US$1 billion at global level.*" in European Commission, *Pharmaceutical Sector Inquiry, Preliminary Report (DG Competition Staff Working Paper)*, 28 November 2008, page 17 (pdf, 1.95 MB).

[19] Whitney, Jake (February 2006). "Pharmaceutical Sales 101: Me-Too Drugs". *Guernica*. Retrieved 31 July 2008.

[20] Finkelstein, Temin "Reasonable Rx: Solving the Drug Price Crisis" 11 January 2008

[21] "Chemical & Engineering News: Top Pharmaceuticals: Prontosil".

[22] Dowling HF (June 1972). "Frustration and foundation. Management of pneumonia before antibiotics". *JAMA*. **220** (10): 1341–5. PMID 4553966. doi:10.1001/jama.1972.03200100053011.

[23] Tone, Andrea and Elizabeth Watkins, *Medicating Modern America: Prescription Drugs in History*. New York and London, New York University, 2007. Print.

[24] Griffith, F. Ll. *The Petrie Papyri: Hieratic Papyri from Kahun and Gurob*

[25] The Kahun Gynaecological Papyrus

[26] H. F. J. Horstmanshoff, Marten Stol, Cornelis Tilburg (2004), *Magic and Rationality in Ancient Near Eastern and Graeco-Roman Medicine*, p. 99, Brill Publishers, ISBN 90-04-13666-5.

[27] See Atharvaveda XIX.34.9

[28] Kenneth G. Zysk, *Asceticism and Healing in Ancient India: Medicine in the Buddhist Monastery*, Oxford University Press, rev. ed. (1998) ISBN 0-19-505956-5.

[29] Heinrich von Staden, *Herophilus: The Art of Medicine in Early Alexandria* (Cambridge: Cambridge University Press, 1989), pp. 1-26.

[30] Everett, Nicholas; Gabra, Martino (2014-08-08). "The pharmacology of medieval sedatives: the "Great Rest" of the Antidotarium Nicolai". *Journal of Ethnopharmacology*. **155** (1): 443–449. ISSN 1872-7573. PMID 24905867. doi:10.1016/j.jep.2014.05.048.

[31] Finkelstein S, Temin P (2008). *Reasonable Rx: Solving the drug price crisis*. FT Press.

[32] Miller, AA; Miller, PF (editor) (2011). *Emerging Trends in Antibacterial Discovery: Answering the Call to Arms*. Caister Academic Press. ISBN 978-1-904455-89-9.

[33] Helene S (2010). "EU Compassionate Use Programmes (CUPs): Regulatory Framework and Points to Consider before CUP Implementation". *Pharm Med*. **24** (4): 223–229. doi:10.1007/bf03256820.

[34] Gina Kolata for the *New York Times*. 12 September 1994 F.D.A. Debate on Speedy Access to AIDS Drugs Is Reopening

[35] Phillips, Lisa (4 September 2008). "Contract Law and Ethical Issues Underscore the Latest Lawsuit About Access to Experimental Drugs for Duchenne Muscular Dystrophy" (PDF). *Neurology Today*. **8** (17): 20–21. doi:10.1097/01.nt.0000337676.20893.50.

[36] "Andrea Sloan Faces Pharma Firm With History of Indifference". *Huffington Post*. 26 September 2013. Retrieved 24 December 2013.

[37] "In cancer drug battle, both sides appeal to ethics". *CNN*. 20 September 2013. Retrieved 24 December 2013.

[38] "The Selection and Use of Essential Medicines - WHO Technical Report Series, No. 914: 4. Other outstanding technical issues: 4.2 Description of essential medicines". Apps.who.int. 2016-04-14. Retrieved 2016-06-25.

[39] Stanley P Kowalksy, 2013. Patent Landscape Analysis of Healthcare Innovations

[40] Banta D.H. (2001). "Worldwide Interest in Global Access to Drugs". *Journal of the American Medical Association*. **285** (22): 2844–46. PMID 11401589. doi:10.1001/jama.285.22.2844.

[41] Ferreira L (2002). "Access to Affordable HIV/AIDS Drugs: The Human Rights Obligations of Multinational Pharmaceutical Corporations". *Fordham Law Review*. **71** (3): 1133–79. PMID 12523370.

[42] Barton J.H.; Emanuel E.J. (2005). "The Patents-Based Pharmaceutical Development Process: Rationale, Problems and Potential Reforms". *Journal of the American Medical Association*. **294** (16): 2075–82. PMID 16249422. doi:10.1001/jama.294.16.2075.

[43] "Misguided Policy on Patents". Generic Pharmaceutical Association (GPhA). Retrieved 8 Oct 2015.

[44] Charlotte Harrison Patent watch Nature Reviews Drug Discovery 12, 336–337 (2013)

[45] U.S. EPA. Pharmaceuticals and Personal Care Products. Accessed 16 March 2009.

7.15 External links

- WHO Model List of Essential Medicines
- Medicines in Development | PhRMA
- Informations and Leaflets of approved pharmaceutical drugs | Medikamio
- SuperCYP: Database for Drug-Cytochrome- and Drug-Drug-interactions

Chapter 8

Drug metabolism

This article is about the scientific concept of drug metabolism. For alternative medicine, see Detoxification (alternative medicine).

Drug metabolism is the metabolic breakdown of drugs by living organisms, usually through specialized enzymatic systems. More generally, **xenobiotic metabolism** (from the Greek xenos "stranger" and biotic "related to living beings") is the set of metabolic pathways that modify the chemical structure of xenobiotics, which are compounds foreign to an organism's normal biochemistry, such any drug or poison. These pathways are a form of biotransformation present in all major groups of organisms, and are considered to be of ancient origin. These reactions often act to detoxify poisonous compounds (although in some cases the intermediates in xenobiotic metabolism can themselves cause toxic effects). The study of drug metabolism is called pharmacokinetics.

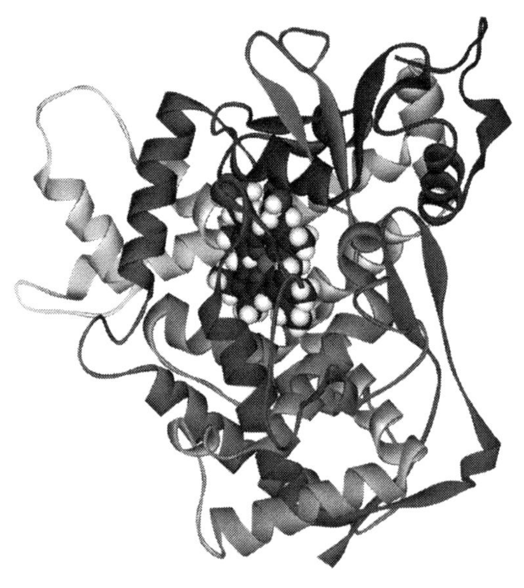

Cytochrome P450 oxidases are important enzymes in xenobiotic metabolism.

The metabolism of pharmaceutical drugs is an important aspect of pharmacology and medicine. For example, the rate of metabolism determines the duration and intensity of a drug's pharmacologic action. Drug metabolism also affects multidrug resistance in infectious diseases and in chemotherapy for cancer, and the actions of some drugs as substrates or inhibitors of enzymes involved in xenobiotic metabolism are a common reason for hazardous drug interactions. These pathways are also important in environmental science, with the xenobiotic metabolism of microorganisms determining whether a pollutant will be broken down during bioremediation, or persist in the environment. The enzymes of xenobiotic metabolism, particularly the glutathione S-transferases are also important in agriculture, since they may produce resistance to pesticides and herbicides.

Drug metabolism is divided into three phases. In phase I, enzymes such as cytochrome P450 oxidases introduce reactive or polar groups into xenobiotics. These modified compounds are then conjugated to polar compounds in phase II reactions. These reactions are catalysed by transferase enzymes such as glutathione S-transferases. Finally, in phase III, the conjugated xenobiotics may be further processed, before being recognised by efflux transporters and pumped out of cells. Drug metabolism often converts lipophilic compounds into hydrophilic products that are more readily excreted.

8.1 Permeability barriers and detoxification

The exact compounds an organism is exposed to will be largely unpredictable, and may differ widely over time; these are major characteristics of xenobiotic toxic stress.[1] The major challenge faced by xenobiotic detoxification systems is that they must be able to remove the almost-limitless number of xenobiotic compounds from the complex mixture of chemicals involved in normal metabolism. The solution that has evolved to address this problem is an elegant combination of physical barriers and low-specificity enzymatic systems.

All organisms use cell membranes as hydrophobic permeability barriers to control access to their internal environment. Polar compounds cannot diffuse across these cell membranes, and the uptake of useful molecules is mediated through transport proteins that specifically select substrates from the extracellular mixture. This selective uptake means that most hydrophilic molecules cannot enter cells, since they are not recognised by any specific transporters.[2] In contrast, the diffusion of hydrophobic compounds across these barriers cannot be controlled, and organisms, therefore, cannot exclude lipid-soluble xenobiotics using membrane barriers.

However, the existence of a permeability barrier means that organisms were able to evolve detoxification systems that exploit the hydrophobicity common to membrane-permeable xenobiotics. These systems therefore solve the specificity problem by possessing such broad substrate specificities that they metabolise almost any non-polar compound.[1] Useful metabolites are excluded since they are polar, and in general contain one or more charged groups.

The detoxification of the reactive by-products of normal metabolism cannot be achieved by the systems outlined above, because these species are derived from normal cellular constituents and usually share their polar characteristics. However, since these compounds are few in number, specific enzymes can recognize and remove them. Examples of these specific detoxification systems are the glyoxalase system, which removes the reactive aldehyde methylglyoxal,[3] and the various antioxidant systems that eliminate reactive oxygen species.[4]

8.2 Phases of detoxification

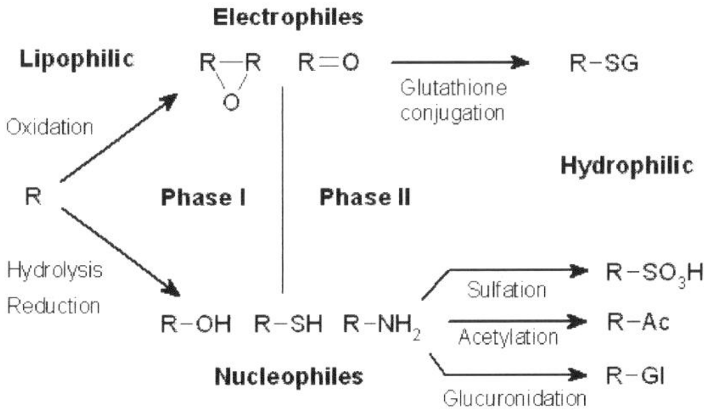

Phases I and II of the metabolism of a lipophilic xenobiotic.

The metabolism of xenobiotics is often divided into three phases:- modification, conjugation, and excretion. These reactions act in concert to detoxify xenobiotics and remove them from cells.

8.2.1 Phase I – modification

In phase I, a variety of enzymes act to introduce reactive and polar groups into their substrates. One of the most common modifications is hydroxylation catalysed by the cytochrome P-450-dependent mixed-function oxidase system. These enzyme complexes act to incorporate an atom of oxygen into nonactivated hydrocarbons, which can result in either the introduction of hydroxyl groups or N-, O- and S-dealkylation of substrates.[5] The reaction mechanism of the P-450 oxidases proceeds through the reduction of cytochrome-bound oxygen and the generation of a highly-reactive oxyferryl species, according to the following scheme:[6]

$$O_2 + NADPH + H^+ + RH \rightarrow NADP^+ + H_2O + ROH$$

Phase I reactions (also termed nonsynthetic reactions) may occur by oxidation, reduction, hydrolysis, cyclization, decyclization, and addition of oxygen or removal of hydrogen, carried out by mixed function oxidases, often in the liver. These oxidative reactions typically involve a cytochrome P450 monooxygenase (often abbreviated CYP), NADPH and oxygen. The classes of pharmaceutical drugs that utilize this method for their metabolism include phenothiazines, paracetamol, and steroids. If the metabolites of phase I reactions are sufficiently polar, they may be readily excreted at this point. However, many phase I products are not eliminated rapidly and undergo a subsequent reaction in which an endogenous substrate combines with the newly incorporated functional group to form a highly polar conjugate.

A common Phase I oxidation involves conversion of a C-H bond to a C-OH. This reaction sometimes converts a pharmacologically inactive compound (a prodrug) to a pharmacologically active one. By the same token, Phase I can turn a nontoxic molecule into a poisonous one (toxification). Simple hydrolysis in the stomach is normally an innocuous reaction, however there are exceptions. For example, phase I metabolism converts acetonitrile to $HOCH_2CN$, which rapidly dissociates into formaldehyde and hydrogen cyanide, both of which are toxic.[7]

Phase I metabolism of drug candidates can be simulated in the laboratory using non-enzyme catalysts.[8] This example of a biomimetic reaction tends to give products that often contains the Phase I metabolites. As an example, the major metabolite of the pharmaceutical trimebutine, desmethyltrimebutine (nor-trimebutine), can be efficiently produced by in vitro oxidation of the commercially available drug. Hydroxylation of an N-methyl group leads to expulsion of a molecule of formaldehyde, while oxidation of the O-methyl groups takes place to a lesser extent.

Oxidation

- Cytochrome P450 monooxygenase system
- Flavin-containing monooxygenase system
- Alcohol dehydrogenase and aldehyde dehydrogenase
- Monoamine oxidase
- Co-oxidation by peroxidases

Reduction

- NADPH-cytochrome P450 reductase

Cytochrome P450 reductase, also known as NADPH: ferrihemoprotein oxidoreductase, NADPH:hemoprotein oxidoreductase, NADPH:P450 oxidoreductase, P450 reductase, POR, CPR, CYPOR, is a membrane-bound enzyme required for electron transfer to cytochrome P450 in the microsome of the eukaryotic cell from a FAD- and FMN-containing enzyme NADPH:cytochrome P450 reductase The general scheme of electron flow in the POR/P450 system is: NADPH → FAD → FMN → P450 → O_2

- Reduced (ferrous) cytochrome P450

During reduction reactions, a chemical can enter *futile cycling*, in which it gains a free-radical electron, then promptly loses it to oxygen (to form a superoxide anion).

Hydrolysis

- Esterases and amidase
- Epoxide hydrolase

8.2.2 Phase II – conjugation

In subsequent phase II reactions, these activated xenobiotic metabolites are conjugated with charged species such as glutathione (GSH), sulfate, glycine, or glucuronic acid. Sites on drugs where conjugation reactions occur include carboxyl (-COOH), hydroxyl (-OH), amino (NH_2), and sulfhydryl (-SH) groups. Products of conjugation reactions have increased molecular weight and tend to be less active than their substrates, unlike Phase I reactions which often produce active metabolites. The addition of large anionic groups (such as GSH) detoxifies reactive electrophiles and produces more polar metabolites that cannot diffuse across membranes, and may, therefore, be actively transported.

These reactions are catalysed by a large group of broad-specificity transferases, which in combination can metabolise almost any hydrophobic compound that contains nucleophilic or electrophilic groups.[1] One of the most important classes of this group is that of the glutathione S-transferases (GSTs).

8.2.3 Phase III – further modification and excretion

After phase II reactions, the xenobiotic conjugates may be further metabolised. A common example is the processing of glutathione conjugates to acetylcysteine (mercapturic acid) conjugates.[10] Here, the γ-glutamate and glycine residues in the glutathione molecule are removed by Gamma-glutamyl transpeptidase and dipeptidases. In the final step, the cystine residue in the conjugate is acetylated.

Conjugates and their metabolites can be excreted from cells in phase III of their metabolism, with the anionic groups acting as affinity tags for a variety of membrane transporters of the multidrug resistance protein (MRP) family.[11] These proteins are members of the family of ATP-binding cassette transporters and can catalyse the ATP-dependent transport of a huge variety of hydrophobic anions,[12] and thus act to remove phase II products to the extracellular medium, where they may be further metabolised or excreted.[13]

8.3 Endogenous toxins

The detoxification of endogenous reactive metabolites such as peroxides and reactive aldehydes often cannot be achieved by the system described above. This is the result of these species' being derived from normal cellular constituents and usually sharing their polar characteristics. However, since these compounds are few in number, it is possible for enzymatic systems to utilize specific molecular recognition to recognize and remove them. The similarity of these molecules to useful metabolites therefore means that different detoxification enzymes are usually required for the metabolism of each group of endogenous toxins. Examples of these specific detoxification systems are the glyoxalase system, which acts to dispose of the reactive aldehyde methylglyoxal, and the various antioxidant systems that remove reactive oxygen species.

8.4 Sites

Quantitatively, the smooth endoplasmic reticulum of the liver cell is the principal organ of drug metabolism, although every biological tissue has some ability to metabolize drugs. Factors responsible for the liver's contribution to drug metabolism include that it is a large organ, that it is the first organ perfused by chemicals absorbed in the gut, and that there are very high concentrations of most drug-metabolizing enzyme systems relative to other organs. If a drug is taken into the GI tract, where it enters hepatic circulation through the portal vein, it becomes well-metabolized and is said to show the *first pass effect*.

Other sites of drug metabolism include epithelial cells of the gastrointestinal tract, lungs, kidneys, and the skin. These sites are usually responsible for localized toxicity reactions.

8.5 Factors that affect drug metabolism

The duration and intensity of pharmacological action of most lipophilic drugs are determined by the rate they are metabolized to inactive products. The Cytochrome P450 monooxygenase system is the most important pathway in this regard. In general, anything that *increases* the rate of metabolism (*e.g.*, enzyme induction) of a pharmacologically active metabolite will *decrease* the duration and intensity of the drug action. The opposite is also true (*e.g.*, enzyme inhibition). However, in cases where an enzyme is responsible for metabolizing a pro-drug into a drug, enzyme induction can speed up this conversion and increase drug levels, potentially causing toxicity.

Various *physiological* and *pathological* factors can also affect drug metabolism. Physiological factors that can influence drug metabolism include age, individual variation (*e.g.*, pharmacogenetics), enterohepatic circulation, nutrition, intestinal flora, or sex differences.

In general, drugs are metabolized more slowly in fetal, neonatal and elderly humans and animals than in adults.

Genetic variation (polymorphism) accounts for some of the variability in the effect of drugs. With N-acetyltransferases (involved in *Phase II* reactions), individual variation creates a group of people who acetylate slowly (*slow acetylators*) and those who acetylate quickly, split roughly 50:50 in the population of Canada. This variation may have dramatic consequences, as the slow acetylators are more prone to dose-dependent toxicity.

Cytochrome P450 monooxygenase system enzymes can also vary across individuals, with deficiencies occurring in 1 – 30% of people, depending on their ethnic background.

Dose, frequency, route of administration, tissue distribution and protein binding of the drug affect its metabolism.

Pathological factors can also influence drug metabolism, including liver, kidney, or heart diseases.

In silico modelling and simulation methods allow drug metabolism to be predicted in virtual patient populations prior to performing clinical studies in human subjects.[14] This can be used to identify individuals most at risk from adverse reaction.

8.6 History

Studies on how people transform the substances that they ingest began in the mid-nineteenth century, with chemists discovering that organic chemicals such as benzaldehyde could be oxidized and conjugated to amino acids in the human body.[15] During the remainder of the nineteenth century, several other basic detoxification reactions were discovered, such as methylation, acetylation, and sulfonation.

In the early twentieth century, work moved on to the investigation of the enzymes and pathways that were responsible for the production of these metabolites. This field became defined as a separate area of study with the publication by Richard Williams of the book *Detoxication mechanisms* in 1947.[16] This modern biochemical research resulted in the identification of glutathione *S*-transferases in 1961,[17] followed by the discovery of cytochrome P450s in 1962,[18] and the realization of their central role in xenobiotic metabolism in 1963.[19][20]

8.7 See also

- Antioxidant
- Biodegradation
- Bioremediation
- Microbial biodegradation

8.8 References

[1] Jakoby WB, Ziegler DM (December 1990). "The enzymes of detoxication". *J. Biol. Chem.* **265** (34): 20715–8. PMID 2249981.

[2] Mizuno N, Niwa T, Yotsumoto Y, Sugiyama Y (September 2003). "Impact of drug transporter studies on drug discovery and development". *Pharmacol. Rev.* **55** (3): 425–61. PMID 12869659. doi:10.1124/pr.55.3.1.

[3] Thornalley PJ (July 1990). "The glyoxalase system: new developments towards functional characterization of a metabolic pathway fundamental to biological life". *Biochem. J.* **269** (1): 1–11. PMC 1131522 ©. PMID 2198020.

[4] Sies H (March 1997). "Oxidative stress: oxidants and antioxidants" (PDF). *Exp. Physiol.* **82** (2): 291–5. PMID 9129943. doi:10.1113/expphysiol.1997.sp004024.

[5] Guengerich FP (June 2001). "Common and uncommon cytochrome P450 reactions related to metabolism and chemical toxicity". *Chem. Res. Toxicol.* **14** (6): 611–50. PMID 11409933. doi:10.1021/tx0002583.

[6] Schlichting I, Berendzen J, Chu K, Stock AM, Maves SA, Benson DE, Sweet RM, Ringe D, Petsko GA, Sligar SG (March 2000). "The catalytic pathway of cytochrome p450cam at atomic resolution". *Science*. **287** (5458): 1615–22. PMID 10698731. doi:10.1126/science.287.5458.1615.

[7] "Acetonitrile (EHC 154, 1993)". *www.inchem.org*. Retrieved 2017-05-03.

[8] Akagah B, Lormier AT, Fournet A, Figadère B (December 2008). "Oxidation of antiparasitic 2-substituted quinolines using metalloporphyrin catalysts: scale-up of a biomimetic reaction for metabolite production of drug candidates". *Org. Biomol. Chem.* **6** (24): 4494–7. PMID 19039354. doi:10.1039/b815963g.

[9] Liston HL, Markowitz JS, DeVane CL (October 2001). "Drug glucuronidation in clinical psychopharmacology". *J Clin Psychopharmacol.* **21** (5): 500–15. PMID 11593076. doi:10.1097/00004714-200110000-00008.

[10] Boyland E, Chasseaud LF (1969). "The role of glutathione and glutathione S-transferases in mercapturic acid biosynthesis". *Adv. Enzymol. Relat. Areas Mol. Biol.* Advances in Enzymology – and Related Areas of Molecular Biology. **32**: 173–219. ISBN 9780470122778. PMID 4892500. doi:10.1002/9780470122778.ch5.

[11] Homolya L, Váradi A, Sarkadi B (2003). "Multidrug resistance-associated proteins: Export pumps for conjugates with glutathione, glucuronate or sulfate". *BioFactors.* **17** (1–4): 103–14. PMID 12897433. doi:10.1002/biof.5520170111.

[12] König J, Nies AT, Cui Y, Leier I, Keppler D (December 1999). "Conjugate export pumps of the multidrug resistance protein (MRP) family: localization, substrate specificity, and MRP2-mediated drug resistance". *Biochim. Biophys. Acta.* **1461** (2): 377–94. PMID 10581368. doi:10.1016/S0005-2736(99)00169-8.

[13] Commandeur JN, Stijntjes GJ, Vermeulen NP (June 1995). "Enzymes and transport systems involved in the formation and disposition of glutathione S-conjugates. Role in bioactivation and detoxication mechanisms of xenobiotics". *Pharmacol. Rev.* **47** (2): 271–330. PMID 7568330.

[14] Rostami-Hodjegan A, Tucker GT (February 2007). "Simulation and prediction of in vivo drug metabolism in human populations from *in vitro* data". *Nat Rev Drug Discov.* **6** (2): 140–8. PMID 17268485. doi:10.1038/nrd2173.

[15] Murphy PJ (June 2001). "Xenobiotic metabolism: a look from the past to the future". *Drug Metab. Dispos.* **29** (6): 779–80. PMID 11353742.

[16] Neuberger A, Smith RL (1983). "Richard Tecwyn Williams: the man, his work, his impact". *Drug Metab. Rev.* **14** (3): 559–607. PMID 6347595. doi:10.3109/03602538308991399.

[17] Booth J, Boyland E, Sims P (June 1961). "An enzyme from rat liver catalysing conjugations with glutathione". *Biochem. J.* **79** (3): 516–24. PMC 1205680 ©. PMID 16748905.

[18] Omura T, Sato R (April 1962). "A new cytochrome in liver microsomes". *J. Biol. Chem.* **237**: 1375–6. PMID 14482007.

[19] Estabrook RW (December 2003). "A passion for P450s (remembrances of the early history of research on cytochrome P450)". *Drug Metab. Dispos.* **31** (12): 1461–73. PMID 14625342. doi:10.1124/dmd.31.12.1461.

[20] Estabrook RW, Cooper DY, Rosenthal O (1963). "The light reversible carbon monoxide inhibition of steroid C-21 hydroxylase system in adrenal cortex". *Biochem Z.* **338**: 741–55. PMID 14087340.

8.9 Further reading

- Parvez H, Reiss C (2001). *Molecular Responses to Xenobiotics*. Elsevier. ISBN 0-345-42277-5.
- Ioannides C (2001). *Enzyme Systems That Metabolise Drugs and Other Xenobiotics*. John Wiley and Sons. ISBN 0-471-89466-4.
- Richardson M (1996). *Environmental Xenobiotics*. Taylor & Francis Ltd. ISBN 0-7484-0399-X.
- Ioannides C (1996). *Cytochromes P450: Metabolic and Toxicological Aspects*. CRC Press Inc. ISBN 0-8493-9224-1.
- Awasthi YC (2006). *Toxicology of Glutathionine S-transferses*. CRC Press Inc. ISBN 0-8493-2983-3.

8.10 External links

- Databases
  - Drug metabolism database
  - Directory of P450-containing Systems

    - University of Minnesota Biocatalysis/Biodegradation Database
    - SPORCalc

- Drug metabolism
    - Small Molecule Drug Metabolism
    - Drug metabolism portal

- Microbial biodegradation
    - Microbial Biodegradation, Bioremediation and Biotransformation

- History
    - History of Xenobiotic Metabolism at the Wayback Machine (archived July 13, 2007)

Chapter 9

Xenobiotic

Not to be confused with xenobiology.

A **xenobiotic** is a foreign chemical substance found within an organism that is not naturally produced by or expected to be present within. It can also cover substances that are present in much higher concentrations than are usual. Specifically, drugs such as antibiotics are xenobiotics in humans because the human body does not produce them itself, nor are they part of a normal food.

Natural compounds can also become xenobiotics if they are taken up by another organism, such as the uptake of natural human hormones by fish found downstream of sewage treatment plant outfalls, or the chemical defenses produced by some organisms as protection against predators.[1]

The term **xenobiotics**, however, is very often used in the context of pollutants such as dioxins and polychlorinated biphenyls and their effect on the biota, because xenobiotics are understood as substances foreign to an entire biological system, i.e. artificial substances, which did not exist in nature before their synthesis by humans. The term xenobiotic is derived from the Greek words ξένος (xenos) = foreigner, stranger and βίος (bios, vios) = life, plus the Greek suffix for adjectives -τικός, -ή, -ό (tic).

Xenobiotics may be grouped as carcinogens, drugs, environmental pollutants, food additives, hydrocarbons, and pesticides.

9.1 Xenobiotic metabolism

The body removes xenobiotics by xenobiotic metabolism. This consists of the deactivation and the excretion of xenobiotics, and happens mostly in the liver. Excretion routes are urine, feces, breath, and sweat. Hepatic enzymes are responsible for the metabolism of xenobiotics by first activating them (oxidation, reduction, hydrolysis and/or hydration of the xenobiotic), and then conjugating the active secondary metabolite with glucuronic acid, sulphuric acid, or glutathione, followed by excretion in bile or urine. An example of a group of enzymes involved in xenobiotic metabolism is hepatic microsomal cytochrome P450. These enzymes that metabolize xenobiotics are very important for the pharmaceutical industry, because they are responsible for the breakdown of medications.

Organisms can also evolve to tolerate xenobiotics. An example is the co-evolution of the production of tetrodotoxin in the rough-skinned newt and the evolution of tetrodotoxin resistance in its predator, the Common Garter Snake. In this predator–prey pair, an evolutionary arms race has produced high levels of toxin in the newt and correspondingly high levels of resistance in the snake.[2] This evolutionary response is based on the snake evolving modified forms of the ion channels that the toxin acts upon, so becoming resistant to its effects.[3]

9.2 Xenobiotics in the environment

Main article: Environmental xenobiotic

Xenobiotic substances are an issue for sewage treatment systems, since they are many in number, and each will present its own problems as to how to remove them (and whether it is worth trying to). It can be dangerous to the health.

Some xenobiotics are resistant to degradation. For example, they may be synthetic organochlorides such as plastics and pesticides, or naturally occurring organic chemicals such as polyaromatic hydrocarbons (PAHs) and some fractions of crude oil and coal. However, it is believed that microorganisms are capable of degrading almost all the different complex and resistant xenobiotics found on the earth.[4] Many xenobiotics produce a variety of biological effects, which is used when they are characterized using bioassay. Before they can be registered for sale in most countries, xenobiotic pesticides must undergo extensive evaluation for risk factors, such as toxicity to humans, ecotoxicity, or persistence in the environment. For example, during the registration process, the herbicide, cloransulam-methyl was found to degrade relatively quickly in soil.[5]

9.3 Inter-species organ transplantation

Main article: Xenotransplantation

The term **xenobiotic** is also used to refer to organs transplanted from one species to another. For example, some researchers hope that hearts and other organs could be transplanted from pigs to humans. Many people die every year whose lives could have been saved if a critical organ had been available for transplant. Kidneys are currently the most commonly transplanted organ. Xenobiotic organs would need to be developed in such a way that they would not be rejected by the immune system.

9.4 See also

Drug metabolism – Xenobiotic metabolism is redirected to the special case: Drug metabolism.

9.5 References

[1] Mansuy D (2013). "Metabolism of xenobiotics: beneficial and adverse effects". *Biol Aujourdhui*. **207** (1): 33–37. PMID 23694723. doi:10.1051/jbio/2013003.

[2] Brodie ED, Ridenhour BJ, Brodie ED (2002). "The evolutionary response of predators to dangerous prey: hotspots and coldspots in the geographic mosaic of coevolution between garter snakes and newts". *Evolution*. **56** (10): 2067–82. PMID 12449493. doi:10.1554/0014-3820(2002)056[2067:teropt]2.0.co;2.

[3] Geffeney S, Brodie ED, Ruben PC, Brodie ED (2002). "Mechanisms of adaptation in a predator–prey arms race: TTX-resistant sodium channels". *Science*. **297** (5585): 1336–9. PMID 12193784. doi:10.1126/science.1074310.

[4] Alexander M. (1999) Biodegradation and Bioremediation, Elsevier Science.

[5] Wolt JD, Smith JK, Sims JK, Duebelbeis DO (1996). "Products and kinetics of cloransulam-methyl aerobic soil metabolism". *J. Agric. Food Chem.* **44**: 324–332. doi:10.1021/jf9503570.

Chapter 10

Biotransformation

Biotransformation is the chemical modification (or modifications) made by an organism on a chemical compound. If this modification ends in mineral compounds like CO_2, NH_4^+, or H_2O, the biotransformation is called mineralisation.

Biotransformation means chemical alteration of chemicals such as nutrients, amino acids, toxins, and drugs in the body. It is also needed to render nonpolar compounds polar so that they are not reabsorbed in renal tubules and are excreted. Biotransformation of xenobiotics can dominate toxicokinetics and the metabolites may reach higher concentrations in organisms than their parent compounds.[1]

10.1 Drug metabolism

Main article: Drug metabolism

The metabolism of a drug or toxin in a body is an example of a biotransformation. The body typically deals with a foreign compound by making it more water-soluble, to increase the rate of its excretion through the urine. There are many different processes that can occur; the pathways of drug metabolism can be divided into:

- phase I
- phase II

Drugs can undergo one of four potential biotransformations: Active Drug to Inactive Metabolite, Active Drug to Active Metabolite, Inactive Drug to Active Metabolite, Active Drug to Toxic Metabolite (biotoxification).

10.1.1 Phase I reaction

- Includes oxidative, reductive, and hydrolytic reactions.
- In these type of reactions, a polar group is either introduced or unmasked, so the drug molecule becomes more water-soluble and can be excreted.
- Reactions are non-synthetic in nature and in general produce a more water-soluble and less active metabolites.
- The majority of metabolites are generated by a common hydroxylating enzyme system known as Cytochrome P450.

10.1.2 Phase II reaction

- These reactions involve covalent attachment of small polar endogenous molecule such as glucuronic acid, sulfate, or glycine to form water-soluble compounds.
- This is also known as a *conjugation reaction.*
- The final compounds have a larger molecular weight.

10.2 Microbial biotransformation

Biotransformation of various pollutants is a sustainable way to clean up contaminated environments.[2] These bioremediation and biotransformation methods harness the naturally occurring, microbial catabolic diversity to degrade, transform or accumulate a huge range of compounds including hydrocarbons (e.g. oil), polychlorinated biphenyls (PCBs), polyaromatic hydrocarbons (PAHs), pharmaceutical substances, radionuclides and metals. Major methodological breakthroughs in recent years have enabled detailed genomic, metagenomic, proteomic, bioinformatic and other high-throughput analyses of environmentally relevant microorganisms providing unprecedented insights into biotransformation and biodegradative pathways and the ability of organisms to adapt to changing environmental conditions.

Biological processes play a major role in the removal of contaminants and pollutants from the environment. Some microorganisms possess an astonishing catabolic versatility to degrade or transform such compounds. New methodological breakthroughs in sequencing, genomics, proteomics, bioinformatics and imaging are producing vast amounts of information. In the field of Environmental Microbiology, genome-based global studies open a new era providing unprecedented *in silico* views of metabolic and regulatory networks, as well as clues to the evolution of biochemical pathways relevant to biotransformation and to the molecular adaptation strategies to changing environmental conditions. Functional genomic and metagenomic approaches are increasing our understanding of the relative importance of different pathways and regulatory networks to carbon flux in particular environments and for particular compounds and they are accelerating the development of bioremediation technologies and biotransformation processes.[2] Also there is other approach of biotransformation called enzymatic biotransformation.

10.3 Oil biodegradation

Petroleum oil is toxic for most life forms and episodic and chronic pollution of the environment by oil causes major ecological perturbations. Marine environments are especially vulnerable, since oil spills of coastal regions and the open sea are poorly containable and mitigation is difficult. In addition to pollution through human activities, millions of tons of petroleum enter the marine environment every year from natural seepages. Despite its toxicity, a considerable fraction of petroleum oil entering marine systems is eliminated by the hydrocarbon-degrading activities of microbial communities, in particular by a remarkable recently discovered group of specialists, the so-called hydrocarbonoclastic bacteria (HCB). *Alcanivorax borkumensis*, a paradigm of HCB and probably the most important global oil degrader, was the first to be subjected to a functional genomic analysis. This analysis has yielded important new insights into its capacity for (i) n-alkane degradation including metabolism, biosurfactant production and biofilm formation, (ii) scavenging of nutrients and cofactors in the oligotrophic marine environment, as well as (iii) coping with various habitat-specific stresses. The understanding thereby gained constitutes a significant advance in efforts towards the design of new knowledge-based strategies for the mitigation of ecological damage caused by oil pollution of marine habitats. HCB also have potential biotechnological applications in the areas of bioplastics and biocatalysis.[3]

10.4 Metabolic engineering and biocatalytic applications

The study of the fate of persistent organic chemicals in the environment has revealed a large reservoir of enzymatic reactions with a large potential in preparative organic synthesis, which has already been exploited for a number of oxygenases on pilot and even on industrial scale. Novel catalysts can be obtained from metagenomic libraries and DNA sequence based approaches. Our increasing capabilities in adapting the catalysts to specific reactions and process requirements by rational and random mutagenesis broadens the scope for application in the fine chemical industry, but also in the field of biodegradation. In many cases, these catalysts need to be exploited in whole cell bioconversions or in fermentations, calling for system-wide approaches to understanding strain physiology and metabolism and rational approaches to the engineering of whole cells as they are increasingly put forward in the area of systems biotechnology and synthetic biology.[4]

10.5 See also

- Biodegradation
- Microbial biodegradation
- Xenobiotic metabolism

10.6 References

[1] Ashauer, R; Hintermeister, A; O'Connor, I; Elumelu, M; et al. (2012). "Significance of Xenobiotic Metabolism for Bioaccumulation Kinetics of Organic Chemicals in Gammarus pulex". *Environ. Sci. Technol.* **46**: 3498–3508. doi:10.1021/es204611h.

[2] Diaz E (editor). (2008). *Microbial Biodegradation: Genomics and Molecular Biology* (1st ed.). Caister Academic Press. ISBN 978-1-904455-17-2.

[3] Martins VAP; et al. (2008). "Genomic Insights into Oil Biodegradation in Marine Systems". *Microbial Biodegradation: Genomics and Molecular Biology*. Caister Academic Press. ISBN 978-1-904455-17-2.

[4] Meyer A and Panke S (2008). "Genomics in Metabolic Engineering and Biocatalytic Applications of the Pollutant Degradation Machinery". *Microbial Biodegradation: Genomics and Molecular Biology*. Caister Academic Press. ISBN 978-1-904455-17-2.

10.7 External links

- Biotransformation of Drugs
- Biodegradation, Bioremediation and Biotransformation

- Microbial Biodegradation
- Biotransformation and Bioaccumulation in freshwater invertebrates
- Ecotoxicology & Models

Chapter 11

Multiple drug resistance

This article is about multiple drug resistance in microorganisms. For multiple drug resistance in tumor/cancer cells, see antineoplastic resistance.

Multiple drug resistance (MDR), **multidrug resistance** or **multiresistance** is antimicrobial resistance shown by a species of microorganism to multiple antimicrobial drugs. The types most threatening to public health are MDR bacteria that resist multiple antibiotics; other types include MDR viruses, fungi, and parasites (resistant to multiple antifungal, antiviral, and antiparasitic drugs of a wide chemical variety).[1] Recognizing different degrees of MDR, the terms **extensively drug resistant (XDR)** and **pandrug-resistant (PDR)** have been introduced. The definitions were published in 2011 in the journal *Clinical Microbiology and Infection* and are openly accessible.[2]

11.1 Common multidrug-resistant organisms (MDROs)

Common multidrug-resistant organisms are usually bacteria:

- Vancomycin-Resistant Enterococci (VRE)
- Methicillin-Resistant *Staphylococcus aureus* (MRSA)
- Extended-spectrum β-lactamase (ESBLs) producing Gram-negative bacteria
- *Klebsiella pneumoniae* carbapenemase (KPC) producing Gram-negatives
- MultiDrug-Resistant gram negative rods (MDR GNR) MDRGN bacteria such as *Enterobacter species, E.coli, Klebsiella pneumoniae, Acinetobacter baumannii, Pseudomonas aeruginosa*

A group of gram-positive and gram-negative bacteria of particular recent importance have been dubbed as the ESKAPE group (*Enterococcus faecium, Staphylococcus aureus, Klebsiella pneumoniae, Acinetobacter baumannii, Pseudomonas aeruginosa* and Enterobacter species).[3]

- Multi-drug-resistant tuberculosis

11.2 Bacterial resistance to antibiotics

Main article: Antibiotic resistance

Various microorganisms have survived for thousands of years by their ability to adapt to antimicrobial agents. They do so via spontaneous mutation or by DNA transfer. This process enables some bacteria to oppose the action of certain antibiotics, rendering the antibiotics ineffective.[4] These microorganisms employ several mechanisms in attaining multi-drug resistance:

- No longer relying on a glycoprotein cell wall
- Enzymatic deactivation of antibiotics
- Decreased cell wall permeability to antibiotics
- Altered target sites of antibiotic
- Efflux mechanisms to remove antibiotics[5]
- Increased mutation rate as a stress response[6]

Many different bacteria now exhibit multi-drug resistance, including staphylococci, enterococci, gonococci, streptococci, salmonella, as well as numerous other gram-negative bacteria and *Mycobacterium tuberculosis*. Antibiotic resistant bacteria are able to transfer copies of DNA that code for a mechanism of resistance to other bacteria even distantly related to them, which then are also able to pass on the resistance genes and so generations of antibiotics resistant bacteria are produced.[7] This process is called horizontal gene transfer.

11.3 Antifungal resistance

Yeasts such as *Candida species* can become resistant under long term treatment with azole preparations, requiring treatment with a different drug class. Scedosporium prolificans infections are almost uniformly fatal because of their resistance to multiple antifungal agents.[8]

11.4 Antiviral resistance

HIV is the prime example of MDR against antivirals, as it mutates rapidly under monotherapy. Influenza virus has become increasingly MDR; first to amantadenes, then to neuraminidase inhibitors such as oseltamivir, (2008-2009: 98.5% of Influenza A tested resistant), also more commonly in immunoincompetent people Cytomegalovirus can become resistant to ganciclovir and foscarnet under treatment, especially in immunosuppressed patients. Herpes simplex virus rarely becomes resistant to acyclovir preparations, mostly in the form of cross-resistance to famciclovir and valacyclovir, usually in immunosuppressed patients.

11.5 Antiparasitic resistance

The prime example for MDR against antiparasitic drugs is malaria. *Plasmodium vivax* has become chloroquine and sulfadoxine-pyrimethamine resistant a few decades ago, and as of 2012 artemisinin-resistant Plasmodium falciparum has emerged in western Cambodia and western Thailand. *Toxoplasma gondii* can also become resistant to artemisinin, as well as atovaquone and sulfadiazine, but is not usually MDR[9] Antihelminthic resistance is mainly reported in the veterinary literature, for example in connection with the practice of livestock drenching[10] and has been recent focus of FDA regulation.

11.6 Preventing the emergence of antimicrobial resistance

To limit the development of antimicrobial resistance, it has been suggested to:

- Use the appropriate antimicrobial for an infection; e.g. no antibiotics for viral infections
- Identify the causative organism whenever possible
- Select an antimicrobial which targets the specific organism, rather than relying on a broad-spectrum antimicrobial
- Complete an appropriate duration of antimicrobial treatment (not too short and not too long)
- Use the correct dose for eradication; subtherapeutic dosing is associated with resistance, as demonstrated in food animals.

The medical community relies on education of its prescribers, and self-regulation in the form of appeals to voluntary antimicrobial stewardship, which at hospitals may take the form of an antimicrobial stewardship program. It has been argued that depending on the cultural context government can aid in educating the public on the importance of restrictive use of antibiotics for human clinical use, but unlike narcotics, there is no regulation of its use anywhere in the world at this time. Antibiotic use has been restricted or regulated for treating animals raised for human consumption with success, in Denmark for example.

Infection prevention is the most efficient strategy of prevention of an infection with a MDR organism within a hospital, because there are few alternatives to antibiotics in the case of an extensively resistant or panresistant infection; if an infection is localized, removal or excision can be attempted (with MDR-TB the lung for example), but in the case of a systemic infection only generic measures like boosting the immune system with immunoglobulins may be possible. The use of bacteriophages (viruses which kill bacteria) has no clinical application at the present time.

It is necessary to develop new antibiotics over time since the selection of resistant bacteria cannot be prevented completely. This means with every application of a specific antibiotic, the survival of a few bacteria which already got a resistance gene against the substance is promoted, and the concerning bacterial population amplifies. Therefore, the resistance gene is farther distributed in the organism and the environment, and a higher percentage of bacteria does no longer respond to a therapy with this specific antibiotic.

11.7 See also

- Drug resistance
- MDRGN bacteria
- Xenobiotic metabolism
- Multidrug tolerance
- NDM1 enzymatic resistance
- Herbicide resistance

11.8 References

[1] Drug Resistance, Multiple at the US National Library of Medicine Medical Subject Headings (MeSH)

[2] A.-P. Magiorakos , A. Srinivasan, R. B. Carey, Y. Carmeli, M. E. Falagas, C. G. Giske, S. Harbarth, J. F. Hinndler *et al.* Multidrug-resistant, extensively drug-resistant and pandrug-resistant bacteria.... Clinical Microbiology and Infection, Vol 8, Iss. 3 first published 27 July 2011 [via Wiley Online Library]. Retrieved 16 August 2014.

[3] Boucher, HW, Talbot GH, Bradley JS, Edwards JE, Gilvert D, Rice LB, Schedul M., Spellberg B., Bartlett J. (1 Jan 2009). "Bad buds, no drugs: no ESKAPE! An update from the Infectious Diseases Society of America". *Clinical Infectious Diseases*. **48** (1): 1–12. doi:10.1086/595011.

[4] Bennett PM (March 2008). "Plasmid encoded antibiotic resistance: acquisition and transfer of antibiotic resistance genes in bacteria". *Br. J. Pharmacol.* 153 Suppl 1: S347–57. PMC 2268074 . PMID 18193080. doi:10.1038/sj.bjp.0707607.

[5] Li XZ, Nikaido H (August 2009). "Efflux-mediated drug resistance in bacteria: an update". *Drugs*. **69** (12): 1555–623. PMC 2847397 . PMID 19678712. doi:10.2165/11317030-000000000-00000.

[6] Stix G (April 2006). "An antibiotic resistance fighter". *Sci. Am.* **294** (4): 80–3. PMID 16596883. doi:10.1038/scientificamerican0406-80.

[7] Hussain, T. Pakistan at the verge of potential epidemics by multi-drug resistant pathogenic bacteria (2015). Adv. Life Sci. 2(2). pp: 46-47

[8] Howden BP, Slavin MA, Schwarer AP, Mijch AM (February 2003). "Successful control of disseminated Scedosporium prolificans infection with a combination of voriconazole and terbinafine". *Eur. J. Clin. Microbiol. Infect. Dis.* **22** (2): 111–3. PMID 12627286. doi:10.1007/s10096-002-0877-z.

[9] Doliwa C, Escotte-Binet S, Aubert D, Velard F, Schmid A, Geers R, Villena I. Induction of sulfadiazine resistance in vitro in Toxoplasma gondii.Exp Parasitol. 2013 Feb;133(2):131-6.

[10] Laurenson YC, Bishop SC, Forbes AB, Kyriazakis I.Modelling the short- and long-term impacts of drenching frequency and targeted selective treatment on the performance of grazing lambs and the emergence of antihelmintic resistance.Parasitology. 2013 Feb 1:1-12.

11.9 Further reading

- Greene HL, Noble JH (2001). *Textbook of primary care medicine*. St. Louis: Mosby. ISBN 0-323-00828-3.

11.10 External links

- BURDEN of Resistance and Disease in European Nations - An EU-Project to estimate the financial burden of antibiotic resistance in European Hospitals
- European Centre of Disease Prevention and Control and (ECDC): Multidrug-resistant, extensively drug-resistant and pandrug-resistant bacteria: An international expert proposal for interim standard definitions for acquired resistance http://www.ecdc.europa.eu/en/activities/diseaseprogrammes/ARHAI/Pages/public_consultation_clinical_microbiology_infection_article.aspx
- State of Connecticut Department of Public Health MDRO information http://www.ct.gov/dph/cwp/view.asp?a=3136&q=424162

APUA or Alliance for the Prudent Use of Antibiotics http://www.tufts.edu/med/apua/about_issue/multi_drug.shtml

Chapter 12

Drug interaction

Certain medicines can interact pharmacologically and affect the activity of other medicines.

A **drug interaction** is a situation in which a substance (usually another drug) affects the activity of a drug when both are administered together. This action can be synergistic (when the drug's effect is increased) or antagonistic (when the drug's effect is decreased) or a new effect can be produced that neither produces on its own. Typically, interactions between drugs come to mind (drug-drug interaction). However, interactions may also exist between drugs and foods (drug-food interactions), as well as drugs and medicinal plants or herbs (drug-plant interactions). People taking antidepressant drugs such as monoamine oxidase inhibitors should not take food containing tyramine as hypertensive crisis may occur (an example of a drug-food interaction). These interactions may occur out of accidental misuse or due to lack of knowledge about the active ingredients involved in the relevant substances.[1]

It is therefore easy to see the importance of these pharmacological interactions in the practice of medicine. If a patient is taking two drugs and one of them increases the effect of the other it is possible that an overdose may occur. The interaction of the two drugs may also increase the risk that side effects will occur. On the other hand, if the action of a drug is reduced it may cease to have any therapeutic use because of under dosage. Notwithstanding the above, on occasion these interactions may be sought in order to obtain an improved therapeutic effect.[2] Examples of this include the use of codeine with paracetamol to increase its analgesic effect. Or the combination of clavulanic acid with amoxicillin in order to overcome bacterial resistance to the antibiotic. It should also be remembered that there are interactions that, from a theoretical standpoint, may occur but in clinical practice have no important repercussions.

The pharmaceutical interactions that are of special interest to the practice of medicine are primarily those that have negative effects for an organism. The risk that a pharmacological interaction will appear increases as a function of the number of drugs administered to a patient at the same time.[3] Over a third (36%) of older adults in the U.S. regularly use 5 or more medications or supplements and 15% are potentially at risk for a major drug-drug interaction.[4] Both the use of medications and subsequent adverse drug interactions have increased significantly between 2005-2011.[4]

It is possible that an interaction will occur between a drug and another substance present in the organism (i.e. foods or alcohol). Or in certain specific situations a drug may even react with itself, such as occurs with dehydration. In other situations, the interaction does not involve any effect on the drug. In certain cases, the presence of a drug in an individual's blood may affect certain types of laboratory analysis (**analytical interference**).

It is also possible for interactions to occur outside an organism before administration of the drugs has taken place. This can occur when two drugs are mixed, for example, in a saline solution prior to intravenous injection. Some classic examples of this type of interaction include that thiopentone and suxamethonium should not be placed in the same syringe and same is true for benzylpenicillin and heparin. These situations will all be discussed under the same heading due to their conceptual similarity.

Drug interactions may be the result of various processes. These processes may include alterations in the pharmacokinetics of the drug, such as alterations in the absorption, distribution, metabolism, and excretion (ADME) of a drug. Alternatively, drug interactions may be the result of the pharmacodynamic properties of the drug, e.g. the co-administration of a receptor antagonist and an agonist for the same receptor.

12.1 Synergy and antagonism

When the interaction causes an increase in the effects of one or both of the drugs the interaction is called a **synergistic effect**. An "additive synergy" occurs when the final effect is equal to the sum of the effects of the two drugs (Although some authors argue that this is not true synergy). When the final effect is much greater than the sum of the two effects this is called **enhanced synergy**. This concept is recognized by the majority of authors,[5] although other authors only refer to synergy when there is an enhanced effect. These authors use the term "additive effect" for additive synergy and they reserve use of the term "synergistic effect" for enhanced synergy.[6] The opposite effect to synergy is termed **antagonism**. Two drugs are antagonistic when their interaction causes a decrease in the effects of one or both of the drugs.

Both synergy and antagonism can both occur during different phases of the interaction of a drug with an organism, with each effect having a different name. For example, when the synergy occurs at a cellular receptor level this is termed agonism, and the substances involved are termed *agonists*. On the other hand, in the case of antagonism the substances involved are known as inverse agonists. The different responses of a receptor to the action of a drug has resulted in a number of classifications, which use terms such as "partial agonist", "competitive agonist" etc. These concepts have fundamental applications in the pharmacodynamics of these interactions. The proliferation of existing classifications at this level, along with the fact that the exact reaction mechanisms for many drugs are not well understood means that it is almost impossible to offer a clear classification for these concepts. It is even likely that many authors would misapply any given classification.[5]

12.2 Underlying factors

It is possible to take advantage of positive drug interactions. However, the negative interactions are usually of more interest because of their pathological significance and also because they are often unexpected and may even go undiagnosed. By studying the conditions that favour the appearance of interactions it should be possible to prevent them or at least diagnose them in time. The factors or conditions that predispose or favor the appearance of interactions include:[5]

- Old age: factors relating to how human physiology changes with age may affect the interaction of drugs. For example, liver metabolism, kidney function, nerve transmission or the functioning of bone marrow all decrease with age. In addition, in old age there is a sensory decrease that increases the chances of errors being made in the administration of drugs.[7]
- Polypharmacy: The more drugs a patient takes the more likely it will be that some of them will interact.[8]
- Genetic factors: Genes synthesize enzymes that metabolize drugs. Some races have genotypic variations that could decrease or increase the activity of these enzymes. The consequence of this would, on occasions, be a greater predisposition towards drug interactions and therefore a greater predisposition for adverse effects to occur. This is seen in genotype variations in the isozymes of cytochrome P450.
- Hepatic or renal diseases: The blood concentrations of drugs that are metabolized in the liver and / or eliminated by the kidneys may be altered if either of these organs is not functioning correctly. If this is the case an increase in blood concentration is normally seen.[8]
- Serious diseases that could worsen if the dose of the medicine is reduced.
- Drug dependent factors:[9]
    - Narrow therapeutic index: Where the difference between the effective dose and the toxic dose is small.[n. 1] The drug digoxin is an example of this type of drug.
    - Steep dose-response curve: Small changes in the dosage of a drug produce large changes in the drug's concentration in the patient's blood plasma.
    - Saturable hepatic metabolism: In addition to dose effects the capacity to metabolize the drug is greatly decreased

12.3 Analytical interference

The detection of laboratory parameters is based on physicochemical reactions between the substance being measured and reagents designed for this purpose. These reactions can be altered by the presence of drugs giving rise to an over estimation or an underestimation of the real results. Levels of cholesterol and other blood lipids can be overestimated as a consequence of the presence in the blood of some psychotropic drugs. These overestimates should not be confused with the action of other drugs that actually increase blood cholesterol levels due to an interaction with its metabolism. Most experts consider that these are not true interactions, so they will not be dealt with further in this discussion.[10]

These chemical reactions are also known as **pharmacological incompatibilities**. The reactions occur when two or more drugs are mixed outside the body of the organism for the purpose of joint administration.[2] Usually the interaction is antagonistic and it almost always affects both drugs. Examples of these types of interactions include the mixing of penicillins and aminoglycosides in the same serum bottle, which causes the formation of an insoluble precipitate, or the mixing of ciprofloxacin with furosemide. The interaction of some drugs with the transport medium can also be included here. This means that certain drugs cannot be administered in plastic bottles because they bind with the bottle's walls, reducing the drug's concentration in solution.

Many authors do not consider them to be interactions in the strictest sense of the word. An example is the database of the General Council of Official Pharmacists Colleges of Spain (Consejo General de Colegios Oficiales de Farmacéuticos de España),[11] that does not include them among the 90,000 registered interactions.

12.4 Pharmacodynamic interactions

The change in an organism's response on administration of a drug is an important factor in pharmacodynamic interactions. These changes are extraordinarily difficult to classify given the wide variety of modes of action that exist and the fact that many drugs can cause their effect through a number of different mechanisms. This wide diversity also means that, in all but the most obvious cases, it is important to investigate and understand these mechanisms. The well-founded suspicion exists that there are more unknown interactions than known ones.

Pharmacodynamic interactions can occur on:

1. **Pharmacological receptors**:[12] Receptor interactions are the most easily defined, but they are also the most common. From a pharmacodynamic perspective, two drugs can be considered to be:

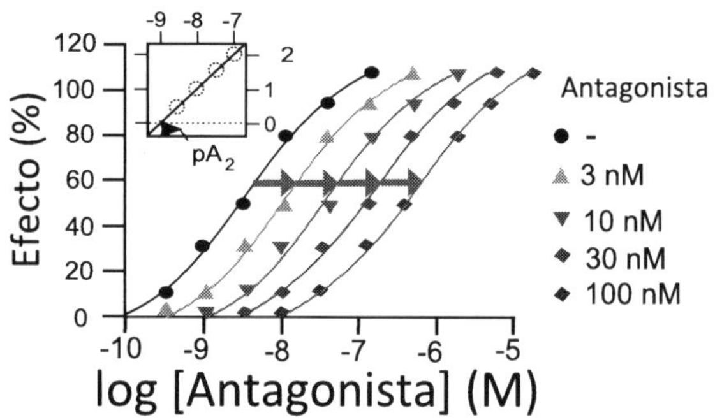

Effects of the competitive inhibition of an agonist by increases in the concentration of an antagonist. A drugs potency can be affected (the response curve shifted to the right) by the presence of an antagonistic interaction. pA_2 known as the Schild representation, a mathematical model of the agonist:antagonist relationship or vice versa.

   (a) **Homodynamic**, if they act on the same receptor. They, in turn can be:

      i. **Pure agonists**, if they bind to the main locus of the receptor, causing a similar effect to that of the main drug.

      ii. **Partial agonists** if, on binding to one of the receptor's secondary loci, they have the same effect as the main drug, but with a lower intensity.

      iii. **Antagonists**, if they bind directly to the receptor's main locus but their effect is opposite to that of the main drug. These include:

         A. **Competitive antagonists**, if they compete with the main drug to bind with the receptor. The amount of antagonist or main drug that binds with the receptor will depend on the concentrations of each one in the plasma.

         B. **Uncompetitive antagonists**, when the antagonist binds to the receptor irreversibly and is not released until the receptor is saturated. In principle the quantity of antagonist and agonist that binds to the receptor will depend on their concentrations. However, the presence of the antagonist will cause the main drug to be released from the receptor regardless of the main drug's concentration, therefore all the receptors will eventually become occupied by the antagonist.

   (b) **Heterodynamic competitors**, if they act on distinct receptors.

2. **Signal transduction mechanisms**: these are molecular processes that commence after the interaction of the drug with the receptor.[13] For example, it is known that hypoglycaemia (low blood glucose) in an organism produces a release of catecholamines, which trigger compensation mechanisms thereby increasing blood glucose levels. The release of catecholamines also triggers a series of symptoms, which allows the organism to recognise what is happening and which act as a stimulant for preventative action (eating sugars). Should a patient be taking a drug such as insulin, which reduces glycaemia, and also be taking another drug such as certain beta-blockers for heart disease, then the beta-blockers will act to block the adrenaline receptors. This will block the reaction triggered by the catecholamines should a hypoglycaemic episode occur. Therefore, the body will not adopt corrective mechanisms and there will be an increased risk of a serious reaction resulting from the ingestion of both drugs at the same time.

3. **Antagonic physiological systems**:[13] Imagine a drug **A** that acts on a certain organ. This effect will increase with increasing concentrations of physiological substance **S** in the organism. Now imagine a drug **B** that acts on another organ, which increases the amount of substance **S**. If both drugs are taken simultaneously it is possible that drug **A** could cause an adverse reaction in the organism as its effect will be indirectly increased by the action of drug **B**. An actual example of this interaction is found in the concomitant use of digoxin and furosemide. The former acts on cardiac fibres and its effect is increased if there are low levels of potassium (K) in blood plasma. Furosemide is a diuretic that lowers arterial tension but favours the loss of K^+. This could lead to hypokalemia (low levels of potassium in the blood), which could increase the toxicity of digoxin.

12.5 Pharmacokinetic interactions

Modifications in the effect of a drug are caused by differences in the absorption, transport, distribution, metabolization or excretion of one or both of the drugs compared with the expected behaviour of each drug when taken individually. These changes are basically modifications in the concentration of the drugs. In this respect two drugs can be **homergic** if they have the same effect in the organism and **heterergic** if their effects are different.

12.5.1 Absorption interactions

Changes in motility

Some drugs, such as the prokinetic agents increase the speed with which a substance passes through the intestines. If a drug is present in the digestive tract's absorption zone for less time its blood concentration will decrease. The opposite will occur with drugs that decrease intestinal motility.

- pH: Drugs can be present in either ionised or non-ionised form, depending on their **pKa** (pH at which the drug reaches equilibrium between its ionised and non-ionised form).[14] The non-ionized forms of drugs are usually easier to absorb, because they will not be repelled by the lipidic bylayer of the cell, most of them can be absorbed by passive diffusion, unless they are too big or too polarized (like glucose or vancomicyn), in which case they may have or not specific and non specific transporters distributed on the entire intestine internal surface, that carries drugs inside the body. Obviously increasing the absorption of a drug will increase its bioavailability, so, changing the drug's state between ionized or not, can be useful or not for certain drugs.

Certain drugs require an acid stomach pH for absorption. Others require the basic pH of the intestines. Any modification in the pH could change this absorption. In the case of the antacids, an increase in pH can inhibit the absorption of other drugs such as zalcitabine (absorption can be decreased by 25%), tipranavir (25%) and amprenavir (up to 35%). However, this occurs less often than an increase in pH causes an increase in absorption. Such as occurs when cimetidine is taken with didanosine. In this case a gap of two to four hours between taking the two drugs is usually sufficient to avoid the interaction.[15]

- Drug solubility: The absorption of some drugs can be drastically reduced if they are administered together with food with a high fat content. This is the case for oral anticoagulants and avocado.
- Formation of non-absorbable complexes:
  - Chelation: The presence of di- or trivalent cations can cause the chelation of certain drugs, making them harder to absorb. This interaction frequently occurs between drugs such as tetracycline or the fluoroquinolones and dairy products (due to the presence of Ca^{++}).
  - Binding with proteins. Some drugs such as sucralfate binds to proteins, especially if they have a high bioavailability. For this reason its

administration is contraindicated in enteral feeding.[16]

    - Finally, another possibility is that the drug is retained in the intestinal lumen forming large complexes that impede its absorption. This can occur with cholestyramine if it is associated with sulfamethoxazol, thyroxine, warfarin or digoxin.

- Acting on the P-glycoprotein of the enterocytes: This appears to be one of the mechanisms promoted by the consumption of grapefruit juice in increasing the bioavailability of various drugs, regardless of its demonstrated inhibitory activity on first pass metabolism.[17]

12.5.2 Transport and distribution interactions

The main interaction mechanism is competition for plasma protein transport. In these cases the drug that arrives first binds with the plasma protein, leaving the other drug dissolved in the plasma, which modifies its concentration. The organism has mechanisms to counteract these situations (by, for example, increasing plasma clearance), which means that they are not usually clinically relevant. However, these situations should be taken into account if there other associated problems are present such as when the method of excretion is affected.[18]

12.5.3 Metabolism interactions

Many drug interactions are due to alterations in drug metabolism.[19] Further, human drug-metabolizing enzymes are typically activated through the engagement of nuclear receptors.[19] One notable system involved in metabolic drug interactions is the enzyme system comprising the cytochrome P450 oxidases.

CYP450

Cytochrome P450 is a very large family of haemoproteins (hemoproteins) that are characterized by their enzymatic activity and their role in the metabolism of a large number of drugs.[20] Of the various families that are present in human beings the most interesting in this respect are the 1, 2 and 3, and the most important enzymes are CYP1A2, CYP2C9, CYP2C19, CYP2D6, CYP2E1 and CYP3A4.[21] The majority of the enzymes are also involved in the metabolism of endogenous substances, such as steroids or sex hormones, which is also important should there be interference with these substances. As a result of these interactions the function of the enzymes can either be stimulated (enzyme induction) or inhibited (enzyme inhibition).

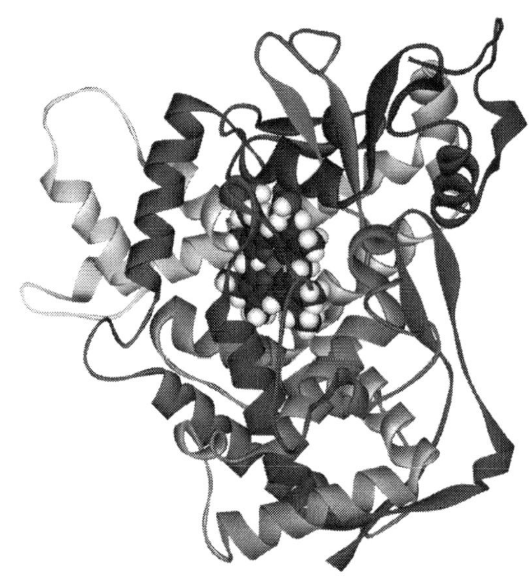

Diagram of cytochrome P450 isoenzyme 2C9 with the haem group in the centre of the enzyme.

Enzymatic inhibition

If drug A is metabolized by a cytochrome P450 enzyme and drug B inhibits or decreases the enzyme's activity, then drug A will remain with high levels in the plasma for longer as its inactivation is slower. As a result, enzymatic inhibition will cause an increase in the drug's effect. This can cause a wide range of adverse reactions.

It is possible that this can occasionally lead to a paradoxical situation, where the enzymatic inhibition causes a decrease in the drug's effect: if the metabolism of drug A gives rise to product A_2, which actually produces the effect of the drug. If the metabolism of drug A is inhibited by drug B the concentration of A_2 that is present in the blood will decrease, as will the final effect of the drug.

Enzymatic induction

If drug A is metabolized by a cytochrome P450 enzyme and drug B induces or increases the enzyme's activity, then blood plasma concentrations of drug A will quickly fall as its inactivation will take place more rapidly. As a result, enzymatic induction will cause a decrease in the drug's effect.

As in the previous case it is possible to find paradoxical situations where an active metabolite causes the drug's effect.

In this case the increase in active metabolite A_2 (following the previous example) produces an increase in the drug's effect.

It can often occur that a patient is taking two drugs that are enzymatic inductors, one inductor and the other inhibitor or both inhibitors, which greatly complicates the control of an individual's medication and the avoidance of possible adverse reactions.

An example of this is shown in the following table for the CYP1A2 enzyme, which is the most common enzyme found in the human liver. The table shows the substrates (drugs metabolized by this enzyme) and the inductors and inhibitors of its activity:[21]

St John's wort can act as an enzyme inductor.

Enzyme CYP3A4 is the enzyme that the greatest number of drugs use as a substrate. Over 100 drugs depend on its metabolism for their activity and many others act on the enzyme as inductors or inhibitors.

Some foods also act as inductors or inhibitors of enzymatic activity. The following table shows the most common:

Grapefruit juice can act as an enzyme inhibitor.

Any study of pharmacological interactions between particular medicines should also discuss the likely interactions of some medicinal plants. The effects caused by medicinal plants should be considered in the same way as those of medicines as their interaction with the organism gives rise to a pharmacological response. Other drugs can modify this response and also the plants can give rise to changes in the effects of other active ingredients. There is little data available regarding interactions involving medicinal plants for the following reasons:

1. **False sense of security** regarding medicinal plants. The interaction between a medicinal plant and a drug is usually overlooked due to a belief in the "safety of medicinal plants."
2. **Variability of composition**, both qualitative and quantitative. The composition of a plant-based drug is often subject to wide variations due to a number of factors such as seasonal differences in concentrations, soil type, climatic changes or the existence of different varieties or chemical races within the same plant species that have variable compositions of the active ingredient. On occasions an interaction can be due to just one active ingredient, but this can be absent in some chemical varieties or it can be present in low concentrations, which will not cause an interaction. Counter interactions can even occur. This occurs, for instance, with ginseng, the *Panax ginseng* variety increases the Prothrombin time, while the *Panax quinquefolius* variety decreases it.[24]
3. **Absence of use in at-risk groups**, such as hospitalized and polypharmacy patients, who tend to have the majority of drug interactions.
4. **Limited consumption** of medicinal plants has given rise to a lack of interest in this area.[25]

They are usually included in the category of foods as they are usually taken as a tea or food supplement. However, medicinal plants are increasingly being taken in a manner more often associated with conventional medicines: pills, tablets, capsules, etc.

12.5.4 Excretion interactions

Renal excretion

Only the free fraction of a drug that is dissolved in the blood plasma can be removed through the kidney. Therefore,

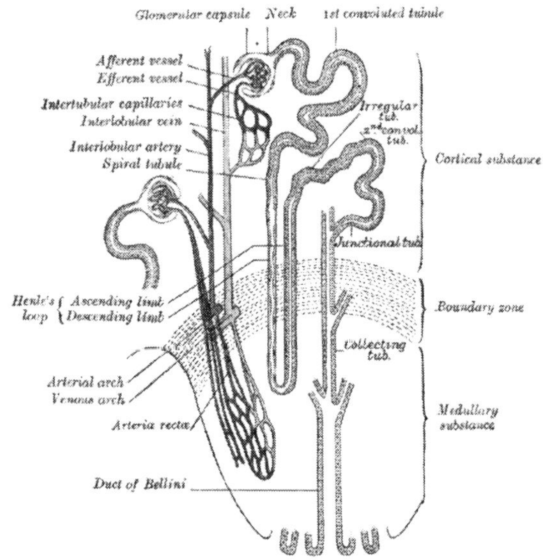

Human kidney nephron.

drugs that are tightly bound to proteins are not available for renal excretion, as long as they are not metabolized when they may be eliminated as metabolites.[26] Creatinine clearance is used as a measure of kidney functioning but it is only useful in cases where the drug is excreted in an unaltered form in the urine. The excretion of drugs from the kidney's nephrons has the same properties as that of any other organic solute: passive filtration, reabsorption and active secretion. In the latter phase the secretion of drugs is an active process that is subject to conditions relating to the saturability of the transported molecule and competition between substrates. Therefore, these are key sites where interactions between drugs could occur. Filtration depends on a number of factors including the pH of the urine, it having been shown that the drugs that act as weak bases are increasingly excreted as the pH of the urine becomes more acidic, and the inverse is true for weak acids. This mechanism is of great use when treating intoxications (by making the urine more acidic or more alkali) and it is also used by some drugs and herbal products to produce their interactive effect.

Bile excretion

Bile excretion is different from kidney excretion as it is always involves energy expenditure in active transport across the epithelium of the bile duct against a concentration gradient. This transport system can also be saturated if the plasma concentrations of the drug are high. Bile excretion of drugs mainly takes place where their molecular weight is greater than 300 and they contain both polar and lipophilic groups. The glucuronidation of the drug in the kidney also facilitates bile excretion. Substances with similar physicochemical properties can block the receptor, which is important in assessing interactions. A drug excreted in the bile duct can occasionally be reabsorbed by the intestines (in the entero-hepatic circuit), which can also lead to interactions with other drugs.

12.6 Epidemiology

Among US adults older than 55, 4% are taking medication and or supplements that put them at risk of a major drug interaction.[28] Potential drug-drug interactions have increased over time[29] and are more common in the low educated elderly even after controlling for age, sex, place of residence, and comorbidity.[30]

12.7 See also

- Deprescribing
- Drug-Interactions.eu
- Cytochrome P450
- Classification of Pharmaco-Therapeutic Referrals
- Pharmacokinetics
- Pharmacodynamics
- Polypharmacy
- Drug interactions can be checked for free online with interaction checkers (note that not all drug interaction checkers provide the same results, and only a drug information expert, such as a pharmacist, should interpret results or provide advice on managing drug interactions)
    - Multi-Drug Interaction Checker by Medscape
    - Drug Interactions Checker by Drugs.com

12.8 Notes

[1] The term effective dose is generally understood to mean the minimum amount of a drug that is need to produce the required effect. The toxic dose is the minimum amount of a drug that will produce a damaging effect.

12.9 References

[1] " *National Prescribing Service*, 2009. Available at http://nps.org.au/news_and_media/media_releases/repository/Forget_the_colour_shape_or_brand__its

[2] María Soledad Fernández Alfonso, Mariano Ruiz Gayo. *Fundamentos de Farmacología Básica y Clínica*. page 232. ISBN 84-8004-689-9

[3] Tannenbaum C, Sheehan NL (July 2014). "Understanding and preventing drug-drug and drug-gene interactions". *Expert Review of Clinical Pharmacology*. **7** (4): 533–44. PMC 4894065. PMID 24745854. doi:10.1586/17512433.2014.910111.

[4] Qato DM, Wilder J, Schumm LP, Gillet V, Alexander GC (April 2016). "Changes in Prescription and Over-the-Counter Medication and Dietary Supplement Use Among Older Adults in the United States, 2005 vs 2011". *JAMA Internal Medicine*. **176** (4): 473–82. PMID 26998708. doi:10.1001/jamainternmed.2015.8581.

[5] Baños Díez, J. E.; March Pujol, M (2002). *Farmacología ocular* (in Spanish) (2da ed.). Edicions UPC. p. 87. ISBN 8483016478. Retrieved 23 May 2009.

[6] Suárez Zuzunaga, A. *Justificación Farmacológica para las Asociaciones Analgésicas* Available on [www.spmed.org.pe/portal/images/stories/biliotecavirtual/.../dolor/justificacion_farmacolgica_para_las_asociaciones_analgesicas.ppt]

[7] Merle L, Laroche ML, Dantoine T, Charmes JP (2005). "Predicting and Preventing Adverse Drug Reactions in the Very Old". *Drugs and Aging*. **22** (5): 375–392. doi:10.2165/00002512-200522050-00003.

[8] García Morillo, J.S. *Optimización del tratamiento de enfermos pluripatológicos en atención primaria* UCAMI HHUU Virgen del Rocio. Sevilla. Spain. Available for members of SEMI at: ponencias de la II Reunión de Paciente Pluripatológico y Edad Avanzada

[9] Castells Molina,S.; Castells,S. y Hernández Pérez, M. *Farmacología en enfermería* Published by Elsevier Spain, 2007 ISBN 84-8174-993-1, 9788481749939 Available from

[10] Gago Bádenas, F. *Curso de Farmacología General. Tema 7.- Interacciones farmacológicas*. en

[11] *Panorama Actual del Medicamento*, number 245, July–August 2001, pages. 583–590

[12] S Gonzalez. "Interacciones Farmacológicas" (in Spanish). Retrieved 1 January 2009.

[13] *Curso de Farmacología Clínica Aplicada*, in El Médico Interactivo

[14] Malgor — Valsecia, *Farmacología general: Farmacocinética*.Cap. 2. en "Archived copy" (PDF). Archived from the original (PDF) on 2012-09-07. Retrieved 2012-03-20. Revised 25 September 2008

[15] Alicia Gutierrez Valanvia y Luis F. López-Cortés *Interacciones farmacológicas entre fármacos antirretrovirales y fármacos usados para ciertos transtornos gastrointestinales.* on accessed 24 September 2008

[16] Marduga Sanz, Mariano. *Interacciones de los alimentos con los medicamentos*. on

[17] Tatro,DS. *Update: Drug interaction with grapefruit juice.* Druglink, 2004. 8 (5), page 35ss

[18] Valsecia, Mabel en

[19] Elizabeth Lipp (2008-06-15). "Tackling Drug-Interaction Issues Early On". *Genetic Engineering & Biotechnology News*. Mary Ann Liebert, Inc. pp. 14, 16, 18, 20. Retrieved 2008-07-06. (subtitle) Researchers explore a number of strategies to better predict drug responses in the clinic

[20] IUPAC, *Compendium of Chemical Terminology*, 2nd ed. (the "Gold Book") (1997). Online corrected version: (2006–) "cytochrome P450". Danielson PB (December 2002). "The cytochrome P450 superfamily: biochemistry, evolution and drug metabolism in humans". *Current Drug Metabolism*. **3** (6): 561–97. PMID 12369887. doi:10.2174/1389200023337054.

[21] Nelson D (2003). Cytochrome P450s in humans Archived July 10, 2009, at the Wayback Machine.. Consulted 9 May 2005.

[22] Bailey DG, Malcolm J, Arnold O, Spence JD (August 1998). "Grapefruit juice-drug interactions". *British Journal of Clinical Pharmacology*. **46** (2): 101–10. PMC 1873672. PMID 9723817. doi:10.1046/j.1365-2125.1998.00764.x. Comment in: Mouly S, Paine MF (August 2001). "Effect of grapefruit juice on the disposition of omeprazole". *British Journal of Clinical Pharmacology*. **52** (2): 216–7. PMC 2014525. PMID 11488783. doi:10.1111/j.1365-2125.1978.00999.pp.x.

[23] Covarrubias-Gómez, A.; et al. (January–March 2005). "¿Qué se auto-administra su paciente?: Interacciones farmacológicas de la medicina herbal". *Revista Mexicana de Anestesiología*. **28** (1): 32–42.

[24] J. C. Tres *Interacción entre fármacos y plantas medicinales*. on Archived April 15, 2012, at the Wayback Machine.

[25] Zaragozá F, Ladero M, Rabasco AM et al. *Plantas Medicinales (Fitoterapia Práctica)*. Second Edition, 2001.

[26] Gago Bádenas, F. *Curso de Farmacología General. Tema 6.- Excreción de los fármacos*. en

[27] , *Farmacología general: Farmacocinética*.Cap. 2. en "Archived copy" (PDF). Archived from the original (PDF) on 2012-09-07. Retrieved 2012-03-20. Revised 25 September 2008

[28] Qato DM, Alexander GC, Conti RM, Johnson M, Schumm P, Lindau ST (December 2008). "Use of prescription and over-the-counter medications and dietary supplements among older adults in the United States". *Jama*. **300** (24): 2867–78. PMC 2702513 . PMID 19109115. doi:10.1001/jama.2008.892.

[29] Haider SI, Johnell K, Thorslund M, Fastbom J (December 2007). "Trends in polypharmacy and potential drug-drug interactions across educational groups in elderly patients in Sweden for the period 1992 - 2002". *International Journal of Clinical Pharmacology and Therapeutics*. **45** (12): 643–53. PMID 18184532. doi:10.5414/cpp45643.

[30] Haider SI, Johnell K, Weitoft GR, Thorslund M, Fastbom J (January 2009). "The influence of educational level on polypharmacy and inappropriate drug use: a register-based study of more than 600,000 older people". *Journal of the American Geriatrics Society*. **57** (1): 62–9. PMID 19054196. doi:10.1111/j.1532-5415.2008.02040.x.

12.10 Bibliography

MA Cos. *Interacciones de fármacos y sus implicancias clínicas.* In: *Farmacología Humana.* Chap. 10, pp. 165–176. (J. Flórez y col. Eds). Masson SA, Barcelona. 1997.

12.11 External links

- Drug Interactions: What You Should Know. U.S. Food and Drug Administration, Center for Drug Evaluation and Research
- On MedlinePlus there is information for non-professionals about drug interactions
- Medfacts Pocket Guide of Drug Interactions, Nephrology Pharmacy Associates
- Cytochrome P450 table maintained by the Indiana University School of Medicine
- Drugs.com Drug Interaction Checker
- EHealthme.com drug interactions checker based on 30 million reports from FDA and community
- 11.500 possible interactions of 5.000 substances
- Drug-Interactions.eu. Drug interactions tool for multiple and non-drug interactions. Data from over 2500 drug labels and 5000 trials, 10000 active ingredients (drugs, metabolites, herbal and not traditional medicine products, excipients, solvents, drugs of abuse, etc.), 11000 constants, 7000 AUC ratio values, 300 enzymes.
- "What really happens when you mix medications?". TED (talks). Retrieved April 2, 2016.

Chapter 13

Bioremediation

Bioremediation is a waste management technique that involves the use of organisms to neutralize pollutants from a contaminated site.[1] According to the United States EPA, bioremediation is a "treatment that uses naturally occurring organisms to break down hazardous substances into less toxic or non toxic substances". Technologies can be generally classified as *in situ* or *ex situ*. In situ *bioremediation involves treating the contaminated material at the site, while* ex situ *involves the removal of the contaminated material to be treated elsewhere. Some examples of bioremediation related technologies are phytoremediation, bioventing, bioleaching, landfarming, bioreactor, composting, bioaugmentation, rhizofiltration, and biostimulation.*

Bioremediation may occur on its own (natural attenuation or intrinsic bioremediation) or may only effectively occur through the addition of fertilizers, oxygen, etc., that help in enhancing the growth of the pollution-eating microbes within the medium (biostimulation). For example, the US Army Corps of Engineers demonstrated that windrowing and aeration of petroleum-contaminated soils enhanced bioremediation using the technique of landfarming.[2] Depleted soil nitrogen status may encourage biodegradation of some nitrogenous organic chemicals,[3] and soil materials with a high capacity to adsorb pollutants may slow down biodegradation owing to limited bioavailability of the chemicals to microbes.[4] Recent advancements have also proven successful via the addition of matched microbe strains to the medium to enhance the resident microbe population's ability to break down contaminants. Microorganisms used to perform the function of bioremediation are known as **bioremediators**.

However, not all contaminants are easily treated by bioremediation using microorganisms. For example, heavy metals such as cadmium and lead are not readily absorbed or captured by microorganisms. A recent experiment, however, suggests that fish bones have some success absorbing lead from contaminated soil.[5][6] Bone char has been shown to bioremediate small amounts of cadmium, copper, and zinc.[7] A recent experiment suggests that the removals of pollutants (nitrate, silicate, chromium and sulphide) from tannery wastewater were studied in batch experiments using marine microalgae.[8] The assimilation of metals such as mercury into the food chain may worsen matters. Phytoremediation is useful in these circumstances because natural plants or transgenic plants are able to bioaccumulate these toxins in their above-ground parts, which are then harvested for removal.[9] The heavy metals in the harvested biomass may be further concentrated by incineration or even recycled for industrial use. Some damaged artifacts at museums contain microbes which could be specified as bio remediating agents.[10] In contrast to this situation, other contaminants, such as aromatic hydrocarbons as are common in petroleum, are relatively simple targets for microbial degradation, and some soils may even have some capacity to autoremediate, as it were, owing to the presence of autochthonous microbial communities capable of degrading these compounds.[11]

The elimination of a wide range of pollutants and wastes from the environment requires increasing our understanding of the relative importance of different pathways and regulatory networks to carbon flux in particular environments and for particular compounds, and they will certainly accelerate the development of bioremediation technologies and biotransformation processes.[12]

13.1 Genetic engineering approaches

The use of genetic engineering to create organisms specifically designed for bioremediation has great potential.[13] The bacterium *Deinococcus radiodurans* (the most radioresistant organism known) has been modified to consume and digest toluene and ionic mercury from highly radioactive nuclear waste.[14] Releasing genetically augmented organisms into the environment may be problematic as tracking them can be difficult; bioluminescence genes from other species may be inserted to make this easier.[15] :135

13.2 Mycoremediation

Mycoremediation is a form of bioremediation in which fungi are used to decontaminate the area.

One of the primary roles of fungi in the ecosystem is decomposition, which is performed by the mycelium. The mycelium secretes extracellular enzymes and acids that break down lignin and cellulose, the two main building blocks of plant fiber. These are organic compounds composed of long chains of carbon and hydrogen, structurally similar to many organic pollutants. The key to mycoremediation is determining the right fungal species to target a specific pollutant. Certain strains have been reported to successfully degrade the nerve gases VX and sarin.

In one conducted experiment, a plot of soil contaminated with diesel oil was inoculated with mycelia of oyster mushrooms; traditional bioremediation techniques (bacteria) were used on control plots. After four weeks, more than 95% of many of the PAH (polycyclic aromatic hydrocarbons) had been reduced to non-toxic components in the mycelial-inoculated plots. It appears that the natural microbial community participates with the fungi to break down contaminants, eventually into carbon dioxide and water. Wood-degrading fungi are particularly effective in breaking down aromatic pollutants (toxic components of petroleum), as well as chlorinated compounds (certain persistent pesticides; Battelle, 2000).

Fungi can break down some hydrocarbons, especially if these are relatively simple molecules. They require a warm temperature, a slightly acidic pH of 4 to 5, and oxygen.[16] Two species of the Ecuadorian fungus *Pestalotiopsis* are capable of consuming polyurethane in aerobic and anaerobic conditions such as found at the bottom of landfills.[17]

Mycofiltration is a similar process, using fungal mycelia to filter toxic waste and microorganisms from water in soil. VTT Technical Research Centre of Finland reported an 80% recovery of gold from electronic waste using mycofiltration techniques.

13.3 Advantages

There are a number of cost/efficiency advantages to bioremediation, which can be employed in areas that are inaccessible without excavation.[18] For example, hydrocarbon spills (specifically, petrol spills) or certain chlorinated solvents may contaminate groundwater, and introducing the appropriate electron acceptor or electron donor amendment, as appropriate, may significantly reduce contaminant concentrations after a long time allowing for acclimation. This is typically much less expensive than excavation followed by disposal elsewhere, incineration or other *ex situ* treatment strategies, and reduces or eliminates the need for "pump and treat", a practice common at sites where hydrocarbons have contaminated clean groundwater. Using archaea for bioremediation of hydrocarbons also has the advantage of breaking down contaminants at the molecular level, as opposed to simply chemically dispersing the contaminant.[19]

13.4 Monitoring bioremediation

The process of bioremediation can be monitored indirectly by measuring the *Oxidation Reduction Potential* or redox in soil and groundwater, together with pH, temperature, oxygen content, electron acceptor/donor concentrations, and concentration of breakdown products (e.g. carbon dioxide). This table shows the (decreasing) biological breakdown rate as function of the redox potential.

This, by itself and at a single site, gives little information about the process of remediation.

1. It is necessary to sample enough points on and around the contaminated site to be able to determine contours of equal redox potential. Contouring is usually done using specialised software, e.g. using Kriging interpolation.
2. If all the measurements of redox potential show that electron acceptors have been used up, it is in effect an indicator for total microbial activity. Chemical analysis is also required to determine when the levels of contaminants and their breakdown products have been reduced to below regulatory limits.
3. Chemical analysis should also be carried out for assessing transformations in inorganic contaminants (e.g. heavy metals, radionuclides). Unlike organic pollutants, inorganic pollutants cannot be degraded[20] and remediation processes can both increase and decrease their solubility and bio-availability. An increase in heavy metal mobility can occur, even in reductive conditions, during *in-situ* bioremediation.[21]

13.5 See also

- Biodegradation
- Bioleaching
- Biosurfactant
- Chelation

- Dutch standards
- Folkewall
- List of environment topics
- Living machines
- Green wall
- *Mega Borg* Oil Spill
- Microbial biodegradation
- Phytoremediation
- *Pseudomonas putida* (used for degrading oil)
- Restoration ecology
- US Microbics
- Xenocatabolism

13.6 References

[1] "Environmental Inquiry - Bioremediation".

[2] Mann, D. K., T. M. Hurt, E. Malkos, J. Sims, S. Twait and G. Wachter. 1996. Onsite treatment of petroleum, oil, and lubricant (POL)-contaminated soils at Illinois Corps of Engineers lake sites. US Army Corps of Engineers Technical Report No. A862603 (71pages).

[3] Sims, G.K. (2006). "Nitrogen Starvation Promotes Biodegradation of N-Heterocyclic Compounds in Soil". *Soil Biology & Biochemistry*. **38**: 2478–2480. doi:10.1016/j.soilbio.2006.01.006.

[4] O'Loughlin, E. J; Traina, S. J.; Sims, G. K. (2000). "Effects of sorption on the biodegradation of 2-methylpyridine in aqueous suspensions of reference clay minerals". *Environ. Toxicol. and Chem.* **19**: 2168–2174. doi:10.1002/etc.5620190904.

[5] Kris S. Freeman (January 2012). "Remediating Soil Lead with Fishbones". *Environmental Health Perspectives*. **120**: A20–1. PMC 3261960 ට. PMID 22214821. doi:10.1289/ehp.120-a20a.

[6] "Battling lead contamination, one fish bone at a time". *Coast Guard Compass*. July 9, 2012.

[7] Huan Jing Ke Xue (February 2007). "Chemical fixation of metals in soil using bone char and assessment of the soil genotoxicity". *Huan Jing Ke Xue*. **28**: 232–7. PMID 17489175.

[8] Adam s. "marine Biology and oceanography".

[9] Meagher, RB (2000). "Phytoremediation of toxic elemental and organic pollutants". *Current Opinion in Plant Biology*. **3** (2): 153–162. PMID 10712958. doi:10.1016/S1369-5266(99)00054-0.

[10] Francesca Cappitelli; Claudia Sorlini (2008). "Microorganisms Attack Synthetic Polymers in Items Representing Our Cultural Heritage". *Applied and Environmental Microbiology*. **74**: 564–9. PMC 2227722 ට. PMID 18065627. doi:10.1128/AEM.01768-07.

[11] Olapade, OA; Ronk, AJ (2014). "Isolation, Characterization and Community Diversity of Indigenous Putative Toluene-Degrading Bacterial Populations with Catechol-2,3-Dioxygenase Genes in Contaminated Soils". *Microbial Ecology*. **69**: 59–65. PMID 25052383. doi:10.1007/s00248-014-0466-6.

[12] Diaz E (editor). (2008). *Microbial Biodegradation: Genomics and Molecular Biology* (1st ed.). Caister Academic Press. ISBN 1-904455-17-4. http://www.horizonpress.com/biod.

[13] Lovley, DR (2003). "Cleaning up with genomics: applying molecular biology to bioremediation". *Nature Reviews Microbiology*. **1** (1): 35–44. PMID 15040178. doi:10.1038/nrmicro731.

[14] Brim H, McFarlan SC, Fredrickson JK, Minton KW, Zhai M, Wackett LP, Daly MJ (2000). "Engineering Deinococcus radiodurans for metal remediation in radioactive mixed waste environments". *Nature Biotechnology*. **18** (1): 85–90. PMID 10625398. doi:10.1038/71986.

[15] Robert L. Irvine; Subhas K. Sikdar. *Bioremediation Technologies: Principles and Practice*.

[16] Singh, Harbhajan (2006). *Mycoremediation: fungal bioremediation*. New York: Wiley-Interscience. ISBN 0-471-75501-X.

[17] "Biodegradation of Polyester Polyurethane by Endophytic Fungi". *Applied and Environmental Microbiology*. July 2011.

[18] "Why Bioremediation". *JRW Bioremediation*. Retrieved 2016-05-02.

[19] "Archaea Effectiveness, Benefits - Akaya". *Akaya*. Retrieved 2015-09-10.

[20] "Developments in Bioremediation of Soils and Sediments Polluted with Metals and Radionuclides – 1. Microbial Processes and Mechanisms Affecting Bioremediation of Metal Contamination and Influencing Metal Toxicity and Transport". *Reviews in Environmental Science and Bio/Technology*. **4**: 115–156. August 2005. doi:10.1007/s11157-005-2169-4.

[21] "Bioremediation of contaminated marine sediments can enhance metal mobility due to changes of bacterial diversity". *Water Research*. **68**: 637–650. January 2015. doi:10.1016/j.watres.2014.10.035.

13.7 External links

- Phytoremediation, hosted by the Missouri Botanical Garden
- Toxic cadmium ions removal by isolated fungal strain from e-waste recycling facility (Kumar et al., 2012)
- Removal of Cu2+ Ions from Aqueous Solutions Using Copper Resistant Bacteria (Rajeshkumar and Kartic 2011)
- Field Demonstrations of Mycoremediation for Removal of Fecal Coliform Bacteria and Nutrients in the Dungeness Watershed, Washington (Thomas, S. *et al.* 2009)
- Evaluation of Isolated Fungal Strain from e-waste Recycling Facility for Effective Sorption of Toxic Heavy Metal Pb (II) Ions and Fungal Protein Molecular Characterization- a Mycoremediation Approach (Rajeshkumar, 2011)

Chapter 14

Cytochrome P450

Cytochromes P450 (**CYPs**) are proteins of the superfamily containing heme as a cofactor and, therefore, are hemoproteins. CYPs use a variety of small and large molecules as substrates in enzymatic reactions. They are, in general, the terminal oxidase enzymes in electron transfer chains, broadly categorized as P450-containing systems. The term *P450* is derived from the spectrophotometric peak at the wavelength of the absorption maximum of the enzyme (450 nm) when it is in the reduced state and complexed with carbon monoxide.

CYP enzymes have been identified in all kingdoms of life: animals, plants, fungi, protists, bacteria, archaea, and even in viruses.[1] However, they are not omnipresent; for example, they have not been found in *Escherichia coli*.[2][3] More than 200,000 distinct CYP proteins are known.[4]

Most CYPs require a protein partner to deliver one or more electrons to reduce the iron (and eventually molecular oxygen). Based on the nature of the electron transfer proteins, CYPs can be classified into several groups:[5]

- **Microsomal P450 systems**, in which electrons are transferred from NADPH via cytochrome P450 reductase (variously CPR, POR, or CYPOR). Cytochrome b5 (cyb5) can also contribute reducing power to this system after being reduced by cytochrome b5 reductase (CYB5R).
- **Mitochondrial P450 systems**, which employ adrenodoxin reductase and adrenodoxin to transfer electrons from NADPH to P450.
- **Bacterial P450 systems**, which employ a ferredoxin reductase and a ferredoxin to transfer electrons to P450.
- **CYB5R/cyb5/P450 systems**, in which both electrons required by the CYP come from cytochrome b5.
- **FMN/Fd/P450 systems**, originally found in *Rhodococcus* species, in which a FMN-domain-containing reductase is fused to the CYP.
- **P450 only** systems, which do not require external reducing power. Notable ones include thromboxane synthase (CYP5), prostacyclin synthase (CYP8), and CYP74A (allene oxide synthase).

The most common reaction catalyzed by cytochromes P450 is a monooxygenase reaction, e.g., insertion of one atom of oxygen into the aliphatic position of an organic substrate (RH) while the other oxygen atom is reduced to water:

$$RH + O_2 + NADPH + H^+ \rightarrow ROH + H_2O + NADP^+$$

Many hydroxylation reactions (insertion of hydroxyl groups) use CYP enzymes.

14.1 Nomenclature

Genes encoding CYP enzymes, and the enzymes themselves, are designated with the root symbol **CYP** for the superfamily, followed by a number indicating the gene family, a capital letter indicating the subfamily, and another numeral for the individual gene. The convention is to *italicise* the name when referring to the gene. For example, *CYP2E1* is the gene that encodes the enzyme CYP2E1—one of the enzymes involved in paracetamol (acetaminophen) metabolism. The **CYP** nomenclature is the official naming convention, although occasionally **CYP450** or **CYP**$_{450}$ is used synonymously. However, some gene or enzyme names for CYPs may differ from this nomenclature, denoting the catalytic activity and the name of the compound used as substrate. Examples include CYP5A1, thromboxane A_2 synthase, abbreviated to TBXAS1 (**T**hrom**B**o**X**ane **A**$_2$ **S**ynthase **1**), and CYP51A1, lanosterol 14-α-demethylase, sometimes unofficially abbreviated to LDM according to its substrate (**L**anosterol) and activity (**D**e**M**ethylation).[6]

The current nomenclature guidelines suggest that members of new CYP families share at least 40% amino acid identity, while members of subfamilies must share at least 55%

amino acid identity. There are nomenclature committees that assign and track both base gene names (Cytochrome P450 Homepage) and allele names (CYP Allele Nomenclature Committee).

Main article: P450-containing systems

14.2 Mechanism

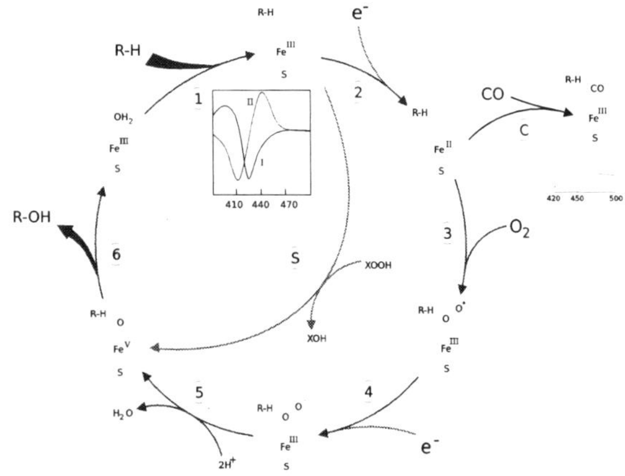

The "Fe(V) intermediate" at the bottom left is a simplification: it is an Fe(IV) with a radical heme ligand.

14.2.1 Structure

The active site of cytochrome P450 contains a heme-iron center. The iron is tethered to the protein via a cysteine thiolate ligand. This cysteine and several flanking residues are highly conserved in known CYPs and have the formal PROSITE signature consensus pattern [FW] - [SGNH] - x - [GD] - {F} - [RKHPT] - {P} - C - [LIVMFAP] - [GAD].[7] Because of the vast variety of reactions catalyzed by CYPs, the activities and properties of the many CYPs differ in many aspects. In general, the P450 catalytic cycle proceeds as follows:

14.2.2 Catalytic cycle

1. Substrate binds in proximity to the heme group, on the side opposite to the axial thiolate. Substrate binding induces a change in the conformation of the active site, often displacing a water molecule from the distal axial coordination position of the heme iron,[8] and changing the state of the heme iron from low-spin to high-spin.[9]
2. Substrate binding induces electron transfer from NAD(P)H via cytochrome P450 reductase or another associated reductase.[10]
3. Molecular oxygen binds to the resulting ferrous heme center at the distal axial coordination position, initially giving a dioxygen adduct not unlike oxy-myoglobin.
4. A second electron is transferred, from either cytochrome P450 reductase, ferredoxins, or cytochrome b5, reducing the Fe-O_2 adduct to give a short-lived peroxo state.
5. The peroxo group formed in step 4 is rapidly protonated twice, releasing one molecule of water and forming the highly reactive species referred to as **P450 Compound 1** (or just Compound I). This highly reactive intermediate was isolated in 2010,[11] P450 Compound 1 is an iron(IV) oxo (or ferryl) species with an additional oxidizing equivalent delocalized over the porphyrin and thiolate ligands. Evidence for the alternative perferryl iron(V)-oxo [8] is lacking.[11]
6. Depending on the substrate and enzyme involved, P450 enzymes can catalyze any of a wide variety of reactions. A hypothetical hydroxylation is shown in this illustration. After the product has been released from the active site, the enzyme returns to its original state, with a water molecule returning to occupy the distal coordination position of the iron nucleus.

Oxygen rebound mechanism utilized by cytochrome P450 for conversion of hydrocarbons to alcohols via the action of "compound I", an iron(IV) oxide bound to a radical heme.

1. An alternative route for mono-oxygenation is via the "peroxide shunt" (path "S" in figure). This pathway entails oxidation of the ferric-substrate complex with oxygen-atom donors such as peroxides and hypochlorites.[12] A hypothetical peroxide "XOOH" is shown in the diagram.

14.2.3 Spectroscopy

Binding of substrate is reflected in the spectral properties of the enzyme, with an increase in absorbance at 390 nm and a decrease at 420 nm. This can be measured by difference spectrometry and is referred to as the "type I" difference spectrum (see inset graph in figure). Some substrates cause an opposite change in spectral properties, a "reverse type I" spectrum, by processes that are as yet unclear. Inhibitors and certain substrates that bind directly to the heme iron give rise to the type II difference spectrum, with a maximum at 430 nm and a minimum at 390 nm (see inset graph in figure). If no reducing equivalents are available, this complex may remain stable, allowing the degree of binding to be determined from absorbance measurements *in vitro*[12] C: If carbon monoxide (CO) binds to reduced P450, the catalytic cycle is interrupted. This reaction yields the classic CO difference spectrum with a maximum at 450 nm.

14.3 P450s in humans

Human CYPs are primarily membrane-associated proteins[13] located either in the inner membrane of mitochondria or in the endoplasmic reticulum of cells. CYPs metabolize thousands of endogenous and exogenous chemicals. Some CYPs metabolize only one (or a very few) substrates, such as *CYP19* (aromatase), while others may metabolize multiple substrates. Both of these characteristics account for their central importance in medicine. Cytochrome P450 enzymes are present in most tissues of the body, and play important roles in hormone synthesis and breakdown (including estrogen and testosterone synthesis and metabolism), cholesterol synthesis, and vitamin D metabolism. Cytochrome P450 enzymes also function to metabolize potentially toxic compounds, including drugs and products of endogenous metabolism such as bilirubin, principally in the liver.

The Human Genome Project has identified 57 human genes coding for the various cytochrome P450 enzymes.[14]

14.3.1 Drug metabolism

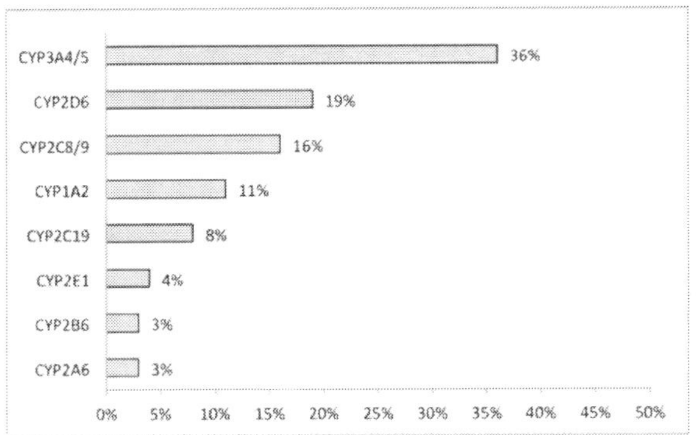

Proportion of antifungal drugs metabolized by different families of CYPs.[15]

Further information: Drug metabolism

CYPs are the major enzymes involved in drug metabolism, accounting for about 75% of the total metabolism.[16] Most drugs undergo deactivation by CYPs, either directly or by facilitated excretion from the body. Also, many substances are bioactivated by CYPs to form their active compounds.

Drug interaction

Many drugs may increase or decrease the activity of various CYP isozymes either by inducing the biosynthesis of an isozyme (enzyme induction) or by directly inhibiting the activity of the CYP (enzyme inhibition). This is a major source of adverse drug interactions, since changes in CYP enzyme activity may affect the metabolism and clearance of various drugs. For example, if one drug inhibits the CYP-mediated metabolism of another drug, the second drug may accumulate within the body to toxic levels. Hence, these drug interactions may necessitate dosage adjustments or choosing drugs that do not interact with the CYP system. Such drug interactions are especially important to take into account when using drugs of vital importance to the patient, drugs with important side-effects and drugs with small therapeutic windows, but any drug may be subject to an altered plasma concentration due to altered drug metabolism.

A classical example includes anti-epileptic drugs. Phenytoin, for example, induces CYP1A2, CYP2C9, CYP2C19, and CYP3A4. Substrates for the latter may be drugs with critical dosage, like amiodarone or carbamazepine, whose blood plasma concentration may either increase because of enzyme inhibition in the former, or decrease because of enzyme induction in the latter.

Interaction of other substances

Naturally occurring compounds may also induce or inhibit CYP activity. For example, bioactive compounds found in grapefruit juice and some other fruit juices, including bergamottin, dihydroxybergamottin, and paradicin-A, have been found to inhibit CYP3A4-mediated metabolism of certain medications, leading to increased bioavailability and, thus, the strong possibility of overdosing.[17] Because of this risk, avoiding grapefruit juice and fresh grapefruits entirely while on drugs is usually advised.[18]

Other examples:

- Saint-John's wort, a common herbal remedy induces CYP3A4, but also inhibits CYP1A1, CYP1B1, and CYP2D6.[19][20]
- Tobacco smoking induces CYP1A2 (example

CYP1A2 substrates are clozapine, olanzapine, and fluvoxamine)[21]

- At relatively high concentrations, starfruit juice has also been shown to inhibit CYP2A6 and other CYPs.[22] Watercress is also a known inhibitor of the cytochrome P450 CYP2E1, which may result in altered drug metabolism for individuals on certain medications (e.g., chlorzoxazone).[23]

- Tributyltin has been found to inhibit the function of Cytochrome P450, leading to masculinization of mollusks.[24]

- Goldenseal, with its two notable alkaloids berberine and hydrastine, has been shown to alter P450-marker enzymatic activities (involving CYP2C9, CYP2D6, and CYP3A4).[25]

14.3.2 Other specific CYP functions

Steroid hormones

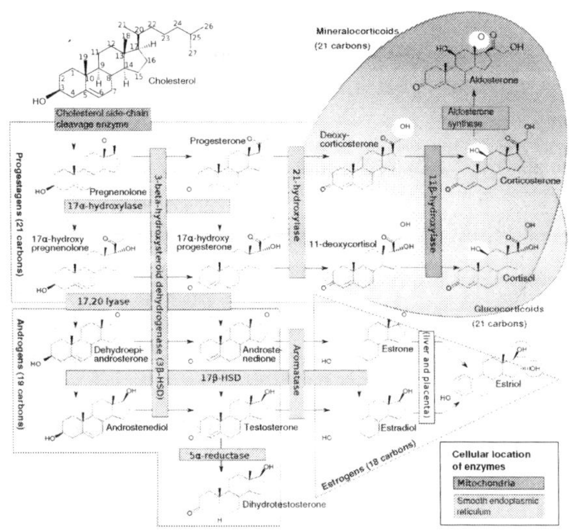

Steroidogenesis, showing many of the enzyme activities that are performed by cytochrome P450 enzymes.[26] *HSD: Hydroxysteroid dehydrogenase.*

A subset of cytochrome P450 enzymes play important roles in the synthesis of steroid hormones (steroidogenesis) by the adrenals, gonads, and peripheral tissue:

- CYP11A1 (also known as P450scc or P450c11a1) in adrenal mitochondria affects "the activity formerly known as 20,22-desmolase" (steroid 20α-hydroxylase, steroid 22-hydroxylase, cholesterol side-chain scission).

- CYP11B1 (encoding the protein P450c11β) found in the inner mitochondrial membrane of adrenal cortex has steroid 11β-hydroxylase, steroid 18-hydroxylase, and steroid 18-methyloxidase activities.

- CYP11B2 (encoding the protein P450c11AS), found only in the mitochondria of the adrenal zona glomerulosa, has steroid 11β-hydroxylase, steroid 18-hydroxylase, and steroid 18-methyloxidase activities.

- CYP17A1, in endoplasmic reticulum of adrenal cortex has steroid 17α-hydroxylase and 17,20-lyase activities.

- CYP21A1 (P450c21) in adrenal cortex conducts 21-hydroxylase activity.

- CYP19A (P450arom, aromatase) in endoplasmic reticulum of gonads, brain, adipose tissue, and elsewhere catalyzes aromatization of androgens to estrogens.

Polyunsaturated fatty acids and eicosanoids

Certain cytochrome P450 enzymes are critical in metabolizing polyunstaturated fatty acids (PUFAs) to biologically active, intercellular cell signaling molecules (eicosanoids) and/or metabolize biologically active metabolites of the PUFA to less active or inactive products. These CYPs possess Cytochrome P450 omega hydroxylase and/or epoxygenase enzyme activity.

- CYP1A1, CYP1A2, and CYP2E1 metabolize endogenous PUFAs to signaling molecules: they metabolize arachidonic acid (i.e. AA) to 19-hydroxyeicosatetraenoic acid (i.e. 19-HETE; see 20-Hydroxyeicosatetraenoic acid); eicosapentaenoic acid (i.e. EPA) to epoxyeicosatetraenoic acids (i.e. EEQs); and docosahexaenoic acid (i.e. DHA) to epoxydocosapentaenoic acids (i.e. EDPs).

- CYP2C8, CYP2C9, CYP2C18, CYP2C19, and CYP2J2 metabolize endogenous PUFAs to signaling molecules: they metabolize AA to epoxyeicosatetraenoic acids (i.e. EETs); EPA to EEQs; and DHA to EDPs.

- CYP2S1 metabolizes PUFA to signaling molecules: it metabolizes AA to EETs ad EPA to EEQs.

- CYP3A4 metabolizes AA to EET signaling molecules.

- CYP4A11 metabolizes endogenous PUFAs to signaling molecules: it metabolizes AA to 20-HETE and EETs; it also hydroxylates DHA to 22-hydroxy-DHA (i.e. 12-HDHA).

- CYP4F2, CYP4F3A, and CYP4F3B (see CYP4F3 for latter two CYPs) metabolize PUFAs to signaling molecules: they metabolizes AA to 20-HETE. They also metabolize EPA to 19-hydroxyeicosapentaenoic acid (19-HEPE) and 20-hydroxyeicosapentaenoic acid (20-HEPE) as well as metabolize DHA to 22-HDA. They also inactivate or reduce the activity of signaling molecules: they metabolize leukotriene B4 (LTB4) to 20-hydroxy-LTB4, 5-hydroxyeicosatetraenoic acid (5-HETE) to 5,20-diHETE, 5-oxo-eicosatetraenoic acid (5-oxo-ETE) to 5-oxo,20-hydroxy-ETE, 12-hydroxyeicosatetraenoic acid (12-HETE) to 12,20-diHETE, EETs to 20-hydroxy-EETs, and lipoxins to 20-hydroxy products.
- CYP4F8 and CYP4F12 metabolize PUFAs to signaling molecules: they metabolizes EPA to EEQs and DHA to EDPs. They also metabolize AA to 18-hydroxyeicosatetraenoic acid (18-HETE) and 19-HETE.
- CYP4F11 inactivates or reduces the activity of signaling molecules: it metabolizes LTB4 to 20-hydroxy-LTB4, (5-HETE) to 5,20-diHETE, (5-oxo-ETE) to 5-oxo,20-hydroxy-ETE, (12-HETE) to 12,20-diHETE, EETs to 20-hydroxy-EETs, and lipoxins to 20-hydroxy products.
- CYP4F22 ω-hydroxylates extremely long "very long chain fatty acids", i.e. fatty acids that are 28 or more carbons long. The ω-hydroxylation of these special fatty acids is critical to creating and maintaining the skins water barrier function; autosomal recessive inactivating mutations of CYP4F22 are associated with the Lamellar ichthyosis subtype of Congenital ichthyosiform erythrodema in humans.[27]

14.3.3 CYP families in humans

Humans have 57 genes and more than 59 pseudogenes divided among 18 families of cytochrome P450 genes and 43 subfamilies.[28] This is a summary of the genes and of the proteins they encode. See the homepage of the Cytochrome P450 Nomenclature Committee for detailed information.[14]

14.4 P450s in other species

14.4.1 Animals

Many animals have as many or more CYP genes than humans do. Reported numbers range from 35 genes in the sponge *Amphimedon queenslandica* to 235 genes in the cephalochordate *Branchiostoma floridae*.[29] Mice have genes for 101 CYPs, and sea urchins have even more (perhaps as many as 120 genes).[30] Most CYP enzymes are presumed to have monooxygenase activity, as is the case for most mammalian CYPs that have been investigated (except for, e.g., CYP19 and CYP5). Gene and genome sequencing is far outpacing biochemical characterization of enzymatic function, though many genes with close homology to CYPs with known function have been found, giving clues to their functionality.

The classes of CYPs most often investigated in non-human animals are those either involved in development (e.g., retinoic acid or hormone metabolism) or involved in the metabolism of toxic compounds (such as heterocyclic amines or polyaromatic hydrocarbons). Often there are differences in gene regulation or enzyme function of CYPs in related animals that explain observed differences in susceptibility to toxic compounds (ex. canines inability to metabolize xanthines such as caffeine). Some drugs undergo metabolism in both species via different enzymes, resulting in different metabolites, while other drugs are metabolized in one species but excreted unchanged in another species. For this reason, one species's reaction to a substance is not a reliable indication of the substance's effects in humans.

CYPs have been extensively examined in mice, rats, dogs, and less so in zebrafish, in order to facilitate use of these model organisms in drug discovery and toxicology. Recently CYPs have also been discovered in avian species, in particular turkeys, that may turn out to be a great model for cancer research in humans.[31] CYP1A5 and CYP3A37 in turkeys were found to be very similar to the human CYP1A2 and CYP3A4 respectively, in terms of their kinetic properties as well as in the metabolism of aflatoxin B1.[32]

CYPs have also been heavily studied in insects, often to understand pesticide resistance. For example, CYP6G1 is linked to insecticide resistance in DDT-resistant Drosophila melanogaster[33] and CYP6Z1 in the mosquito malaria vector Anopheles gambiae is capable of directly metabolizing DDT.[34]

14.4.2 Microbial

Microbial cytochromes P450 are often soluble enzymes and are involved in diverse metabolic processes. In bacteria the distribution of P450s is very variable with many bacteria having no identified P450s (e.g. E.coli). Some bacteria, predominantly actinomycetes, have numerous P450s (e.g.,[35][36]). Those so far identified are generally involved in either biotransformation of xenobiotic compounds (e.g. CYP105A1 from Streptomyces griseolus metabolizes sulfonylurea herbicides to less toxic derivatives,[37]) or are part of specialised metabolite biosynthetic pathways (e.g. CYP170B1 catalyses production of the sesquiterpenoid albaflavenone in Streptomyces albus,[38]). Although no P450 has yet been shown to be essential in a microbe, the CYP105 family is highly conserved with a representative in every streptomycete genome sequenced so far ([39]). Due to the solubility of bacterial P450 enzymes, they are generally regarded as easier to work with than the predominantly membrane bound eukaryotic P450s. This, combined with the remarkable chemistry they catalyse, has led to many studies using the heterologously expressed proteins in vitro. Few studies have investigated what P450s do in vivo, what the natural substrate(s) are and how P450s contribute to survival of the bacteria in the natural environment.Three examples that have contributed significantly to structural and mechanistic studies are listed here, but many different families exist.

- Cytochrome P450cam (CYP101) originally from *Pseudomonas putida* has been used as a model for many cytochromes P450 and was the first cytochrome P450 three-dimensional protein structure solved by X-ray crystallography. This enzyme is part of a camphor-hydroxylating catalytic cycle consisting of two electron transfer steps from putidaredoxin, a 2Fe-2S cluster-containing protein cofactor.
- Cytochrome P450 eryF (CYP107A1) originally from the actinomycete bacterium *Saccharopolyspora erythraea* is responsible for the biosynthesis of the antibiotic erythromycin by C6-hydroxylation of the macrolide 6-deoxyerythronolide B.
- Cytochrome P450 BM3 (CYP102A1) from the soil bacterium *Bacillus megaterium* catalyzes the NADPH-dependent hydroxylation of several long-chain fatty acids at the ω–1 through ω–3 positions. Unlike almost every other known CYP (except CYP505A1, cytochrome P450 foxy), it constitutes a natural fusion protein between the CYP domain and an electron donating cofactor. Thus, BM3 is potentially very useful in biotechnological applications.[40][41]
- Cytochrome P450 119 (CYP119) isolated from the thermophillic archea *Sulfolobus acidocaldarius* [42] has been used in a variety of mechanistic studies.[11] Because thermophillic enzymes evolved to function at high temperatures, they tend to function more slowly at room temperature (if at all) and are therefore excellent mechanistic models.

14.4.3 Fungi

The commonly used azole class antifungal drugs work by inhibition of the fungal cytochrome P450 14α-demethylase. This interrupts the conversion of lanosterol to ergosterol, a component of the fungal cell membrane. (This is useful only because humans' P450 have a different sensitivity; this is how this class of antifungals work.)[43]

Significant research is ongoing into fungal P450s, as a number of fungi are pathogenic to humans (such as Candida yeast and Aspergillus) and to plants.

Cunninghamella elegans is a candidate for use as a model for mammalian drug metabolism.

14.4.4 Plants

Plant cytochrome P450s are involved in a wide range of biosynthetic reactions and target a diverse range of biomolecules. These reactions lead to various fatty acid conjugates, plant hormones, secondary metabolites, lignins, and a variety of defensive compounds.[44] Plant genome annotations suggest that Cytochrome P450 genes make up as much as 1% of the plant genes. The number and diversity of P450 genes is responsible, in part, for the multitude of bioactive compounds.[45]

14.5 P450s in biotechnology

The remarkable reactivity and substrate promiscuity of P450s have long attracted the attention of chemists.[46] Recent progress towards realizing the potential of using P450s towards difficult oxidations have included: (i) eliminating the need for natural co-factors by replacing them with inexpensive peroxide containing molecules,[47] (ii) exploring the compatibility of p450s with organic solvents,[48] and (iii) the use of small, non-chiral auxiliaries to predictably direct P450 oxidation.

14.6 InterPro subfamilies

InterPro subfamilies:

- Cytochrome P450, B-class InterPro: *IPR002397*
- Cytochrome P450, mitochondrial InterPro: *IPR002399*
- Cytochrome P450, E-class, group I InterPro: *IPR002401*
- Cytochrome P450, E-class, group II InterPro: *IPR002402*
- Cytochrome P450, E-class, group IV InterPro: *IPR002403*

- Aromatase

Clozapine, imipramine, paracetamol, phenacetin Heterocyclic aryl amines Inducible and CYP1A2 5-10% deficient oxidize uroporphyrinogen to uroporphyrin (CYP1A2) in heme metabolism, but they may have additional undiscovered endogenous substrates. are inducible by some polycyclic hydrocarbons, some of which are found in cigarette smoke and charred food.

These enzymes are of interest, because in assays, they can activate compounds to carcinogens. High levels of CYP1A2 have been linked to an increased risk of colon cancer. Since the 1A2 enzyme can be induced by cigarette smoking, this links smoking with colon cancer.[49]

14.7 References

[1] Lamb DC, Lei L, Warrilow AG, Lepesheva GI, Mullins JG, Waterman MR, Kelly SL (August 2009). "The first virally encoded cytochrome p450". *Journal of Virology*. **83** (16): 8266–9. PMC 2715754. PMID 19515774. doi:10.1128/JVI.00289-09.

[2] Roland Sigel; Sigel, Astrid; Sigel, Helmut (2007). *The Ubiquitous Roles of Cytochrome P450 Proteins: Metal Ions in Life Sciences*. New York: Wiley. ISBN 0-470-01672-8.

[3] Danielson PB (December 2002). "The cytochrome P450 superfamily: biochemistry, evolution and drug metabolism in humans". *Current Drug Metabolism*. **3** (6): 561–97. PMID 12369887. doi:10.2174/1389200023337054.

[4] Nelson D. "Cytochrome P450 Homepage". University of Tennessee. Retrieved 2014-11-13.

[5] Hanukoglu, Israel (1996). "Electron Transfer Proteins of Cytochrome P450 Systems". *Advances in Molecular and Cell Biology*. Advances in Molecular and Cell Biology. **14**: 29–56. ISBN 9780762301133. ISSN 1569-2558. doi:10.1016/S1569-2558(08)60339-2.

[6] "NCBI sequence viewer". Retrieved 2007-11-19.

[7] PROSITE consensus pattern for P450

[8] Meunier B, de Visser SP, Shaik S (September 2004). "Mechanism of oxidation reactions catalyzed by cytochrome p450 enzymes". *Chemical Reviews*. **104** (9): 3947–80. PMID 15352783. doi:10.1021/cr020443g.

[9] Poulos TL, Finzel BC, Howard AJ (June 1987). "High-resolution crystal structure of cytochrome P450cam". *Journal of Molecular Biology*. **195** (3): 687–700. PMID 3656428. doi:10.1016/0022-2836(87)90190-2.

[10] Sligar SG, Cinti DL, Gibson GG, Schenkman JB (October 1979). "Spin state control of the hepatic cytochrome P450 redox potential". *Biochemical and Biophysical Research Communications*. **90** (3): 925–32. PMID 228675. doi:10.1016/0006-291X(79)91916-8.

[11] Rittle J, Green MT (November 2010). "Cytochrome P450 compound I: capture, characterization, and C-H bond activation kinetics". *Science*. **330** (6006): 933–7. Bibcode:2010Sci...330..933R. PMID 21071661. doi:10.1126/science.1193478.

[12] Ortiz de Montellano, Paul R.; Paul R. Ortiz de Montellano (2005). *Cytochrome P450: structure, mechanism, and biochemistry* (3rd ed.). New York: Kluwer Academic/Plenum Publishers. ISBN 0-306-48324-6.

[13] Berka K, Hendrychová T, Anzenbacher P, Otyepka M (October 2011). "Membrane position of ibuprofen agrees with suggested access path entrance to cytochrome P450 2C9 active site". *The Journal of Physical Chemistry. A*. **115** (41): 11248–55. PMC 3257864. PMID 21744854. doi:10.1021/jp204488j.

[14] "P450 Table".

[15] doctorfungus > Antifungal Drug Interactions Content Director: Russell E. Lewis, Pharm.D. Retrieved on Jan 23, 2010

[16] Guengerich FP (January 2008). "Cytochrome p450 and chemical toxicology". *Chemical Research in Toxicology*. **21** (1): 70–83. PMID 18052394. doi:10.1021/tx700079z. (Metabolism in this context is the chemical modification or degradation of drugs.)

[17] Bailey DG, Dresser GK (2004). "Interactions between grapefruit juice and cardiovascular drugs". *American Journal of Cardiovascular Drugs*. **4** (5): 281–97. PMID 15449971. doi:10.2165/00129784-200404050-00002.

[18] Zeratsky K (2008-11-06). "Grapefruit juice: Can it cause drug interactions?". *Ask a food & nutrition specialist*. MayoClinic.com. Retrieved 2009-02-09.

[19] Chaudhary A, Willett KL (January 2006). "Inhibition of human cytochrome CYP 1 enzymes by flavonoids of St. John's wort". *Toxicology*. **217** (2–3): 194–205. PMID 16271822. doi:10.1016/j.tox.2005.09.010.

[20] Strandell J, Neil A, Carlin G (February 2004). "An approach to the in vitro evaluation of potential for cytochrome P450 enzyme inhibition from herbals and other natural remedies". *Phytomedicine*. **11** (2–3): 98–104. PMID 15070158. doi:10.1078/0944-7113-00379.

[21] Kroon LA (September 2007). "Drug interactions with smoking". *American Journal of Health-System Pharmacy*. **64** (18): 1917–21. PMID 17823102. doi:10.2146/ajhp060414.

[22] Zhang JW, Liu Y, Cheng J, Li W, Ma H, Liu HT, Sun J, Wang LM, He YQ, Wang Y, Wang ZT, Yang L (2007). "Inhibition of human liver cytochrome P450 by star fruit juice". *Journal of Pharmacy & Pharmaceutical Sciences*. **10** (4): 496–503. PMID 18261370. doi:10.18433/j30593.

[23] Leclercq I, Desager JP, Horsmans Y (August 1998). "Inhibition of chlorzoxazone metabolism, a clinical probe for CYP2E1, by a single ingestion of watercress". *Clinical Pharmacology and Therapeutics*. **64** (2): 144–9. PMID 9728894. doi:10.1016/S0009-9236(98)90147-3.

[24] Walmsley, Simon. "Tributyltin pollution on a global scale. An overview of relevant and recent research: impacts and issues." (PDF). WWF UK.

[25] Chatterjee P, Franklin MR (November 2003). "Human cytochrome p450 inhibition and metabolic-intermediate complex formation by goldenseal extract and its methylenedioxyphenyl components". *Drug Metabolism and Disposition*. **31** (11): 1391–7. PMID 14570772. doi:10.1124/dmd.31.11.1391.

[26] Häggström, Mikael; Richfield, David (2014). "Diagram of the pathways of human steroidogenesis". *WikiJournal of Medicine*. **1** (1). ISSN 2002-4436. doi:10.15347/wjm/2014.005.

[27] Sugiura K, Akiyama M (July 2015). "Update on autosomal recessive congenital ichthyosis: mRNA analysis using hair samples is a powerful tool for genetic diagnosis". *Journal of Dermatological Science*. **79** (1): 4–9. PMID 25982146. doi:10.1016/j.jdermsci.2015.04.009.

[28] Nelson D (2003). Cytochromes P450 in humans. Retrieved May 9, 2005.

[29] Nelson DR, Goldstone JV, Stegeman JJ (February 2013). "The cytochrome P450 genesis locus: the origin and evolution of animal cytochrome P450s". *Philosophical Transactions of the Royal Society of London. Series B, Biological Sciences*. **368** (1612): 20120474. PMC 3538424. PMID 23297357. doi:10.1098/rstb.2012.0474.

[30] Goldstone JV, Hamdoun A, Cole BJ, Howard-Ashby M, Nebert DW, Scally M, Dean M, Epel D, Hahn ME, Stegeman JJ (December 2006). "The chemical defensome: environmental sensing and response genes in the Strongylocentrotus purpuratus genome". *Developmental Biology*. **300** (1): 366–84. PMC 3166225. PMID 17097629. doi:10.1016/j.ydbio.2006.08.066.

[31] Rawal S, Kim JE, Coulombe R (December 2010). "Aflatoxin B1 in poultry: toxicology, metabolism and prevention". *Research in Veterinary Science*. **89** (3): 325–31. PMID 20462619. doi:10.1016/j.rvsc.2010.04.011.

[32] Rawal S, Coulombe RA (August 2011). "Metabolism of aflatoxin B1 in turkey liver microsomes: the relative roles of cytochromes P450 1A5 and 3A37". *Toxicology and Applied Pharmacology*. **254** (3): 349–54. PMID 21616088. doi:10.1016/j.taap.2011.05.010.

[33] McCart C, Ffrench-Constant RH (June 2008). "Dissecting the insecticide-resistance- associated cytochrome P450 gene Cyp6g1". *Pest Management Science*. **64** (6): 639–45. PMID 18338338. doi:10.1002/ps.1567.

[34] Chiu TL, Wen Z, Rupasinghe SG, Schuler MA (July 2008). "Comparative molecular modeling of Anopheles gambiae CYP6Z1, a mosquito P450 capable of metabolizing DDT". *Proceedings of the National Academy of Sciences of the United States of America*. **105** (26): 8855–60. Bibcode:2008PNAS..105.8855C. PMC 2449330. PMID 18577597. doi:10.1073/pnas.0709249105.

[35] McLean KJ, Clift D, Lewis DG, Sabri M, Balding PR, Sutcliffe MJ, Leys D, Munro AW (May 2006). "The preponderance of P450s in the Mycobacterium tuberculosis genome". *Trends in Microbiology*. **14** (5): 220–8. PMID 16581251. doi:10.1016/j.tim.2006.03.002.

[36] Ikeda H, Ishikawa J, Hanamoto A, Shinose M, Kikuchi H, Shiba T, Sakaki Y, Hattori M, Omura S (May 2003). "Complete genome sequence and comparative analysis of the industrial microorganism Streptomyces avermitilis". *Nature Biotechnology*. **21** (5): 526–31. PMID 12692562. doi:10.1038/nbt820.

[37] Leto, O'Keefe (1988). "Identification of constitutive and herbicide inducible cytochromes P-450 in Streptomyces griseolus". *Arch Microbiol*. **149** (5): 406–12. doi:10.1007/BF00425579.

[38] Moody SC, Zhao B, Lei L, Nelson DR, Mullins JG, Waterman MR, Kelly SL, Lamb DC (May 2012). "Investigating conservation of the albaflavenone biosynthetic pathway and CYP170 bifunctionality in streptomycetes". *The FEBS Journal*. **279** (9): 1640–9. PMID 22151149. doi:10.1111/j.1742-4658.2011.08447.x.

[39] Moody SC, Loveridge EJ (December 2014). "CYP105-diverse structures, functions and roles in an intriguing family of enzymes in Streptomyces". *Journal of Applied Microbiology*. **117** (6): 1549–63. PMC 4265290. PMID 25294646. doi:10.1111/jam.12662.

[40] Narhi LO, Fulco AJ (June 1986). "Characterization of a catalytically self-sufficient 119,000-dalton cytochrome P-450 monooxygenase induced by barbiturates in Bacillus megaterium". *The Journal of Biological Chemistry*. **261** (16): 7160–9. PMID 3086309.

[41] Girvan HM, Waltham TN, Neeli R, Collins HF, McLean KJ, Scrutton NS, Leys D, Munro AW (December 2006). "Flavocytochrome P450 BM3 and the origin of CYP102 fusion species". *Biochemical Society Transactions*. **34** (Pt 6): 1173–7. PMID 17073779. doi:10.1042/BST0341173.

[42] Wright RL, Harris K, Solow B, White RH, Kennelly PJ (April 1996). "Cloning of a potential cytochrome P450 from the archaeon Sulfolobus solfataricus". *FEBS Letters*. **384** (3): 235–9. PMID 8617361. doi:10.1016/0014-5793(96)00322-5.

[43] Vanden Bossche H, Marichal P, Gorrens J, Coene MC (September 1990). "Biochemical basis for the activity and selectivity of oral antifungal drugs". *British Journal of Clinical Practice. Supplement*. **71**: 41–6. PMID 2091733.

[44] Schuler MA, Werck-Reichhart D (2003-01-01). "Functional genomics of P450s". *Annual Review of Plant Biology*. **54** (1): 629–67. PMID 14503006. doi:10.1146/annurev.arplant.54.031902.134840.

[45] Mizutani M, Sato F (March 2011). "Unusual P450 reactions in plant secondary metabolism". *Archives of Biochemistry and Biophysics*. P450 Catalysis Mechanisms. **507** (1): 194–203. PMID 20920462. doi:10.1016/j.abb.2010.09.026.

[46] Chefson A, Auclair K (October 2006). "Progress towards the easier use of P450 enzymes". *Molecular bioSystems*. **2** (10): 462–9. PMID 17216026. doi:10.1039/b607001a.

[47] Chefson A, Zhao J, Auclair K (June 2006). "Replacement of natural cofactors by selected hydrogen peroxide donors or organic peroxides results in improved activity for CYP3A4 and CYP2D6". *Chembiochem*. **7** (6): 916–9. PMID 16671126. doi:10.1002/cbic.200600006.

[48] Chefson A, Auclair K (July 2007). "CYP3A4 activity in the presence of organic cosolvents, ionic liquids, or water-immiscible organic solvents". *Chembiochem*. **8** (10): 1189–97. PMID 17526062. doi:10.1002/cbic.200700128.

[49] Petros WP, Younis IR, Ford JN, Weed SA (October 2012). "Effects of tobacco smoking and nicotine on cancer treatment". *Pharmacotherapy*. **32** (10): 920–31. PMC 3499669 🔓. PMID 23033231. doi:10.1002/phar.1117.

14.8 External links

- Degtyarenko K (2009-01-09). "Directory of P450-containing Systems". International Centre for Genetic Engineering and Biotechnology. Retrieved 2009-02-10.
- Estabrook RW (December 2003). "A passion for P450s (rememberances of the early history of research on cytochrome P450)". *Drug Metabolism and Disposition*. **31** (12): 1461–73. PMID 14625342. doi:10.1124/dmd.31.12.1461.
- Feyereisen R (2005-12-19). "The Insect P450 Site". Institut National de la Recherche Agronomique. Retrieved 2009-02-10.
- Flockhart DA (2007). "Cytochrome P450 drug interaction table". Indiana University-Purdue University Indianapolis. Retrieved 2009-02-10.
- Fowler L, Mercer A. "Cytochrome P450 Animated Tutorial". School of Pharmacy, London. Retrieved 2009-02-10.
- Preissner S (2010). "Cytochrome P450 database". Nucleic Acids Research.
- Sim SC (2008-09-04). "Human Cytochrome P450 (CYP) Allele Nomenclature Committee". Karolinska Institutet. Retrieved 2009-02-10.
- Hazai E (2012-02-12). "Cytochrome P450 enzyme-substrate selectivity prediction".
- Performance of P450 inhibition Studies The performance of *in vitro* cytochrome P450 inhibition studies including analysis of the data.
- DDI Regulatory Guidance Request a guide to drug-drug interaction regulatory recommendations.
- Expanding the toolbox of cytochrome P450s through enzyme engineering Video by the Turner Group, University of Manchester, UK

Chapter 15

Transferase

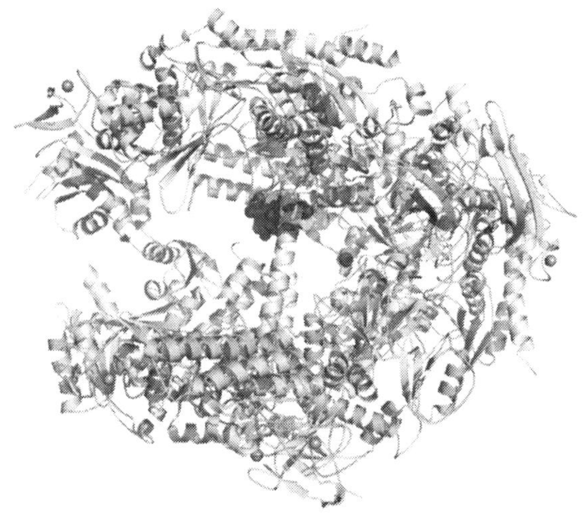

RNA polymerase from Saccharomyces cerevisiae *complexed with α-amanitin (in red). Despite the use of the term "polymerase," RNA polymerases are classified as a form of nucleotidyl transferase.*[1]

A **transferase** is any one of a class of enzymes that enact the transfer of specific functional groups (e.g. a methyl or glycosyl group) from one molecule (called the donor) to another (called the acceptor).[2] They are involved in hundreds of different biochemical pathways throughout biology, and are integral to some of life's most important processes.

Transferases are involved in myriad reactions in the cell. Three examples of these reactions are the activity of coenzyme A (CoA) transferase, which transfers thiol esters,[3] the action of N-acetyltransferase, which is part of the pathway that metabolizes tryptophan,[4] and the regulation of pyruvate dehydrogenase (PDH), which converts pyruvate to acetyl CoA.[5] Transferases are also utilized during translation. In this case, an amino acid chain is the functional group transferred by a peptidyl transferase. The transfer involves the removal of the growing amino acid chain from the tRNA molecule in the A-site of the ribosome and its subsequent addition to the amino acid attached to the tRNA in the P-site.[6]

Mechanistically, an enzyme that catalyzed the following reaction would be a transferase:

$$X group + Y \xrightarrow[transferase]{} X + Y group$$

In the above reaction, X would be the donor, and Y would be the acceptor.[7] "Group" would be the functional group transferred as a result of transferase activity. The donor is often a coenzyme.

15.1 History

Some of the most important discoveries relating to transferases occurred as early as the 1930s. Earliest discoveries of transferase activity occurred in other classifications of enzymes, including Beta-galactosidase, protease, and acid/base phosphatase. Prior to the realization that individual enzymes were capable of such a task, it was believed that two or more enzymes enacted functional group transfers.[8]

Transamination, or the transfer of an amine (or NH_2) group from an amino acid to a keto acid by an aminotransferase (also known as a "transaminase"), was first noted in 1930 by D. M. Needham, after observing the disappearance of glutamic acid added to pigeon breast muscle.[9] This observance was later verified by the discovery of its reaction mechanism by Braunstein and Kritzmann in 1937.[10] Their analysis showed that this reversible reaction could be applied to other tissues.[11] This assertion was validated by Rudolf Schoenheimer's work with radioisotopes as tracers in 1937.[12][13] This in turn would pave the way for the possibility that similar transfers were a primary means of producing most amino acids via amino transfer.[14]

Another such example of early transferase research and later reclassification involved the discovery of uridyl transferase. In 1953, the enzyme UDP-glucose pyrophosphorylase was shown to be a transferase, when it was found that it could reversibly produce UTP and G1P from UDP-glucose and an organic pyrophosphate.[15]

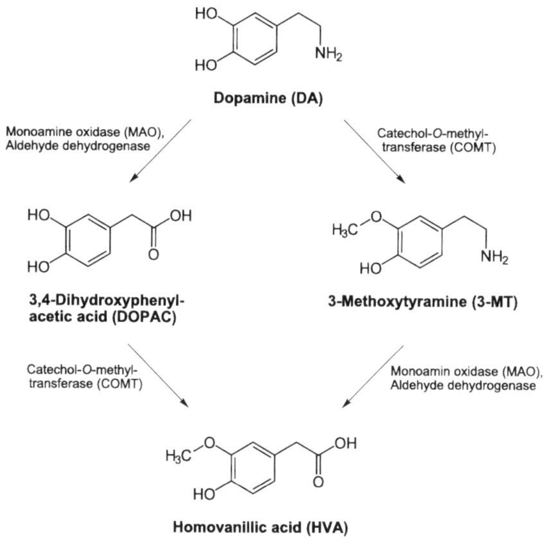

Biodegradation of dopamine via catechol-O-methyltransferase (along with other enzymes). The mechanism for dopamine degradation led to the Nobel Prize in Physiology or Medicine in 1970.

Another example of historical significance relating to transferase is the discovery of the mechanism of catecholamine breakdown by catechol-O-methyltransferase. This discovery was a large part of the reason for Julius Axelrod's 1970 Nobel Prize in Physiology or Medicine (shared with Sir Bernard Katz and Ulf von Euler).[16]

Classification of transferases continues to this day, with new ones being discovered frequently.[17][18] An example of this is Pipe, a sulfotransferase involved in the dorsal-ventral patterning of *Drosophilia*.[19] Initially, the exact mechanism of Pipe was unknown, due to a lack of information on its substrate.[20] Research into Pipe's catalytic activity eliminated the likelihood of it being a heparan sulfate glycosaminoglycan.[21] Further research has shown that Pipe targets the ovarian structures for sulfation.[22] Pipe is currently classified as a *Drosophilia* heparan sulfate 2-O-sulfotransferase.[23]

15.2 Nomenclature

Systematic names of transferases are constructed in the form of "donor:acceptor grouptransferase."[24] For example, methylamine:L-glutamate N-methyltransferase would be the standard naming convention for the transferase methylamine-glutamate N-methyltransferase, where methylamine is the donor, L-glutamate is the acceptor, and methyltransferase is the EC category grouping. This same action by the transferase can be illustrated as follows:

$$\text{methylamine} + \text{L-glutamate} \rightleftharpoons NH_3 + \text{N-methyl-L-glutamate}$$[25]

However, other accepted names are more frequently used for transferases, and are often formed as "acceptor grouptransferase" or "donor grouptransferase." For example, a DNA methyltransferase is a transferase that catalyzes the transfer of a methyl group to a DNA acceptor. In practice, many molecules are not referred to using this terminology due to more prevalent common names.[26] For example, RNA Polymerase is the modern common name for what was formerly known as RNA nucleotidyltransferase, a kind of nucleotidyl transferase that transfers nucleotides to the 3' end of a growing RNA strand.[27] In the EC system of classification, the accepted name for RNA Polymerase is DNA-directed RNA polymerase.[28]

15.3 Classification

Described primarily based on the type of biochemical group transferred, transferases can be divided into ten categories (based on the EC Number classification).[29] These categories comprise over 450 different unique enzymes.[30] In the EC numbering system, transferases have been given a classification of **EC2**. Hydrogen is not considered a functional group when it comes to transferase targets; instead, hydrogen transfer is included under oxidoreductases,[30] due to electron transfer considerations.

15.4 Reactions

15.4.1 EC 2.1: single carbon transferases

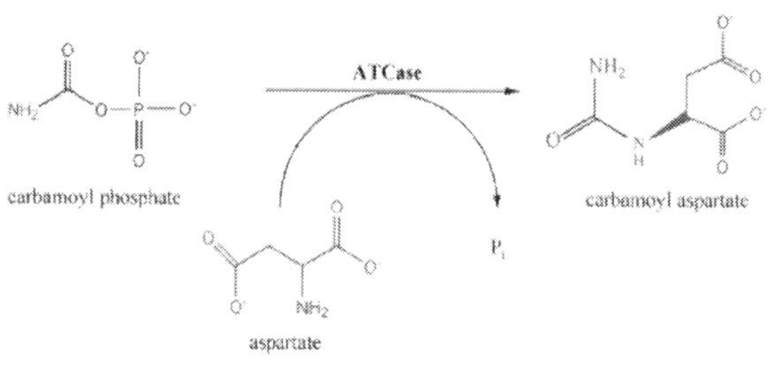

Reaction involving aspartate transcarbamylase.

EC 2.1 includes enzymes that transfer single-carbon groups. This category consists of transfers of methyl, hydroxymethyl, formyl, carboxy, carbamoyl, and amido groups.[31] Carbamoyltransferases, as an example, transfer a carbamoyl group from one molecule to another.[32] Carbamoyl groups follow the formula NH_2CO.[33] In ATCase such a transfer is written as Carbamyl phosphate + L-aspertate → L-carbamyl aspartate + phosphate.[34]

15.4.2 EC 2.2: aldehyde and ketone transferases

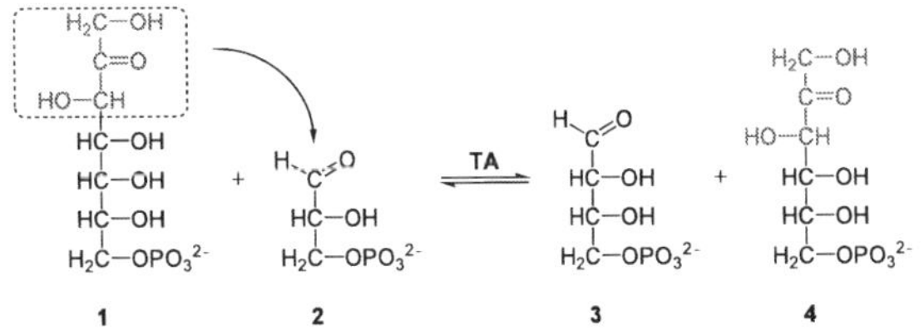

The reaction catalyzed by transaldolase

Enzymes that transfer aldehyde or ketone groups and included in EC 2.2. This category consists of various transketolases and transaldolases.[35] Transaldolase, the namesake of aldehyde transferases, is an important part of the pentose phosphate pathway.[36] The reaction it catalyzes consists of a transfer of a dihydroxyacetone functional group to Glyceraldehyde 3-phosphate (also known as G3P). The reaction is as follows: sedoheptulose 7-phosphate + glyceraldehyde 3-phosphate ⇌ erythrose 4-phosphate + fructose 6-phosphate.[37]

15.4.3 EC 2.3: acyl transferases

Transfer of acyl groups or acyl groups that become alkyl groups during the process of being transferred are key aspects of EC 2.3. Further, this category also differentiates between amino-acyl and non-amino-acyl groups. Peptidyl transferase is a ribozyme that facilitates formation of peptide bonds during translation.[38] As an aminoacyltransferase, it catalyzes the transfer of a peptide to an aminoacyl-tRNA, following this reaction: peptidyl-tRNAA + aminoacyl-tRNAB ⇌ tRNAA + peptidyl aminoacyl-tRNAB.[39]

15.4.4 EC 2.4: glycosyl, hexosyl, and pentosyl transferases

EC 2.4 includes enzymes that transfer glycosyl groups, as well as those that transfer hexose and pentose. Glycosyltransferase is a subcategory of EC 2.4 transferases that is involved in biosynthesis of disaccharides and polysaccharides through transfer of monosaccharides to other molecules.[40] An example of a prominent glycosyltransferase is lactose synthase which is a dimer possessing two protein subunits. Its primary action is to produce lactose from glucose and UDP-galactose.[41] This occurs via the following pathway: UDP-β-D-galactose + D-glucose ⇌ UDP + lactose.[42]

15.4.5 EC 2.5: alkyl and aryl transferases

EC 2.5 relates to enzymes that transfer alkyl or aryl groups, but does not include methyl groups. This is in contrast to functional groups that become alkyl groups when transferred, as those are included in EC 2.3. EC 2.5 currently only possesses one sub-class: Alkyl and aryl transferases.[43] Cysteine synthase, for example, catalyzes the formation of acetic acids and cysteine from O_3-acetyl-L-serine and hydrogen sulfide: O_3-acetyl-L-serine + H_2S ⇌ L-cysteine + acetate.[44]

15.4.6 EC 2.6: nitrogenous transferases

Aspartate aminotransferase can act on several different amino acids

The grouping consistent with transfer of nitrogenous groups is EC 2.6. This includes enzymes like transaminase (also known as "aminotransferase"), and a very small number of oximinotransferases and other nitrogen group transferring enzymes. EC 2.6 previously included amidinotransferase but it has since been reclassified as a subcategory of EC 2.1 (single-carbon transferring enzymes).[45] In the case of aspartate transaminase, which can act on tyrosine, phenylalanine, and tryptophan, it reversibly transfers an amino group from one molecule to the other.[46]

The reaction, for example, follows the following order: L-aspartate +2-oxoglutarate ⇌ oxaloacetate + L-glutamate.[47]

15.4.7 EC 2.7: phosphorus transferases

While EC 2.7 includes enzymes that transfer phosphorus-containing groups, it also includes nuclotidyl transferases as well.[48] Sub-category phosphotransferase is divided up in categories based on the type of group that accepts the transfer.[24] Groups that are classified as phosphate acceptors include: alcohols, carboxy groups, nitrogenous groups, and phosphate groups.[29] Further constituents of this subclass of transferases are various kinases. A prominent kinase is cyclin-dependent kinase (or CDK), which comprises a sub-family of protein kinases. As their name implies, CDKs are heavily dependent on specific cyclin molecules for activation.[49] Once combined, the CDK-cyclin complex is capable of enacting its function within the cell cycle.[50]

The reaction catalyzed by CDK is as follows: ATP + a target protein → ADP + a phosphoprotein.[51]

15.4.8 EC 2.8: sulfur transferases

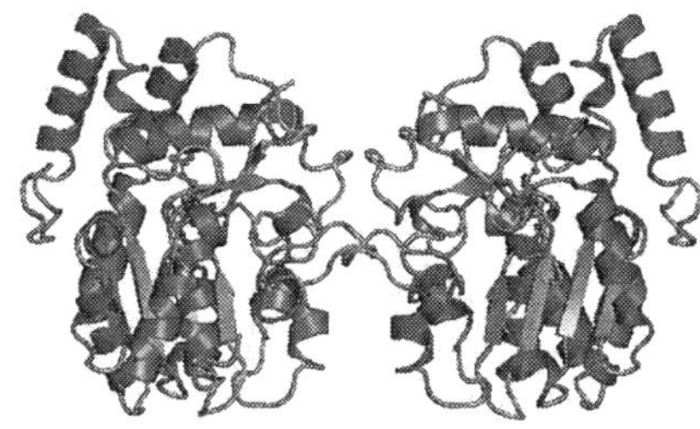

Ribbon diagram of a variant structure of estrogen sulfotransferase (PDB 1aqy EBI)[52]

Transfer of sulfur-containing groups is covered by EC 2.8 and is subdivided into the subcategories of sulfurtransferases, sulfotransferases, and CoA-transferases, as well as enzymes that transfer alkylthio groups.[53] A specific group of sulfotransferases are those that use PAPS as a sulfate group donor.[54] Within this group is alcohol sulfotransferase which has a broad targeting capacity.[55] Due to this, alcohol sulfotransferase is also known by several other names including "hydroxysteroid sulfotransferase," "steroid sulfokinase," and "estrogen sulfotransferase."[56] Decreases in its activity has been linked to human liver disease.[57] This transferase acts via the following reaction: 3'-phosphoadenylyl sulfate + an alcohol ⇌ adenosine 3',5'bisphosphate + an alkyl sulfate.[58]

15.4.9 EC 2.9: selenium transferases

EC 2.9 includes enzymes that transfer selenium-containing groups.[59] This category only contains two transferases, and thus is one of the smallest categories of transferase. Selenocysteine synthase, which was first added to the classification system in 1999, converts seryl-tRNA(Sec UCA) into selenocysteyl-tRNA(Sec UCA).[60]

15.4.10 EC 2.10: metal transferases

The category of EC 2.10 includes enzymes that transfer molybdenum or tungsten-containing groups. However, as of 2011, only one enzyme has been added: molybdopterin molybdotransferase.[61] This enzyme is a component of MoCo biosynthesis in *Escherichia coli*.[62] The reaction it catalyzes is as follows: adenylyl-molybdopterin + molybdate → molybdenum cofactor + AMP.[63]

15.5 Role in histo-blood group

The A and B transferases are the foundation of the human ABO blood group system. Both A and B transferases are glycosyltransferases, meaning they transfer a sugar molecule onto an H-antigen.[64] This allows H-antigen to synthesize the glycoprotein and glycolipid conjugates that are known as the A/B antigens.[64] The full name of A transferase is alpha 1-3-N-acetylgalactosaminyltransferase[65] and its function in the cell is to add N-acetylgalactosamine to H-antigen, creating A-antigen.[66]:55 The full name of B transferase is alpha 1-3-galactosyltransferase,[65] and its function in the cell is to add a galactose molecule to H-antigen, creating B-antigen.[66]

It is possible for *Homo sapiens* to have any of four different blood types: Type A (express A antigens), Type B (express B antigens), Type AB (express both A and B antigens) and Type O (express neither A nor B antigens).[67] The gene for A and B transferases is located on chromosome 9.[68] The gene contains seven exons and six introns[69] and the gene itself is over 18kb long.[70] The alleles for A and B transferases are extremely similar. The resulting enzymes only differ in 4 amino acid residues.[66] The differing residues are located at positions 176, 235, 266, and 268 in the enzymes.[66]:82–83

15.6 Deficiencies

Transferase deficiencies are at the root of many common illnesses. The most common result of a transferase deficiency is a buildup of a cellular product.

15.6.1 SCOT deficiency

Succinyl-CoA:3-ketoacid CoA transferase deficiency (or SCOT deficiency) leads to a buildup of ketones.[71] Ketones are created upon the breakdown of fats in the body and are an important energy source.[72] Inability to utilize ketones leads to intermittent ketoacidosis, which usually

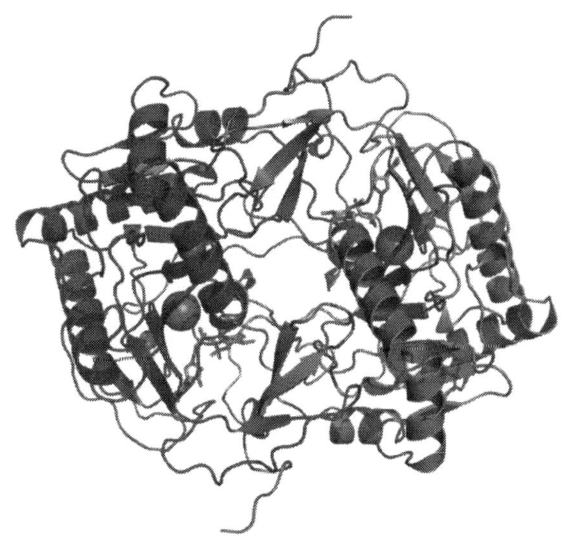

A deficiency of this transferase, E. coli galactose-1-phosphate uridyltransferase is a known cause of galactosemia

first manifests during infancy.[72] Disease sufferers experience nausea, vomiting, inability to feed, and breathing difficulties.[72] In extreme cases, ketoacidosis can lead to coma and death.[72] The deficiency is caused by mutation in the gene OXTC1.[73] Treatments mostly rely on controlling the diet of the patient.[74]

15.6.2 CPT-II deficiency

Carnitine palmitoyltransferase II deficiency (also known as CPT-II deficiency) leads to an excess long chain fatty acids, as the body lacks the ability to transport fatty acids into the mitochondria to be processed as a fuel source.[75] The disease is caused by a defect in the gene CPT2.[76] This deficiency will present in patients in one of three ways: lethal neonatal, severe infantile hepatocardiomuscular, and myopathic form.[76] The myopathic is the least severe form of the deficiency and can manifest at any point in the lifespan of the patient.[76] The other two forms appear in infancy.[76] Common symptoms of the lethal neonatal form and the severe infantile forms are liver failure, heart problems, seizures and death.[76] The myopathic form is characterized by muscle pain and weakness following vigorous exercise.[76] Treatment generally includes dietary modifications and carnitine supplements.[76]

15.6.3 Galactosemia

Galactosemia results from an inability to process galactose, a simple sugar.[77] This deficiency occurs when the gene for galactose-1-phosphate uridylyltransferase (GALT) has any number of mutations, leading to a deficiency in the amount of GALT produced.[78][79] There are two forms of Galactosemia: classic and Duarte.[80] Duarte galactosemia is generally less severe than classic galactosemia and is caused by a deficiency of galactokinase.[81] Galactosemia renders infants unable to process the sugars in breast milk, which leads to vomiting and anorexia within days of birth.[81] Most symptoms of the disease are caused by a buildup of galactose-1-phosphate in the body.[81] Common symptoms include liver failure, sepsis, failure to grow, and mental impairment, among others.[82] Buildup of a second toxic substance, galactitol, occurs in the lenses of the eyes, causing cataracts.[83] Currently, the only available treatment is early diagnosis followed by adherence to a diet devoid of lactose, and prescription of antibiotics for infections that may develop.[84]

15.6.4 Choline acetyltransferase deficiencies

Choline acetyltransferase (also known as ChAT or CAT) is an important enzyme which produces the neurotransmitter acetylcholine.[85] Acetylcholine is involved in many neuropsychic functions such as memory, attention, sleep and arousal.[86][87][88] The enzyme is globular in shape and consists of a single amino acid chain.[89] ChAT functions to transfer an acetyl group from acetyl co-enzyme A to choline in the synapses of nerve cells and exists in two forms: soluble and membrane bound.[89] The ChAT gene is located on chromosome 10.[90]

Alzheimer's disease

Decreased expression of ChAT is one of the hallmarks of Alzheimer's disease.[91] Patients with Alzheimer's disease show a 30 to 90% reduction in activity in several regions of the brain, including the temporal lobe, the parietal lobe and the frontal lobe.[92] However, ChAT deficiency is not believed to be the main cause of this disease.[89]

Amyotrophic lateral sclerosis (ALS or Lou Gehrig's disease)

Patients with ALS show a marked decrease in ChAT activity in motor neurons in the spinal cord and brain.[93] Low levels of ChAT activity are an early indication of the disease and are detectable long before motor neurons begin to die. This can even be detected before the patient is symptomatic.[94]

Huntington's disease

Patients with Huntington's also show a marked decrease in ChAT production.[95] Though the specific cause of the reduced production is not clear, it is believed that the death of medium-sized motor neurons with spiny dendrites leads to the lower levels of ChAT production.[89]

Schizophrenia

Patients with Schizophrenia also exhibit decreased levels of ChAT, localized to the mesopontine tegment of the brain[96] and the nucleus accumbens,[97] which is believed to correlate with the decreased cognitive functioning experienced by these patients.[89]

Sudden infant death syndrome (SIDS)

Recent studies have shown that SIDS infants show decreased levels of ChAT in both the hypothalamus and the striatum.[89] SIDS infants also display fewer neurons capable of producing ChAT in the vagus system.[98] These defects in the medulla could lead to an inability to control essential autonomic functions such as the cardiovascular and respiratory systems.[98]

Congenital myasthenic syndrome (CMS)

CMS is a family of diseases that are characterized by defects in neuromuscular transmission which leads to recurrent bouts of apnea (inability to breathe) that can be fatal.[99] ChAT deficiency is implicated in myasthenia syndromes where the transition problem occurs presynaptically.[100] These syndromes are characterized by the patients' inability to resynthesize acetylcholine.[100]

15.7 Uses in biotechnology

15.7.1 Terminal transferases

Terminal transferases are transferases that can be used to label DNA or to produce plasmid vectors.[101] It accomplishes both of these tasks by adding deoxynucleotides in the form of a template to the downstream end or 3' end of an existing DNA molecule. Terminal transferase is one of the few DNA polymerases that can function without an RNA primer.[101]

15.7.2 Glutathione transferases

The family of glutathione transferases (GST) is extremely diverse, and therefore can be used for a number of biotechnological purposes. Plants use glutathione transferases as a means to segregate toxic metals from the rest of the cell.[102] These glutathione transferases can be used to create biosensors to detect contaminants such as herbicides and insecticides.[103] Glutathione transferases are also used in transgenic plants to increase resistance to both biotic and abiotic stress.[103] Glutathione transferases are currently being explored as targets for anti-cancer medications due to their role in drug resistance.[103] Further, glutathione transferase genes have been investigated due to their ability to prevent oxidative damage and have shown improved resistance in transgenic cultigens.[104]

15.7.3 Rubber transferases

Currently the only available commercial source of natural rubber is the Hevea plant (Hevea brasiliensis). Natural rubber is superior to synthetic rubber in a number of commercial uses.[105] Efforts are being made to produce transgenic plants capable of synthesizing natural rubber, including tobacco and sunflower.[106] These efforts are focused on sequencing the subunits of the rubber transferase enzyme complex in order to transfect these genes into other plants.[106]

15.8 References

[1] "EC 2.7.7 Nucleotidyltransferases". *Enzyme Nomenclature. Recommendations.* Nomenclature Committee of the International Union of Biochemistry and Molecular Biology (NC-IUBMB). Retrieved 14 November 2013.

[2] "Transferase". *Genetics Home Reference.* National Institute of Health. Retrieved 4 November 2013.

[3] Moore SA, Jencks WP (Sep 1982). "Model reactions for CoA transferase involving thiol transfer. Anhydride formation from thiol esters and carboxylic acids". *The Journal of Biological Chemistry*. **257** (18): 10882–92. PMID 6955307.

[4] Wishart D. "Tryptophan Metabolism". *Small Molecule Pathway Database.* Department of Computing Science and Biological Sciences, University of Alberta. Retrieved 4 November 2013.

[5] Herbst EA, MacPherson RE, LeBlanc PJ, Roy BD, Jeoung NH, Harris RA, Peters SJ (Jan 2014). "Pyruvate dehydrogenase kinase-4 contributes to the recirculation of gluconeogenic precursors during postexercise glycogen recovery". *American Journal of Physiology. Regulatory, Integrative and*

Comparative Physiology. **306** (2): R102–7. PMC 3921314. PMID 24305065. doi:10.1152/ajpregu.00150.2013.

[6] Watson, James D. *Molecular Biology of the Gene*. Upper Saddle River, NJ: Pearson, 2013. Print.

[7] Boyce S, Tipton KF (2005). "Enzyme Classification and Nomenclature". *ELS*. ISBN 0470016175. doi:10.1038/npg.els.0003893.

[8] Morton RK (Jul 1953). "Transferase activity of hydrolytic enzymes". *Nature*. **172** (4367): 65–8. PMID 13072573. doi:10.1038/172065a0.

[9] Cohen PP (Sep 1939). "Transamination in pigeon breast muscle". *The Biochemical Journal*. **33** (9): 1478–87. PMC 1264599. PMID 16747057. doi:10.1042/bj0331478.

[10] Snell EE, Jenkins WT (December 1959). "The mechanism of the transamination reaction". *Journal of Cellular and Comparative Physiology*. **54** (S1): 161–177. doi:10.1002/jcp.1030540413.

[11] Braunstein AE, Kritzmann MG (1937). "Formation and Breakdown of Amino-acids by Inter-molecular Transfer of the Amino Group". *Nature*. **140** (3542): 503–504. doi:10.1038/140503b0.

[12] Schoenheimer R (1949). *The Dynamic State of Body Constituents*. Hafner Publishing Co Ltd. ISBN 978-0-02-851800-8.

[13] Guggenheim KY (Nov 1991). "Rudolf Schoenheimer and the concept of the dynamic state of body constituents". *The Journal of Nutrition*. **121** (11): 1701–4. PMID 1941176.

[14] Hird FJ, Rowsell EV (Sep 1950). "Additional transaminations by insoluble particle preparations of rat liver". *Nature*. **166** (4221): 517–8. PMID 14780123. doi:10.1038/166517a0.

[15] Munch-Petersen A, Kalckar HM, Cutolo E, Smith EE (Dec 1953). "Uridyl transferases and the formation of uridine triphosphate; enzymic production of uridine triphosphate: uridine diphosphoglucose pyrophosphorolysis". *Nature*. **172** (4388): 1036–7. PMID 13111246. doi:10.1038/1721036a0.

[16] "Physiology or Medicine 1970 - Press Release". *Nobelprize.org*. Nobel Media AB. Retrieved 5 November 2013.

[17] Lambalot RH, Gehring AM, Flugel RS, Zuber P, LaCelle M, Marahiel MA, Reid R, Khosla C, Walsh CT (Nov 1996). "A new enzyme superfamily - the phosphopantetheinyl transferases". *Chemistry & Biology*. **3** (11): 923–36. PMID 8939709. doi:10.1016/S1074-5521(96)90181-7.

[18] Wongtrakul J, Pongjaroenkit S, Leelapat P, Nachaiwieng W, Prapanthadara LA, Ketterman AJ (Mar 2010). "Expression and characterization of three new glutathione transferases, an epsilon (AcGSTE2-2), omega (AcGSTO1-1), and theta (AcGSTT1-1) from Anopheles cracens (Diptera: Culicidae), a major Thai malaria vector". *Journal of Medical Entomology*. **47** (2): 162–71. PMID 20380296. doi:10.1603/me09132.

[19] Sen J, Goltz JS, Stevens L, Stein D (Nov 1998). "Spatially restricted expression of pipe in the Drosophila egg chamber defines embryonic dorsal-ventral polarity". *Cell*. **95** (4): 471–81. PMID 9827800. doi:10.1016/s0092-8674(00)81615-3.

[20] Moussian B, Roth S (Nov 2005). "Dorsoventral axis formation in the Drosophila embryo shaping and transducing a morphogen gradient". *Current Biology*. **15** (21): R887–99. PMID 16271864. doi:10.1016/j.cub.2005.10.026.

[21] Zhu X, Sen J, Stevens L, Goltz JS, Stein D (Sep 2005). "Drosophila pipe protein activity in the ovary and the embryonic salivary gland does not require heparan sulfate glycosaminoglycans". *Development*. **132** (17): 3813–22. PMID 16049108. doi:10.1242/dev.01962.

[22] Zhang Z, Stevens LM, Stein D (Jul 2009). "Sulfation of eggshell components by Pipe defines dorsal-ventral polarity in the Drosophila embryo". *Current Biology*. **19** (14): 1200–5. PMC 2733793. PMID 19540119. doi:10.1016/j.cub.2009.05.050.

[23] Xu D, Song D, Pedersen LC, Liu J (Mar 2007). "Mutational study of heparan sulfate 2-O-sulfotransferase and chondroitin sulfate 2-O-sulfotransferase". *The Journal of Biological Chemistry*. **282** (11): 8356–67. PMID 17227754. doi:10.1074/jbc.M608062200.

[24] "EC 2 Introduction". *School of Biological & Chemical Sciences at Queen Mary, University of London*. Nomenclature Committee of the International Union of Biochemistry and Molecular Biology (NC-IUBMB). Retrieved 5 November 2013.

[25] Shaw WV, Tsai L, Stadtman ER (Feb 1966). "The enzymatic synthesis of N-methylglutamic acid". *The Journal of Biological Chemistry*. **241** (4): 935–45. PMID 5905132.

[26] Lower S. "Naming Chemical Substances". *Chem1 General Chemistry Virtual Textbook*. Retrieved 13 November 2013.

[27] Hausmann R. *To grasp the essence of life: a history of molecular biology*. Dordrecht: Springer. pp. 198–199. ISBN 978-90-481-6205-5.

[28] "EC 2.7.7.6". *IUBMB Enzyme Nomenclature*. Nomenclature Committee of the International Union of Biochemistry and Molecular Biology (NC-IUBMB). Retrieved 12 November 2013.

[29] "EC2 Transferase Nomenclature". *School of Biological & Chemical Sciences at Queen Mary, University of London*. Nomenclature Committee of the International Union of Biochemistry and Molecular Biology (NC-IUBMB). Retrieved 4 November 2013.

[30] "Transferase". *Encyclopædia Britannica*. Encyclopædia Britannica, Inc. Retrieved 28 July 2016.

[31] "EC 2.1.3: Carboxy- and Carbamoyltransferases". *School of Biological & Chemical Sciences at Queen Mary, University of London.* Nomenclature Committee of the International Union of Biochemistry and Molecular Biology (NC-IUBMB). Retrieved 25 November 2013.

[32] "carbamoyltransferase". *The Free Dictionary.* Farlex, Inc. Retrieved 25 November 2013.

[33] "carbamoyl group (CHEBI:23004)". *ChEBI: The database and ontology of Chemical Entities of Biological Interest.* European Molecular Biology Laboratory. Retrieved 25 November 2013.

[34] Reichard P, Hanshoff G (1956). "Aspartate Carbamyl Transferase from Escherichia Coli" (PDF). *Acta Chemica Scandinavica*: 548–566.

[35] "ENZYME class 2.2.1". *ExPASy: Bioinformatics Resource Portal.* Swiss Institute of Bioinformatics. Retrieved 25 November 2013.

[36] "Pentose Phosphate Pathway". *Molecular Biochemistry II Notes.* The Biochemistry and Biophysics Program at Renssalaer Polytechnic Institute. Retrieved 25 November 2013.

[37] "EC 2.2.1.2 Transaldolase". *Enzyme Structures Database.* European Molecular Biology Laboratory. Retrieved 25 November 2013.

[38] Voorhees RM, Weixlbaumer A, Loakes D, Kelley AC, Ramakrishnan V (May 2009). "Insights into substrate stabilization from snapshots of the peptidyl transferase center of the intact 70S ribosome". *Nature Structural & Molecular Biology.* **16** (5): 528–33. PMC 2679717 ⓐ. PMID 19363482. doi:10.1038/nsmb.1577.

[39] "ENZYME entry: EC 2.3.2.12". *ExPASy: Bioinformatics Resource Portal.* Swiss Institute of Bioinformatics. Retrieved 26 November 2013.

[40] "Keyword Glycosyltransferase". *UniProt.* UniProt Consortium. Retrieved 26 November 2013.

[41] Fitzgerald DK, Brodbeck U, Kiyosawa I, Mawal R, Colvin B, Ebner KE (Apr 1970). "Alpha-lactalbumin and the lactose synthetase reaction". *The Journal of Biological Chemistry.* **245** (8): 2103–8. PMID 5440844.

[42] "ENZYME entry: EC 2.4.1.22". *ExPASy: Bioinformatics Resource Portal.* Swiss Institute of Bioinformatics. Retrieved 26 November 2013.

[43] "EC 2.5". *IntEnz.* European Molecular Biology Laboratory. Retrieved 26 November 2013.

[44] Qabazard B, Ahmed S, Li L, Arlt VM, Moore PK, Stürzenbaum SR (2013). "C. elegans aging is modulated by hydrogen sulfide and the sulfhydrylase/cysteine synthase cysl-2". *PLOS ONE.* **8** (11): e80135. PMC 3832670 ⓐ. PMID 24260346. doi:10.1371/journal.pone.0080135.

[45] "EC 2.6.2". *IUBMB Enzyme Nomenclatur.* Nomenclature Committee of the International Union of Biochemistry and Molecular Biology (NC-IUBMB). Retrieved 28 November 2013.

[46] Kirsch JF, Eichele G, Ford GC, Vincent MG, Jansonius JN, Gehring H, Christen P (Apr 1984). "Mechanism of action of aspartate aminotransferase proposed on the basis of its spatial structure". *Journal of Molecular Biology.* **174** (3): 497–525. PMID 6143829. doi:10.1016/0022-2836(84)90333-4.

[47] "Enzyme entry:2.6.1.1". *ExPASy: Bioinformatics Resource Portal.* Swiss Institute of Bioinformatics. Retrieved 28 November 2013.

[48] "EC 2.7". *School of Biological & Chemical Sciences at Queen Mary, University of London.* Nomenclature Committee of the International Union of Biochemistry and Molecular Biology (NC-IUBMB). Retrieved 4 December 2013.

[49] Yee A, Wu L, Liu L, Kobayashi R, Xiong Y, Hall FL (Jan 1996). "Biochemical characterization of the human cyclin-dependent protein kinase activating kinase. Identification of p35 as a novel regulatory subunit". *The Journal of Biological Chemistry.* **271** (1): 471–7. PMID 8550604.

[50] Lewis R (2008). *Human genetics : concepts and applications* (8th ed.). Boston: McGraw-Hill/Higher Education. p. 32. ISBN 978-0-07-299539-8.

[51] "ENZYME Entry: EC 2.7.11.22". *ExPASy: Bioinformatics Resource Portal.* Swiss Institute of Bioinformatics. Retrieved 4 December 2013.

[52] "1aqy Summary". *Protein Data Bank in Europe Bringing Structure to Biology.* The European Bioinformatics Institute. Retrieved 11 December 2013.

[53] "EC 2.8 Transferring Sulfur-Containing Groups". *School of Biological & Chemical Sciences at Queen Mary, University of London.* Nomenclature Committee of the International Union of Biochemistry and Molecular Biology (NC-IUBMB). Retrieved 11 December 2013.

[54] Negishi M, Pedersen LG, Petrotchenko E, Shevtsov S, Gorokhov A, Kakuta Y, Pedersen LC (Jun 2001). "Structure and function of sulfotransferases". *Archives of Biochemistry and Biophysics.* **390** (2): 149–57. PMID 11396917. doi:10.1006/abbi.2001.2368.

[55] "EC 2.8 Transferring Sulfur-Containing Groups". *School of Biological & Chemical Sciences at Queen Mary, University of London.* Nomenclature Committee of the International Union of Biochemistry and Molecular Biology (NC-IUBMB). Retrieved 11 December 2013.

[56] "Enzyme 2.8.2.2". *Kegg: DBGET.* Kyoto University Bioinformatics Center. Retrieved 11 December 2013.

[57] Ou Z, Shi X, Gilroy RK, Kirisci L, Romkes M, Lynch C, Wang H, Xu M, Jiang M, Ren S, Gramignoli R, Strom SC,

Huang M, Xie W (Jan 2013). "Regulation of the human hydroxysteroid sulfotransferase (SULT2A1) by RORα and RORγ and its potential relevance to human liver diseases". *Molecular Endocrinology*. **27** (1): 106–15. PMC 3545217. PMID 23211525. doi:10.1210/me.2012-1145.

[58] Sekura RD, Marcus CJ, Lyon ES, Jakoby WB (May 1979). "Assay of sulfotransferases". *Analytical Biochemistry*. **95** (1): 82–6. PMID 495970. doi:10.1016/0003-2697(79)90188-x.

[59] "EC 2.9.1". *School of Biological & Chemical Sciences at Queen Mary, University of London*. Nomenclature Committee of the International Union of Biochemistry and Molecular Biology (NC-IUBMB). Retrieved 11 December 2013.

[60] Forchhammer K, Böck A (Apr 1991). "Selenocysteine synthase from Escherichia coli. Analysis of the reaction sequence". *The Journal of Biological Chemistry*. **266** (10): 6324–8. PMID 2007585.

[61] "EC 2.10.1". *School of Biological & Chemical Sciences at Queen Mary, University of London*. Nomenclature Committee of the International Union of Biochemistry and Molecular Biology (NC-IUBMB). Retrieved 11 December 2013.

[62] Nichols JD, Xiang S, Schindelin H, Rajagopalan KV (Jan 2007). "Mutational analysis of Escherichia coli MoeA: two functional activities map to the active site cleft". *Biochemistry*. **46** (1): 78–86. PMC 1868504. PMID 17198377. doi:10.1021/bi061551q.

[63] Wünschiers R, Jahn M, Jahn D, Schomburg I, Peifer S, Heinzle E, Burtscher H, Garbe J, Steen A, Schobert M, Oesterhelt D, Wachtveitl J, Chang A (2010). "Chapter 3: Metabolism". In Michal G, Schomburg D. *Biochemical Pathways: an Atlas of Biochemistry and Molecular Biology* (2nd ed.). Oxford: Wiley-Blackwell. p. 140. ISBN 9780470146842. doi:10.1002/9781118657072.ch3.

[64] Nishida C, Tomita T, Nishiyama M, Suzuki R, Hara M, Itoh Y, Ogawa H, Okumura K, Nishiyama C (2011). "B-transferase with a Pro234Ser substitution acquires AB-transferase activity". *Bioscience, Biotechnology, and Biochemistry*. **75** (8): 1570–5. PMID 21821934. doi:10.1271/bbb.110276.

[65] "ABO ABO blood group (transferase A, alpha 1-3-N-acetylgalactosaminyltransferase; transferase B, alpha 1-3-galactosyltransferase) [Homo sapiens (human)]". NCBI. Retrieved 2 December 2013.

[66] Datta SP, Smith GH, Campbell PN (2000). *Oxford Dictionary of Biochemistry and Molecular Biology* (Rev. ed.). Oxford: Oxford Univ. Press. ISBN 978-0-19-850673-7.

[67] O'Neil D. "ABO Blood Groups". *Human Blood: An Introduction to Its Components and Types*. Behavioral Sciences Department, Palomar College. Retrieved 2 December 2013.

[68] "ABO Blood Group (Transferase A, Alpha 1-3-N-Acetylgalactosaminyltransferase;Transferase B, Alpha 1-3-Galactosyltransferase)". *GeneCards: The Human Gene Compendium*. Weizmann Institute of Science. Retrieved 2 December 2013.

[69] Moran, Lawrence. "Human ABO Gene". Retrieved 2 December 2013.

[70] Kidd, Kenneth. "ABO blood group (transferase A, alpha 1-3-N-acetylgalactosaminyltransferase; transferase B, alpha 1-3-galactosyltransferase)". Retrieved 2 December 2013.

[71] "Succinyl-CoA:3-ketoacid CoA transferase deficiency". *Genetics Home Reference*. National Institute of Health. Retrieved 4 November 2013.

[72] "SUCCINYL-CoA:3-OXOACID CoA TRANSFERASE DEFICIENCY". OMIM. Retrieved 22 November 2013.

[73] "SCOT deficiency". NIH. Retrieved 22 November 2013.

[74] "Succinyl-CoA 3-Oxoacid Transferase Deficiency" (PDF). Climb National Information Centre. Retrieved 22 November 2013.

[75] "Carnitine plamitoyltransferase I deficiency". *Genetics Home Reference*. National Institute of Health. Retrieved 4 November 2013.

[76] Weiser, Thomas. "Carnitine Palmitoyltransferase II Deficiency". NIH. Retrieved 22 November 2013.

[77] "Galactosemia". *Genetics Home Reference*. National Institute of Health. Retrieved 4 November 2013.

[78] Dobrowolski SF, Banas RA, Suzow JG, Berkley M, Naylor EW (Feb 2003). "Analysis of common mutations in the galactose-1-phosphate uridyl transferase gene: new assays to increase the sensitivity and specificity of newborn screening for galactosemia". *The Journal of Molecular Diagnostics*. **5** (1): 42–7. PMC 1907369. PMID 12552079. doi:10.1016/S1525-1578(10)60450-3.

[79] Murphy M, McHugh B, Tighe O, Mayne P, O'Neill C, Naughten E, Croke DT (Jul 1999). "Genetic basis of transferase-deficient galactosaemia in Ireland and the population history of the Irish Travellers". *European Journal of Human Genetics*. **7** (5): 549–54. PMID 10439960. doi:10.1038/sj.ejhg.5200327.

[80] Mahmood U, Imran M, Naik SI, Cheema HA, Saeed A, Arshad M, Mahmood S (Nov 2012). "Detection of common mutations in the GALT gene through ARMS". *Gene*. **509** (2): 291–4. PMID 22963887. doi:10.1016/j.gene.2012.08.010.

[81] "Galactosemia". NORD. Retrieved 22 November 2013.

[82] Berry GT (2000). "Classic Galactosemia and Clinical Variant Galactosemia". *GeneReviews [Internet]*. PMID 20301691.

[83] Bosch AM (Aug 2006). "Classical galactosaemia revisited". *Journal of Inherited Metabolic Disease*. **29** (4): 516–25. PMID 16838075. doi:10.1007/s10545-006-0382-0.

[84] Karadag N, Zenciroglu A, Eminoglu FT, Dilli D, Karagol BS, Kundak A, Dursun A, Hakan N, Okumus N (2013). "Literature review and outcome of classic galactosemia diagnosed in the neonatal period". *Clinical Laboratory*. **59** (9–10): 1139–46. PMID 24273939. doi:10.7754/clin.lab.2013.121235.

[85] Strauss WL, Kemper RR, Jayakar P, Kong CF, Hersh LB, Hilt DC, Rabin M (Feb 1991). "Human choline acetyltransferase gene maps to region 10q11-q22.2 by in situ hybridization". *Genomics*. **9** (2): 396–8. PMID 1840566. doi:10.1016/0888-7543(91)90273-H.

[86] Braida D, Ponzoni L, Martucci R, Sparatore F, Gotti C, Sala M (May 2014). "Role of neuronal nicotinic acetylcholine receptors (nAChRs) on learning and memory in zebrafish". *Psychopharmacology*. **231** (9): 1975–85. PMID 24311357. doi:10.1007/s00213-013-3340-1.

[87] Stone TW (Sep 1972). "Cholinergic mechanisms in the rat somatosensory cerebral cortex". *The Journal of Physiology*. **225** (2): 485–99. PMC 1331117 . PMID 5074408. doi:10.1113/jphysiol.1972.sp009951.

[88] Guzman MS, De Jaeger X, Drangova M, Prado MA, Gros R, Prado VF (Mar 2013). "Mice with selective elimination of striatal acetylcholine release are lean, show altered energy homeostasis and changed sleep/wake cycle". *Journal of Neurochemistry*. **124** (5): 658–69. PMID 23240572. doi:10.1111/jnc.12128.

[89] Oda Y (Nov 1999). "Choline acetyltransferase: the structure, distribution and pathologic changes in the central nervous system" (PDF). *Pathology International*. **49** (11): 921–37. PMID 10594838. doi:10.1046/j.1440-1827.1999.00977.x.

[90] "Choline O-Acetyltransferase". *GeneCards: The Human Gene Compendium*. Weizmann Institute of Science. Retrieved 5 December 2013.

[91] Szigeti C, Bencsik N, Simonka AJ, Legradi A, Kasa P, Gulya K (May 2013). "Long-term effects of selective immunolesions of cholinergic neurons of the nucleus basalis magnocellularis on the ascending cholinergic pathways in the rat: a model for Alzheimer's disease". *Brain Research Bulletin*. **94**: 9–16. PMID 23357177. doi:10.1016/j.brainresbull.2013.01.007.

[92] González-Castañeda RE, Sánchez-González VJ, Flores-Soto M, Vázquez-Camacho G, Macías-Islas MA, Ortiz GG (Mar 2013). "Neural restrictive silencer factor and choline acetyltransferase expression in cerebral tissue of Alzheimer's Disease patients: A pilot study". *Genetics and Molecular Biology*. **36** (1): 28–36. PMC 3615522 . PMID 23569405. doi:10.1590/S1415-47572013000100005.

[93] Rowland LP, Shneider NA (May 2001). "Amyotrophic lateral sclerosis". *The New England Journal of Medicine*. **344** (22): 1688–700. PMID 11386269. doi:10.1056/NEJM200105313442207.

[94] Casas C, Herrando-Grabulosa M, Manzano R, Mancuso R, Osta R, Navarro X (Mar 2013). "Early presymptomatic cholinergic dysfunction in a murine model of amyotrophic lateral sclerosis". *Brain and Behavior*. **3** (2): 145–58. PMC 3607155 . PMID 23531559. doi:10.1002/brb3.104.

[95] Smith R, Chung H, Rundquist S, Maat-Schieman ML, Colgan L, Englund E, Liu YJ, Roos RA, Faull RL, Brundin P, Li JY (Nov 2006). "Cholinergic neuronal defect without cell loss in Huntington's disease". *Human Molecular Genetics*. **15** (21): 3119–31. PMID 16987871. doi:10.1093/hmg/ddl252.

[96] Karson CN, Casanova MF, Kleinman JE, Griffin WS (Mar 1993). "Choline acetyltransferase in schizophrenia". *The American Journal of Psychiatry*. **150** (3): 454–9. PMID 8434662. doi:10.1176/ajp.150.3.454.

[97] Mancama D, Mata I, Kerwin RW, Arranz MJ (Oct 2007). "Choline acetyltransferase variants and their influence in schizophrenia and olanzapine response". *American Journal of Medical Genetics Part B*. **144B** (7): 849–53. PMID 17503482. doi:10.1002/ajmg.b.30468.

[98] Mallard C, Tolcos M, Leditschke J, Campbell P, Rees S (Mar 1999). "Reduction in choline acetyltransferase immunoreactivity but not muscarinic-m2 receptor immunoreactivity in the brainstem of SIDS infants". *Journal of Neuropathology and Experimental Neurology*. **58** (3): 255–64. PMID 10197817. doi:10.1097/00005072-199903000-00005.

[99] Engel AG, Shen XM, Selcen D, Sine S (Dec 2012). "New horizons for congenital myasthenic syndromes". *Annals of the New York Academy of Sciences*. **1275**: 54–62. PMC 3546605 . PMID 23278578. doi:10.1111/j.1749-6632.2012.06803.x.

[100] Maselli RA, Chen D, Mo D, Bowe C, Fenton G, Wollmann RL (Feb 2003). "Choline acetyltransferase mutations in myasthenic syndrome due to deficient acetylcholine resynthesis". *Muscle & Nerve*. **27** (2): 180–7. PMID 12548525. doi:10.1002/mus.10300.

[101] Bowen, R. "Terminal Transferase". *Biotechnology and Genetic Engineering*. Colorado State University. Retrieved 10 November 2013.

[102] Kumar B, Singh-Pareek SL, Sopory SK (2008). "Chapter 23: Glutathione Homeostasis and Abiotic Stresses in Plants: Physiological, Biochemical and Molecular Approaches". In Kumar A, Sopory S. *Recent advances in plant biotechnology and its applications : Prof. Dr. Karl-Hermann Neumann commemorative volume*. New Delhi: I.K. International Pub. House. ISBN 9788189866099.

[103] Chronopoulou EG, Labrou NE (2009). "Glutathione transferases: emerging multidisciplinary tools in red and green biotechnology". *Recent Patents on Biotechnology*. **3** (3): 211–23. PMID 19747150. doi:10.2174/187220809789389135.

[104] Sytykiewicz H (2011). "Expression patterns of glutathione transferase gene (GstI) in maize seedlings under juglone-induced oxidative stress". *International Journal of Molecular Sciences*. **12** (11): 7982–95. PMC 3233451 . PMID 22174645. doi:10.3390/ijms12117982.

[105] Shintani D. "What is Rubber?". *Elastomics*. University of Nevada, Reno. Retrieved 23 November 2013.

[106] "Development of Domestic Natural Rubber-Producing Industrial Crops Through Biotechnology". USDA. Retrieved 23 November 2013.

Chapter 16

Glutathione S-transferase

Glutathione *S*-transferases (**GSTs**), previously known as **ligandins**, comprise a family of eukaryotic and prokaryotic phase II metabolic isozymes best known for their ability to catalyze the conjugation of the reduced form of glutathione (GSH) to xenobiotic substrates for the purpose of detoxification. The GST family consists of three superfamilies: the cytosolic, mitochondrial, and microsomal—also known as MAPEG—proteins.[1][2][3] Members of the GST superfamily are extremely diverse in amino acid sequence, and a large fraction of the sequences deposited in public databases are of unknown function.[4] The Enzyme Function Initiative (EFI) is using GSTs as a model superfamily to identify new GST functions.

GSTs can constitute up to 10% of cytosolic protein in some mammalian organs.[5][6] GSTs catalyse the conjugation of GSH — via a sulfhydryl group — to electrophilic centers on a wide variety of substrates in order to make the compounds more water-soluble.[7][8] This activity detoxifies endogenous compounds such as peroxidised lipids and enables the breakdown of xenobiotics. GSTs may also bind toxins and function as transport proteins, which gave rise to the early term for GSTs, *ligandin*.[9][10]

16.1 Classification

Protein sequence and structure are important additional classification criteria for the three superfamilies (cytosolic, mitochondrial, and MAPEG) of GSTs: while classes from the cytosolic superfamily of GSTs possess more than 40% sequence homology, those from other classes may have less than 25%. Cytosolic GSTs are divided into 13 classes based upon their structure: alpha, beta, delta, epsilon, zeta, theta, mu, nu, pi, sigma, tau, phi, and omega. Mitochondrial GSTs are in class kappa. The MAPEG superfamily of microsomal GSTs consists of subgroups designated I-IV, between which amino acid sequences share less than 20% identity. Human cytosolic GSTs belong to the alpha, zeta, theta, mu, pi, sigma, and omega classes, while six isozymes belonging to classes I, II, and IV of the MAPEG superfamily are known to exist.[8][11][12]

16.1.1 Nomenclature

Standardized GST nomenclature first proposed in 1992 identifies the species to which the isozyme of interest belongs with a lower-case initial (e.g., "h" for human), which precedes the abbreviation GST. The isozyme class is subsequently identified with an upper-case letter (e.g., "A" for alpha), followed by an Arabic numeral representing the class subfamily (or subunit). Because both mitochondrial and cytosolic GSTs exist as dimers, and only heterodimers form between members of the same class, the second subfamily component of the enzyme dimer is denoted with a hyphen, followed by an additional Arabic numeral.[11][12] Therefore, if a human glutathione *S*-transferase is a homodimer in the pi-class subfamily 1, its name will be written as "hGSTP1-1."

The early nomenclature for GSTs referred to them as "Y" proteins, referring to their separation in the "Y" fraction (as opposed to the "X and Z" fractions) using Sephadex G75 chromatography.[13] As GST sub-units were identified they were referred to as Ya, Yp, etc. with if necessary, a number identifying the monomer isoform (e.g. Yb1). Litwack *et al* proposed the term "Ligandin" to cover the proteins previously known as "Y" proteins.[10]

In clinical chemistry and toxicology, the terms alpha GST, mu GST, and pi GST are most commonly used.

16.2 Structure

The glutathione binding site, or "G-site," is located in the thioredoxin-like domain of both cytosolic and mitochondrial GSTs. The region containing the greatest amount of variability between the assorted classes is that of helix $\alpha 2$, where one of three different amino acid residues interacts with the glycine residue of glutathione. Two subgroups of cytosolic GSTs have been characterized based upon their interaction with glutathione: the Y-GST group, which uses a tyrosine residue to activate glutathione, and the S/C-GST, which instead uses serine or cysteine residues.[8][15]

"GST proteins are globular proteins with an N-terminal mixed helical and beta-strand domain and an all-helical C-terminal domain . "

The porcine pi-class enzyme pGTSP1-1 was the first GST to have its structure determined, and it is representative of other members of the cytosolic GST superfamily, which contain a thioredoxin-like N-terminal domain as well as a C-terminal domain consisting of alpha helices.[8][16]

Mammalian cytosolic GSTs are dimeric, with both subunits being from the same class of GSTs, although not necessarily identical. The monomers are approximately 25 kDa in size.[11][17] They are active over a wide variety of substrates with considerable overlap.[18] The following table lists all GST enzymes of each class known to exist in *Homo sapiens*, as found in the UniProtKB/Swiss-Prot database.

16.3 Function

The activity of GSTs is dependent upon a steady supply of GSH from the synthetic enzymes gamma-glutamylcysteine synthetase and glutathione synthetase, as well as the action of specific transporters to remove conjugates of GSH from the cell. The primary role of GSTs is to detoxify xenobiotics by catalyzing the nucleophilic attack by GSH on electrophilic carbon, sulfur, or nitrogen atoms of said nonpolar xenobiotic substrates, thereby preventing their interaction with crucial cellular proteins and nucleic acids.[12][19] Specifically, the function of GSTs in this role is twofold: to bind both the substrate at the enzyme's hydrophobic H-site and GSH at the adjacent, hydrophilic G-site, which together form the active site of the enzyme; and subsequently to activate the thiol group of GSH, enabling the nucleophilic attack upon the substrate.[11] The glutathione molecule binds in a cleft between N and C-terminal domains - the catalytically important residues are proposed to reside in the N-terminal domain.[20] Both subunits of the GST dimer, whether hetero- or homodimeric in nature, contain a single nonsubstrate binding site, as well as a GSH-binding site. In heterodimeric GST complexes such as those formed by the cytosolic mu and alpha classes, however, the cleft between the two subunits is home to an additional high-affinity nonsubstrate xenobiotic binding site, which may account for the enzymes' ability to form heterodimers.[19][21]

The compounds targeted in this manner by GSTs encompass a diverse range of environmental or otherwise exogenous toxins, including chemotherapeutic agents and other drugs, pesticides, herbicides, carcinogens, and variably-derived epoxides; indeed, GSTs are responsible for the conjugation of $\beta_1-8,9$-epoxide, a reactive intermediate formed from aflatoxin B_1, which is a crucial means of protection against the toxin in rodents. The detoxification reactions comprise the first four steps of mercapturic acid synthesis,[19] with the conjugation to GSH serving to make the substrates more soluble and allowing them to be removed from the cell by transporters such as multidrug resistance-associated protein 1 (MRP1).[8] After export, the conjugation products are converted into mercapturic acids and excreted via the urine or bile.[12]

Most mammalian isoenzymes have affinity for the substrate 1-chloro-2,4-dinitrobenzene, and spectrophotometric assays utilising this substrate are commonly used to report GST activity.[22] However, some endogenous compounds, e.g., bilirubin, can inhibit the activity of GSTs. In mammals, GST isoforms have cell specific distributions (e.g., alpha GST in hepatocytes and pi GST in the biliary tract of the human liver).[23]

16.4 Role in cell signaling

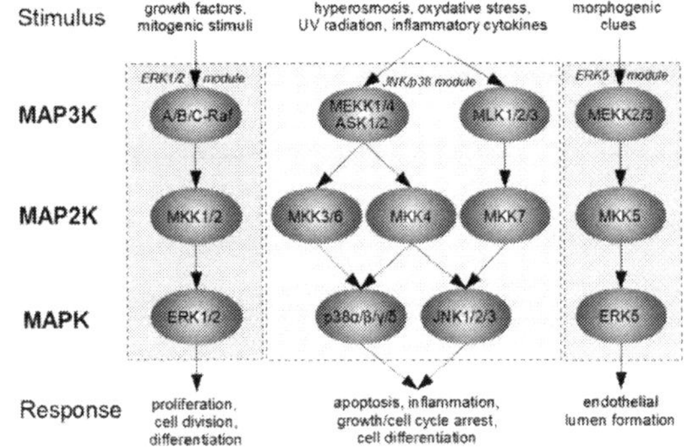

A simplified overview of MAPK pathways in mammals, organised into three main signaling modules (ERK1/2, JNK/p38 and ERK5).

Although best known for their ability to conjugate xenobiotics to GSH and thereby detoxify cellular environments, GSTs are also capable of binding nonsubstrate ligands, with important cell signaling implications. Several GST isozymes from various classes have been shown to inhibit the function of a kinase involved in the MAPK pathway that regulates cell proliferation and death, preventing the kinase from carrying out its role in facilitating the signaling cascade.[24]

Cytosolic GSTP1-1, a well-characterized isozyme of the mammalian GST family, is expressed primarily in heart, lung, and brain tissues; in fact, it is the most common GST expressed outside the liver.[24][25] Based on its overexpression in a majority of human tumor cell lines and prevalence in chemotherapeutic-resistant tumors, GSTP1-1 is thought to play a role in the development of cancer and its potential resistance to drug treatment. Further evidence for this comes from the knowledge that GSTP can selectively inhibit C-jun phosphorylation by JNK, preventing apoptosis.[24] During times of low cellular stress, a complex forms through direct protein–protein interactions between GSTP and the C-terminus of JNK, effectively preventing the action of JNK and thus its induction of the JNK pathway. Cellular oxidative stress causes the dissociation of the complex, oligomerization of GSTP, and induction of the JNK pathway, resulting in apoptosis.[26] The connection between GSTP inhibition of the pro-apoptotic JNK pathway and the isozyme's overexpression in drug-resistant tumor cells may itself account for the tumor cells' ability to escape apoptosis mediated by drugs that are not substrates of GSTP.[24]

Like GSTP, GSTM1 is involved in regulating apoptotic pathways through direct protein–protein interactions, although it acts on ASK1, which is upstream of JNK. The mechanism and result are similar to that of GSTP and JNK, in that GSTM1 sequesters ASK1 through complex formation and prevents its induction of the pro-apoptotic p38 and JNK portions of the MAPK signaling cascade. Like GSTP, GSTM1 interacts with its partner in the absence of oxidative stress, although ASK1 is also involved in heat shock response, which is likewise prevented during ASK1 sequestration. The fact that high levels of GST are associated with resistance to apoptosis induced by a range of substances, including chemotherapeutic agents, supports its putative role in MAPK signaling prevention.[26]

16.4.1 Implications in cancer development

There is a growing body of evidence supporting the role of GST, particularly GSTP, in cancer development and chemotherapeutic resistance. The link between GSTP and cancer is most obvious in the overexpression of GSTP in many cancers, but it is also supported by the fact that the transformed phenotype of tumor cells is associated with aberrantly regulated kinase signaling pathways and cellular addiction to overexpressed proteins. That most anti-cancer drugs are poor substrates for GSTP indicates that the role of elevated GSTP in many tumor cell lines is not to detoxify the compounds, but must have another purpose; this hypothesis is also given credence by the common finding of GSTP overexpression in tumor cell lines that are not drug resistant.[27]

16.5 Clinical significance

In addition to their roles in cancer development and chemotherapeutic drug resistance, GSTs are implicated in a variety of diseases by virtue of their involvement with GSH. Although the evidence is minimal for the influence of GST polymorphisms of the alpha, mu, pi, and theta classes on susceptibility to various types of cancer, numerous studies have implicated such genotypic variations in asthma, atherosclerosis, allergies, and other inflammatory diseases.[19]

Because diabetes is a disease that involves oxidative damage, and GSH metabolism is dysfunctional in diabetic patients, GSTs may represent a potential target for diabetic drug treatment. In addition, insulin administration is known to result in increased GST gene expression through the PI3K/AKT/mTOR pathway and reduced intracellular oxidative stress, while glucagon decreases such gene expression.[28]

Omega-class GST (GSTO) genes, in particular, are associated with neurological diseases such as Alzheimer's, Parkinson's, and amyotrophic lateral sclerosis; again, oxidative stress is believed to be the culprit, with decreased GSTO gene expression resulting in a lowered age of onset for the diseases.[29]

16.5.1 Release of GSTs as an indication of organ damage

The high intracellular concentrations of GSTs coupled with their cell-specific cellular distribution allows them to function as biomarkers for localising and monitoring injury to defined cell types. For example, hepatocytes contain high levels of alpha GST and serum alpha GST has been found to be an indicator of hepatocyte injury in transplantation, toxicity and viral infections.[30][31][32]

Similarly, in humans, renal proximal tubular cells contain high concentrations of alpha GST, while distal tubular cells contain pi GST.[33] This specific distribution enables the measurement of urinary GSTs to be used to quantify and localise renal tubular injury in transplantation, nephrotoxicity and ischaemic injury.[34]

In rodent pre-clinical studies, urinary and serum alpha GST have been shown to be sensitive and specific indicators of renal proximal tubular and hepatocyte necrosis respectively.[35][36]

16.6 GST-tags and the GST pull-down assay

GST can be added to a protein of interest to purify it from solution in a process known as a pull-down assay. This is accomplished by inserting the GST DNA coding sequence next to that which codes for the protein of interest. Thus, after transcription and translation, the GST protein and the protein of interest will be expressed together as a fusion protein. Because the GST protein has a strong binding affinity for GSH, beads coated with the compound can be added to the protein mixture; as a result, the protein of interest attached to the GST will stick to the beads, isolating the protein from the rest of those in solution. The beads are recovered and washed with free GST to detach the protein of interest from the beads, resulting in a purified protein. This technique can be used to elucidate direct protein–protein interactions. A drawback of this assay is that the protein of interest is attached to GST, altering its native state.[37][38]

A GST-tag is often used to separate and purify proteins that contain the GST-fusion protein. The tag is 220 amino acids (roughly 26 KDa) in size,[39] which, compared to tags such as the Myc-tag or the FLAG-tag, is quite large. It can be fused to either the N-terminus or C-terminus of a protein. However, many commercially available sources of GST-tagged plasmids include a thrombin domain for cleavage of the GST tag during protein purification.[37][40]

16.7 See also

- Maltose-binding protein
- Glutathione S-transferase, C-terminal domain
- Bacterial glutathione transferase
- GSTP1
- Glutathione S-transferase Mu 1
- Affinity chromatography

16.8 References

[1] PDB: 1R5A; Udomsinprasert R, Pongjaroenkit S, Wongsantichon J, Oakley AJ, Prapanthadara LA, Wilce MC, Ketterman AJ (June 2005). "Identification, characterization and structure of a new Delta class glutathione transferase isoenzyme". *Biochem. J.* **388** (Pt 3): 763–71. PMC 1183455. PMID 15717864. doi:10.1042/BJ20042015.

[2] Sheehan D, Meade G, Foley VM, Dowd CA (November 2001). "Structure, function and evolution of glutathione transferases: implications for classification of non-mammalian members of an ancient enzyme superfamily". *Biochem. J.* **360** (Pt 1): 1–16. PMC 1222196. PMID 11695986. doi:10.1042/0264-6021:3600001.

[3] Allocati N, Federici L, Masulli M, Di Ilio C (January 2009). "Glutathione transferases in bacteria". *FEBS J.* **276** (1): 58–75. PMID 19016852. doi:10.1111/j.1742-4658.2008.06743.x.

[4] Atkinson, HJ; Babbitt, PC (Nov 24, 2009). "Glutathione transferases are structural and functional outliers in the thioredoxin fold.". *Biochemistry*. **48** (46): 11108–16. PMC 2778357. PMID 19842715. doi:10.1021/bi901180v.

[5] Boyer TD (March 1989). "The glutathione S-transferases: an update". *Hepatology*. **9** (3): 486–96. PMID 2646197. doi:10.1002/hep.1840090324.

[6] Mukanganyama S, Bezabih M, Robert M, et al. (August 2011). "The evaluation of novel natural products as inhibitors of human glutathione transferase P1-1". *J Enzyme Inhib Med Chem*. **26** (4): 460–7. PMID 21028940. doi:10.3109/14756366.2010.526769.

[7] Douglas KT (1987). "Mechanism of action of glutathione-dependent enzymes". *Adv. Enzymol. Relat. Areas Mol. Biol.* **59**: 103–67. PMID 2880477.

[8] Oakley A (May 2011). "Glutathione transferases: a structural perspective". *Drug Metab. Rev.* **43** (2): 138–51. PMID 21428697. doi:10.3109/03602532.2011.558093.

[9] Leaver MJ, George SG (1998). "A piscine glutathione S-transferase which efficiently conjugates the end-products of lipid peroxidation". *Marine Environmental Research*. **46** (1–5): 71–74. doi:10.1016/S0141-1136(97)00071-8.

[10] Litwack G, Ketterer B, Arias IM (December 1971). "Ligandin: a hepatic protein which binds steroids, bilirubin, carcinogens and a number of exogenous organic anions". *Nature*. **234** (5330): 466–7. PMID 4944188. doi:10.1038/234466a0.

[11] Eaton DL, Bammler TK (June 1999). "Concise review of the glutathione S-transferases and their significance to toxicology". *Toxicol. Sci.* **49** (2): 156–64. PMID 10416260. doi:10.1093/toxsci/49.2.156.

[12] Josephy PD (2010). "Genetic variations in human glutathione transferase enzymes: significance for pharmacology and toxicology". *Hum Genomics Proteomics*. **2010**: 876940. PMC 2958679. PMID 20981235. doi:10.4061/2010/876940.

[13] Levi, A.J. (1969). "Two hepatic cytoplasmic protein fractions, Y and Z, and their possible role in the hepatic uptake of bilirubin, sulfobromophthalein, and other anions.". *J. Clin. Invest.* **48** (November): 2156–2167. PMC 297469. PMID 4980931. doi:10.1172/JCI106182.

[14] PDB: 2GST; Ji X, Johnson WW, Sesay MA, Dickert L, Prasad SM, Ammon HL, Armstrong RN, Gilliland GL (February 1994). "Structure and function of the xenobiotic substrate binding site of a glutathione S-transferase as revealed by X-ray crystallographic analysis of product complexes with the diastereomers of 9-(S-glutathionyl)-10-hydroxy-9,10-dihydrophenanthrene". *Biochemistry*. **33** (5): 1043–52. PMID 8110735. doi:10.1021/bi00171a002.

[15] Atkinson HJ, Babbitt PC (November 2009). "Glutathione transferases are structural and functional outliers in the thioredoxin fold". *Biochemistry*. **48** (46): 11108–16. PMC 2778357. PMID 19842715. doi:10.1021/bi901180v.

[16] Park AK, Moon JH, Jang EH, Park H, Ahn IY, Lee KS, et al. (2013). "The structure of a shellfish specific GST class glutathione S-transferase from antarctic bivalve Laternula elliptica reveals novel active site architecture.". *Proteins*. **81** (3): 531–7. PMID 23152139. doi:10.1002/prot.24208.

[17] Landi S (October 2000). "Mammalian class theta GST and differential susceptibility to carcinogens: a review". *Mutat. Res.* **463** (3): 247–83. PMID 11018744. doi:10.1016/s1383-5742(00)00050-8.

[18] Raza H (November 2011). "Dual localization of glutathione S-transferase in the cytosol and mitochondria: implications in oxidative stress, toxicity and disease". *FEBS J.* **278** (22): 4243–51. PMC 3204177. PMID 21929724. doi:10.1111/j.1742-4658.2011.08358.x.

[19] Hayes JD, Flanagan JU, Jowsey IR (2005). "Glutathione transferases". *Annu. Rev. Pharmacol. Toxicol.* **45**: 51–88. PMID 15822171. doi:10.1146/annurev.pharmtox.45.120403.095857.

[20] Nishida M, Harada S, Noguchi S, Satow Y, Inoue H, Takahashi K (1998). "Three-dimensional structure of Escherichia coli glutathione S-transferase complexed with glutathione sulfonate: catalytic roles of Cys10 and His106.". *J Mol Biol.* **281** (1): 135–47. PMID 9680481. doi:10.1006/jmbi.1998.1927.

[21] Vargo MA, Colman RF (January 2001). "Affinity labeling of rat glutathione S-transferase isozyme 1-1 by 17beta -iodoacetoxy-estradiol-3-sulfate". *J. Biol. Chem.* **276** (3): 2031–6. PMID 11031273. doi:10.1074/jbc.M008212200.

[22] Habig WH, Pabst MJ, Fleischner G, Gatmaitan Z, Arias IM, Jakoby WB (October 1974). "The Identity of Glutathione S-Transferase B with Ligandin, a Major Binding Protein of Liver". *Proc. Natl. Acad. Sci. U.S.A.* **71** (10): 3879–82. PMC 434288. PMID 4139704. doi:10.1073/pnas.71.10.3879.

[23] Beckett GJ, Hayes JD (1987). "Glutathione S-transferase measurements and liver disease in man". *Journal of Clinical Biochemistry and Nutrition*. **2**: 1–24. doi:10.3164/jcbn.2.1.

[24] Laborde E (September 2010). "Glutathione transferases as mediators of signaling pathways involved in cell proliferation and cell death". *Cell Death Differ.* **17** (9): 1373–80. PMID 20596078. doi:10.1038/cdd.2010.80.

[25] Adler V, Yin Z, Fuchs SY, et al. (March 1999). "Regulation of JNK signaling by GSTp". *EMBO J.* **18** (5): 1321–34. PMC 1171222. PMID 10064598. doi:10.1093/emboj/18.5.1321.

[26] Townsend DM, Tew KD (October 2003). "The role of glutathione-S-transferase in anti-cancer drug resistance". *Oncogene*. **22** (47): 7369–75. PMID 14576844. doi:10.1038/sj.onc.1206940.

[27] Tew KD, Manevich Y, Grek C, Xiong Y, Uys J, Townsend DM (July 2011). "The role of glutathione S-transferase P in signaling pathways and S-glutathionylation in cancer". *Free Radic. Biol. Med.* **51** (2): 299–313. PMC 3125017. PMID 21558000. doi:10.1016/j.freeradbiomed.2011.04.013.

[28] Franco R, Schoneveld OJ, Pappa A, Panayiotidis MI (2007). "The central role of glutathione in the pathophysiology of human diseases". *Arch. Physiol. Biochem.* **113** (4–5): 234–58. PMID 18158646. doi:10.1080/13813450701661198.

[29] Board PG (May 2011). "The omega-class glutathione transferases: structure, function, and genetics". *Drug Metab. Rev.* **43** (2): 226–35. PMID 21495794. doi:10.3109/03602532.2011.561353.

[30] Beckett GJ, Chapman BJ, Dyson EH, Hayes JD (January 1985). "Plasma glutathione S-transferase measurements after paracetamol overdose: evidence for early hepatocellular damage". *Gut*. **26** (1): 26–31. PMC 1432412. PMID 3965363. doi:10.1136/gut.26.1.26.

[31] Hughes VF, Trull AK, Gimson A, Friend PJ, Jamieson N, Duncan A, Wight DG, Prevost AT, Alexander GJ (November 1997). "Randomized trial to evaluate the clinical benefits of serum alpha-glutathione S-transferase concentration monitoring after liver transplantation". *Transplantation*. **64** (10): 1446–52. PMID 9392310. doi:10.1097/00007890-199711270-00013.

[32] Loguercio C, Caporaso N, Tuccillo C, Morisco F, Del Vecchio Blanco G, Del Vecchio Blanco C (March 1998). "Alpha-glutathione transferases in HCV-related chronic hepatitis: a new predictive index of response to interferon therapy?". *J. Hepatol.* **28** (3): 390–5. PMID 9551675. doi:10.1016/s0168-8278(98)80311-5.

[33] Harrison DJ, Kharbanda R, Cunningham DS, McLellan LI, Hayes JD (June 1989). "Distribution of glutathione S-transferase isoenzymes in human kidney: basis for possible markers of renal injury". *J. Clin. Pathol.* **42** (6): 624–8. PMC 1141991. PMID 2738168. doi:10.1136/jcp.42.6.624.

[34] Sundberg AG, Appelkvist EL, Bäckman L, Dallner G (1994). "Urinary pi-class glutathione transferase as an indicator of tubular damage in the human kidney". *Nephron*. **67** (3): 308–16. PMID 7936021. doi:10.1159/000187985.

[35] Harpur, E; Ennulat, D; Hoffman, D; Betton, G; Gautier, JC; Riefke, B; Bounous, D; Schuster, K; Beushausen, S; Guffroy, M; Shaw, M; Lock, E; Pettit, S; HESI Committee on Biomarkers of, Nephrotoxicity (August 2011). "Biological qualification of biomarkers of chemical-induced renal toxicity in two strains of male rat.". *Toxicological Sciences*. **122** (2): 235–52. PMID 21593213. doi:10.1093/toxsci/kfr112.

[36] Bailey, WJ; Holder, D; Patel, H; Devlin, P; Gonzalez, RJ; Hamilton, V; Muniappa, N; Hamlin, DM; Thomas, CE; Sistare, FD; Glaab, WE (December 2012). "A performance evaluation of three drug-induced liver injury biomarkers in the rat: alpha-glutathione S-transferase, arginase 1, and 4-hydroxyphenyl-pyruvate dioxygenase.". *Toxicological Sciences*. **130** (2): 229–44. PMID 22872058. doi:10.1093/toxsci/kfs243.

[37] Benard V, Bokoch GM (2002). "Assay of Cdc42, Rac, and Rho GTPase activation by affinity methods". *Meth. Enzymol.* **345**: 349–59. PMID 11665618. doi:10.1016/s0076-6879(02)45028-8.

[38] Ren L, Chang E, Makky K, Haas AL, Kaboord B, Walid Qoronfleh M (November 2003). "Glutathione S-transferase pull-down assays using dehydrated immobilized glutathione resin". *Anal. Biochem.* **322** (2): 164–9. PMID 14596823. doi:10.1016/j.ab.2003.07.023.

[39] Long F, Cho W, Ishii Y (September 2011). "Expression and purification of 15N- and 13C-isotope labeled 40-residue human Alzheimer's β-amyloid peptide for NMR-based structural analysis". *Protein Expr. Purif.* **79** (1): 16–24. PMC 3134129 . PMID 21640828. doi:10.1016/j.pep.2011.05.012.

[40] Tinta T, Christiansen LS, Konrad A, et al. (June 2012). "Deoxyribonucleoside kinases in two aquatic bacteria with high specificity for thymidine and deoxyadenosine". *FEMS Microbiol. Lett.* **331** (2): 120–7. PMID 22462611. doi:10.1111/j.1574-6968.2012.02565.x.

16.9 External links

- Overview of Glutathione-S-Transferases
- UMich Orientation of Proteins in Membranes *families/superfamily-199* - MAPEG (Eicosanoid and Glutathione metabolism proteins) family
- Glutathione S-Transferase at the US National Library of Medicine Medical Subject Headings (MeSH)
- EC 2.5.1.18
- Preparation of GST Fusion Proteins
- GST Gene Fusion System Handbook

Chapter 17

Efflux (microbiology)

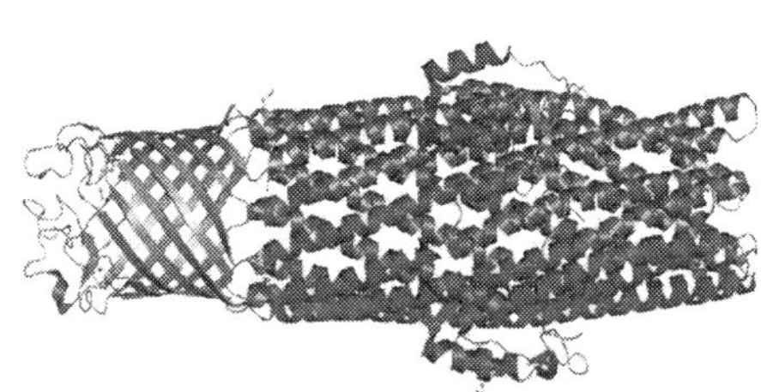

Protein TolC, the outer membrane component of a tripartite efflux pump in Escherichia coli.

Active efflux is a mechanism responsible for moving compounds, like neurotransmitters, toxic substances, and antibiotics, out of the cell; this is considered to be a vital part of xenobiotic metabolism. This mechanism is important in medicine as it can contribute to bacterial antibiotic resistance.

Efflux systems function via an energy-dependent mechanism (active transport) to pump out unwanted toxic substances through specific efflux pumps. Some efflux systems are drug-specific, whereas others may accommodate multiple drugs with small multidrug resistance (SMR)[1] transporters.[2]

17.1 Bacteria

17.1.1 Bacterial efflux pumps

Efflux pumps are proteinaceous transporters localized in the cytoplasmic membrane of all kinds of cells. They are active transporters, meaning that they require a source of chemical energy to perform their function. Some are primary active transporters utilizing adenosine triphosphate hydrolysis as a source of energy, whereas others are secondary active transporters (uniporters, symporters, or antiporters) in which transport is coupled to an electrochemical potential difference created by pumping hydrogen or sodium ions from or to the outside of the cell.
Bacterial efflux transporters are classified into five major superfamilies, based on their amino acid sequence and the energy source used to export their substrates:

1. The major facilitator superfamily (MFS)
2. The ATP-binding cassette superfamily (ABC)
3. The small multidrug resistance family (SMR)
4. The resistance-nodulation-cell division superfamily (RND)
5. The Multi antimicrobial extrusion protein family (MATE).

Of these, only the ABC superfamily are primary transporters, the rest being secondary transporters utilizing proton or sodium gradient as a source of energy. Whereas MFS dominates in Gram positive bacteria, the RND family was once thought to be unique to Gram negative bacteria. They have since been found in all major Kingdoms.

17.1.2 Function

Although antibiotics are the most clinically important substrates of efflux systems, it is probable that most efflux pumps have other natural physiological functions. Examples include:

- The *E. coli* AcrAB efflux system, which has a physiologic role of pumping out bile acids and fatty acids to lower their toxicity.[3]

- The MFS family Ptr pump in *Streptomyces pristinaespiralis* appears to be an autoimmunity pump for this organism when it turns on production of pristinamycins I and II.
- The AcrAB–TolC system in *E. coli* is suspected to have a role in the transport of the calcium-channel components in the *E. coli* membrane.[4]
- The MtrCDE system plays a protective role by providing resistance to faecal lipids in rectal isolates of *Neisseria gonorrhoeae*.[5]
- The AcrAB efflux system of *Erwinia amylovora* is important for this organism's virulence, plant (host) colonization, and resistance to plant toxins.
- The MexXY component of the MexXY-OprM multidrug efflux system of *P. aeruginosa* is inducible by antibiotics that target ribosomes via the PA5471 gene product.[6]

The ability of efflux systems to recognize a large number of compounds other than their natural substrates is probably because substrate recognition is based on physicochemical properties, such as hydrophobicity, aromaticity and ionizable character rather than on defined chemical properties, as in classical enzyme-substrate or ligand-receptor recognition. Because most antibiotics are amphiphilic molecules - possessing both hydrophilic and hydrophobic characters - they are easily recognized by many efflux pumps.

17.1.3 Impact on antimicrobial resistance

The impact of efflux mechanisms on antimicrobial resistance is large; this is usually attributed to the following:

- The genetic elements encoding efflux pumps may be encoded on chromosomes and/or plasmids, thus contributing to both intrinsic (natural) and acquired resistance respectively. As an intrinsic mechanism of resistance, efflux pump genes can survive a hostile environment (for example in the presence of antibiotics) which allows for the selection of mutants that over-express these genes. Being located on transportable genetic elements as plasmids or transposons is also advantageous for the microorganisms as it allows for the easy spread of efflux genes between distant species.
- Antibiotics can act as inducers and regulators of the expression of some efflux pumps.[6]
- Expression of several efflux pumps in a given bacterial species may lead to a broad spectrum of resistance when considering the shared substrates of some multi-drug efflux pumps, where one efflux pump may confer resistance to a wide range of antimicrobials.

17.2 Eukaryotes

In eukaryotic cells, the existence of efflux pumps has been known since the discovery of P-glycoprotein in 1976 by Juliano and Ling.[7] Efflux pumps are one of the major causes of anticancer drug resistance in eukaryotic cells. They include monocarboxylate transporters (MCTs), multiple drug resistance proteins (MDRs)- also referred as P-glycoprotein, multidrug resistance-associated proteins (MRPs), peptide transporters (PEPTs), and Na+ phosphate transporters (NPTs). These transporters are distributed along particular portions of the renal proximal tubule, intestine, liver, blood–brain barrier, and other portions of the brain.

17.3 Efflux inhibitors

Several trials are currently being conducted to develop drugs that can be co-administered with antibiotics to act as inhibitors for the efflux-mediated extrusion of antibiotics. As yet, no efflux inhibitor has been approved for therapeutic use, but some are being used to determine the prevalence of efflux pumps in clinical isolates and in cell biology research. Verapamil, for example, is used to block P-glycoprotein-mediated efflux of DNA-binding fluorophores, thereby facilitating fluorescent cell sorting for DNA content. Various natural products have been shown to inhibit bacterial efflux pumps including the carotenoids capsanthin and capsorubin,[8] the flavonoids rotenone and chrysin,[8] and the alkaloid lysergol.[9] Some nanoparticles, for example zinc oxide, also inhibit bacterial efflux pumps.[10]

17.4 See also

- Antibiotic resistance

17.5 References

[1] Bay, Denice C.; Turner, Raymond J. (2016). *Small Multidrug Resistance Efflux Pumps*. Switzerland: Springer International Publishing. p. 45. ISBN 978-3-319-39658-3.

[2] Sun, Jingjing; Deng, Ziqing; Yan, Aixin (2014). "Bacterial multidrug efflux pumps: Mechanisms, physiology and

pharmacological exploitations". *Biochemical and Biophysical Research Communications*. **453** (2): 254–67. PMID 24878531. doi:10.1016/j.bbrc.2014.05.090.

[3] Okusu, H; Ma, D; Nikaido, H (1996). "AcrAB efflux pump plays a major role in the antibiotic resistance phenotype of Escherichia coli multiple-antibiotic-resistance (Mar) mutants". *Journal of bacteriology*. **178** (1): 306–8. PMC 177656 . PMID 8550435.

[4] Du, Dijun; Wang, Zhao; James, Nathan R.; Voss, Jarrod E.; Klimont, Ewa; Ohene-Agyei, Thelma; Venter, Henrietta; Chiu, Wah; Luisi, Ben F. (2014). "Structure of the AcrAB–TolC multidrug efflux pump". *Nature*. **509** (7501): 512–5. Bibcode:2014Natur.509..512D. PMC 4361902 . PMID 24747401. doi:10.1038/nature13205.

[5] Rouquette, Corinne; Harmon, Jennifer B.; Shafer, William M. (1999). "Induction of the mtrCDE-encoded efflux pump system of Neisseria gonorrhoeae requires MtrA, an AraC-like protein". *Molecular Microbiology*. **33** (3): 651–8. PMID 10417654. doi:10.1046/j.1365-2958.1999.01517.x.

[6] Morita, Y.; Sobel, M. L.; Poole, K. (2006). "Antibiotic Inducibility of the MexXY Multidrug Efflux System of Pseudomonas aeruginosa: Involvement of the Antibiotic-Inducible PA5471 Gene Product". *Journal of Bacteriology*. **188** (5): 1847–55. PMC 1426571 . PMID 16484195. doi:10.1128/JB.188.5.1847-1855.2006.

[7] Juliano, R.L.; Ling, V. (1976). "A surface glycoprotein modulating drug permeability in Chinese hamster ovary cell mutants". *Biochimica et Biophysica Acta (BBA) - Biomembranes*. **455** (1): 152–62. PMID 990323. doi:10.1016/0005-2736(76)90160-7.

[8] Wang, Q.; Michalak, K.; Wesolowska, O.; Deli, J.; Molnar, P.; Hohmann, J.; Molnar, J.; Engi, H. (2010). "Reversal of Multidrug Resistance by Natural Substances from Plants". *Current Topics in Medicinal Chemistry*. **10** (17): 1757–68. PMID 20645919. doi:10.2174/156802610792928103.

[9] Cushnie, T.P. Tim; Cushnie, Benjamart; Lamb, Andrew J. (2014). "Alkaloids: An overview of their antibacterial, antibiotic-enhancing and antivirulence activities". *International Journal of Antimicrobial Agents*. **44** (5): 377–86. PMID 25130096. doi:10.1016/j.ijantimicag.2014.06.001.

[10] Banoee, Maryam; Seif, Sepideh; Nazari, Zeinab E.; Jafari-Fesharaki, Parisa; Shahverdi, Hamid R.; Moballegh, Ali; Moghaddam, Kamyar M.; Shahverdi, Ahmad R. (2010). "ZnO nanoparticles enhanced antibacterial activity of ciprofloxacin against Staphylococcus aureus and Escherichia coli". *Journal of Biomedical Materials Research Part B: Applied Biomaterials*. **93B** (2): 557–61. PMID 20225250. doi:10.1002/jbm.b.31615.

Chapter 18

Antioxidant

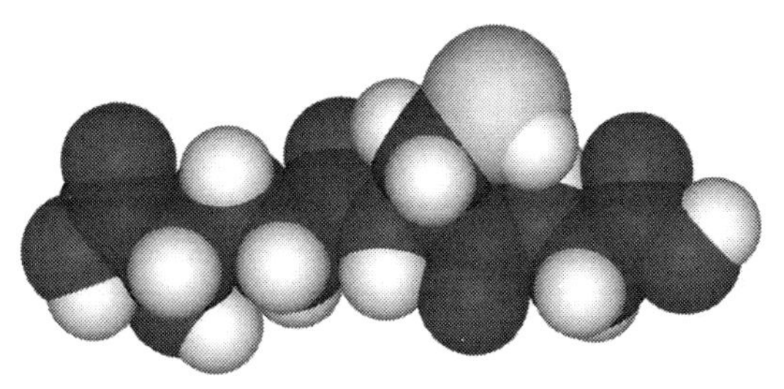

Model of the antioxidant metabolite glutathione. The yellow sphere is the redox-active sulfur atom that provides antioxidant activity, while the red, blue, white, and dark grey spheres represent oxygen, nitrogen, hydrogen, and carbon atoms, respectively.

An **antioxidant** is a molecule that inhibits the oxidation of other molecules. Oxidation is a chemical reaction that can produce free radicals, leading to chain reactions that may damage cells. Antioxidants such as thiols or ascorbic acid (vitamin C) terminate these chain reactions. The term "antioxidant" is mainly used for two different groups of substances: industrial chemicals which are added to products to prevent oxidation, and natural chemicals found in foods and body tissue which are said to have beneficial health effects.

To balance the oxidative state, plants and animals maintain complex systems of overlapping antioxidants, such as glutathione and enzymes (e.g., catalase and superoxide dismutase) produced internally or the dietary antioxidants, vitamin A, vitamin C, and vitamin E.

Antioxidant dietary supplements do not improve health nor are they effective in preventing diseases as shown by randomized clinical trials including supplements of beta-carotene, vitamin A, and vitamin E singly or in different combinations having no effect on mortality rate[1][2] or cancer risk.[3][4] Supplementation with selenium or vitamin E does not reduce the risk of cardiovascular disease.[5][6] Oxidative stress can be considered as either a cause or consequence of some diseases, an area of research stimulating drug development for antioxidant compounds for use as potential therapies.

Industrial antioxidants have diverse uses, such as food and cosmetics preservatives and inhibitors of rubber or gasoline deterioration.[7]

18.1 Health effects

18.1.1 Relation to diet

Although certain levels of antioxidant vitamins in the diet are required for good health, there is considerable debate on whether antioxidant-rich foods or supplements have anti-disease activity. Moreover, if they are actually beneficial, it is unknown which antioxidant(s) are needed from the diet and in what amounts beyond typical dietary intake.[8][9][10] Some authors dispute the hypothesis that antioxidant vitamins could prevent chronic diseases,[8][11] while others maintain such a possibility is unproved and misguided from the beginning.[12]

Polyphenols, which often have antioxidant properties in vitro, are not necessarily antioxidants in vivo due to extensive metabolism.[13] In many polyphenols, the catechol group acts as electron acceptor and is therefore responsible for the antioxidant activity.[14] However, this catechol group undergoes extensive metabolism upon uptake in the human body, for example by catechol-O-methyl transferase, and is therefore no longer able to act as electron acceptor. Many polyphenols may have non-antioxidant roles in minute concentrations that affect cell-to-cell signaling, receptor sensitivity, inflammatory enzyme activity or gene regulation.[15][16][17]

Although dietary antioxidants have been investigated for potential effects on neurodegenerative diseases such as Alzheimer's disease, Parkinson's disease, and amyotrophic lateral sclerosis,[18][19] these studies have been inconclusive.[20][21][22]

18.1.2 Drug candidates

Tirilazad is an antioxidant steroid derivative that inhibits the lipid peroxidation that is believed to play a key role in neuronal death in stroke and head injury. It demonstrated activity in animal models of stroke,[23] but human trials demonstrated no effect on mortality or other outcomes in subarachnoid haemorrhage[24] and worsened results in ischemic stroke.[25]

Similarly, the designed antioxidant NXY-059 exhibited efficacy in animal models, but failed to improve stroke outcomes in a clinical trial.[26] As of November 2014, other antioxidants are being studied as potential neuroprotectants.[27]

Common pharmaceuticals (and supplements) with antioxidant properties may interfere with the efficacy of certain anticancer medication and radiation.[28][29]

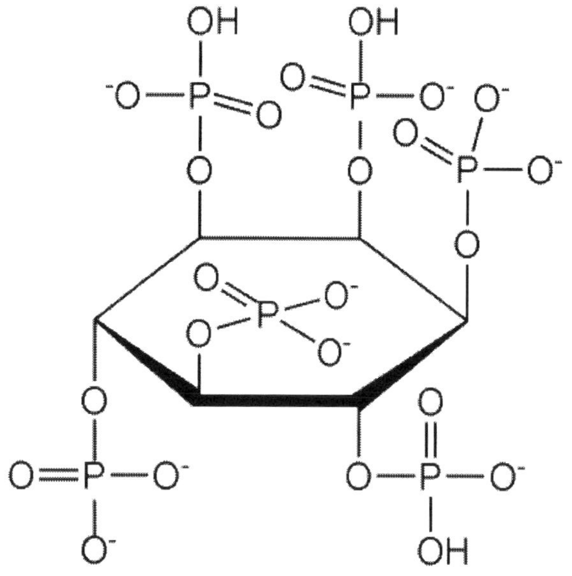

Structure of the metal chelator phytic acid.

18.1.3 Physical exercise

During exercise, oxygen consumption can increase by a factor of more than 10.[30] However, no benefits for physical performance to athletes are seen with vitamin E supplementation[31] and 6 weeks of vitamin E supplementation had no effect on muscle damage in ultramarathon runners.[32] Some research suggests that supplementation with amounts as high as 1000 mg of vitamin C inhibits recovery.[33] Other studies indicated that antioxidant supplementation may attenuate the cardiovascular benefits of exercise.[34]

18.1.4 Adverse effects

See also: Antioxidative stress

Relatively strong reducing acids can have antinutrient effects by binding to dietary minerals such as iron and zinc in the gastrointestinal tract and preventing them from being absorbed.[35] Notable examples are oxalic acid, tannins and phytic acid, which are high in plant-based diets.[36] Calcium and iron deficiencies are not uncommon in diets in developing countries where less meat is eaten and there is high consumption of phytic acid from beans and unleavened whole grain bread.[37]

Nonpolar antioxidants such as eugenol—a major component of oil of cloves—have toxicity limits that can be exceeded with the misuse of undiluted essential oils.[41] Toxicity associated with high doses of water-soluble antioxidants such as ascorbic acid are less of a concern, as these compounds can be excreted rapidly in urine.[42] More seriously, very high doses of some antioxidants may have harmful long-term effects. The beta-carotene and Retinol Efficacy Trial (CARET) study of lung cancer patients found that smokers given supplements containing beta-carotene and vitamin A had increased rates of lung cancer.[43] Subsequent studies confirmed these adverse effects.[44]

These harmful effects may also be seen in non-smokers, as a recent meta-analysis including data from approximately 230,000 patients showed that β-carotene, vitamin A or vitamin E supplementation is associated with increased mortality but saw no significant effect from vitamin C.[45] No health risk was seen when all the randomized controlled studies were examined together, but an increase in mortality was detected when only high-quality and low-bias risk trials were examined separately.[46] As the majority of these low-bias trials dealt with either elderly people, or people with disease, these results may not apply to the general population.[47] This meta-analysis was later repeated and extended by the same authors, with the new analysis published by the Cochrane Collaboration; confirming the previous results.[46] These two publications are consistent with some previous meta-analyzes that also suggested that Vitamin E supplementation increased mortality,[48] and that antioxidant supplements increased the risk of colon cancer.[49] Beta-carotene may also increase lung cancer.[49][50] Overall, the large number of clinical trials carried out on antioxidant supplements suggest that either these products have no effect on health, or that they cause a small increase in mortality in elderly or vulnerable populations.[8][9][45]

While antioxidant supplementation is widely used in attempts to prevent the development of cancer, antioxidants may interfere with cancer treatments,[51] since the environment of cancer cells causes high levels of oxidative stress, making these cells more susceptible to the further oxidative stress induced by treatments. As a result, by reducing the redox stress in cancer cells, antioxidant supplements (and pharmaceuticals) could decrease the effectiveness of radiotherapy and chemotherapy.[28][52][53] On the other hand, other reviews have suggested that antioxidants could reduce side effects or increase survival times.[54][55]

18.2 Oxidative challenge in biology

Further information: Oxidative stress
A paradox in metabolism is that, while the vast majority of complex life on Earth requires oxygen for its existence, oxygen is a highly reactive molecule that damages living organisms by producing reactive oxygen species.[56] Consequently, organisms contain a complex network of antioxidant metabolites and enzymes that work together to prevent oxidative damage to cellular components such as DNA, proteins and lipids.[57][58] In general, antioxidant systems either prevent these reactive species from being formed, or remove them before they can damage vital components of the cell.[56][57] However, reactive oxygen species also have useful cellular functions, such as redox signaling. Thus, the function of antioxidant systems is not to remove oxidants entirely, but instead to keep them at an optimum level.[59]

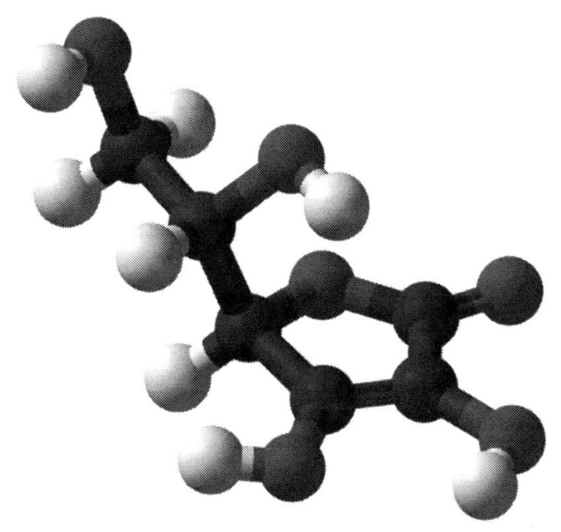

The structure of the antioxidant vitamin ascorbic acid (vitamin C).

The reactive oxygen species produced in cells include hydrogen peroxide (H_2O_2), hypochlorous acid (HClO), and free radicals such as the hydroxyl radical (·OH) and the superoxide anion (O_2^-).[60] The hydroxyl radical is particularly unstable and will react rapidly and non-specifically with most biological molecules. This species is produced from hydrogen peroxide in metal-catalyzed redox reactions such as the Fenton reaction.[61] These oxidants can damage cells by starting chemical chain reactions such as lipid peroxidation, or by oxidizing DNA or proteins.[57] Damage to DNA can cause mutations and possibly cancer, if not reversed by DNA repair mechanisms,[62][63] while damage to proteins causes enzyme inhibition, denaturation and protein degradation.[64]

The use of oxygen as part of the process for generating metabolic energy produces reactive oxygen species.[65] In this process, the superoxide anion is produced as a by-product of several steps in the electron transport chain.[66] Particularly important is the reduction of coenzyme Q in complex III, since a highly reactive free radical is formed as an intermediate ($Q\cdot^-$). This unstable intermediate can lead to electron "leakage", when electrons jump directly to oxygen and form the superoxide anion, instead of moving through the normal series of well-controlled reactions of the electron transport chain.[67] Peroxide is also produced from the oxidation of reduced flavoproteins, such as complex I.[68] However, although these enzymes can produce oxidants, the relative importance of the electron transfer chain to other processes that generate peroxide is unclear.[69][70] In plants, algae, and cyanobacteria, reactive oxygen species are also produced during photosynthesis,[71] particularly under conditions of high light intensity.[72] This effect is partly offset by the involvement of carotenoids in photoinhibition, and in algae and cyanobacteria, by large amount of iodide and selenium,[73] which involves these antioxidants reacting with over-reduced forms of the photosynthetic reaction centres to prevent the production of reactive oxygen species.[74][75]

18.3 Metabolites

Antioxidants are classified into two broad divisions, depending on whether they are soluble in water (hydrophilic) or in lipids (lipophilic). In general, water-soluble antioxidants react with oxidants in the cell cytosol and the blood plasma, while lipid-soluble antioxidants protect cell membranes from lipid peroxidation.[57] These compounds may be synthesized in the body or obtained from the diet.[58] The different antioxidants are present at a wide range of concentrations in body fluids and tissues, with some such as glutathione or ubiquinone mostly present within cells, while others such as uric acid are more evenly distributed (see table below). Some antioxidants are only found in a few organisms and these compounds can be important in pathogens and can be virulence factors.[76]

The relative importance and interactions between these different antioxidants is a very complex question, with the various metabolites and enzyme systems having synergistic and interdependent effects on one another.[77][78] The action of one antioxidant may therefore depend on the proper function of other members of the antioxidant system.[58] The amount of protection provided by any one antioxidant will also depend on its concentration, its reactivity towards the particular reactive oxygen species being considered, and the status of the antioxidants with which it interacts.[58]

Some compounds contribute to antioxidant defense by chelating transition metals and preventing them from catalyzing the production of free radicals in the cell. Particularly important is the ability to sequester iron, which is the function of iron-binding proteins such as transferrin and ferritin.[70] Selenium and zinc are commonly referred to as *antioxidant nutrients*, but these chemical elements have no antioxidant action themselves and are instead required for the activity of some antioxidant enzymes, as is discussed below.

18.3.1 Uric acid

Uric acid is by far the highest concentration antioxidant in human blood. Uric acid (UA) is an antioxidant oxypurine produced from xanthine by the enzyme xanthine oxidase, and is an intermediate product of purine metabolism.[91] In almost all land animals, urate oxidase further catalyzes the oxidation of uric acid to allantoin,[92] but in humans and most higher primates, the urate oxidase gene is nonfunctional, so that UA is not further broken down.[92][93] The evolutionary reasons for this loss of urate conversion to allantoin remain the topic of active speculation.[94][95] The antioxidant effects of uric acid have led researchers to suggest this mutation was beneficial to early primates and humans.[95][96] Studies of high altitude acclimatization support the hypothesis that urate acts as an antioxidant by mitigating the oxidative stress caused by high-altitude hypoxia.[97]

Uric acid has the highest concentration of any blood antioxidant[85] and provides over half of the total antioxidant capacity of human serum.[98] Uric acid's antioxidant activities are also complex, given that it does not react with some oxidants, such as superoxide, but does act against peroxynitrite,[99] peroxides, and hypochlorous acid.[91] Concerns over elevated UA's contribution to gout must be considered as one of many risk factors.[100] By itself, UA-related risk of gout at high levels (415–530 μmol/L) is only 0.5% per year with an increase to 4.5% per year at UA supersaturation levels (535+ μmol/L).[101] Many of these aforementioned studies determined UA's antioxidant actions within normal physiological levels,[97][99] and some found antioxidant activity at levels as high as 285 μmol/L.[102]

18.3.2 Vitamin C

Ascorbic acid or "vitamin C" is a monosaccharide oxidation-reduction (redox) catalyst found in both animals and plants. As one of the enzymes needed to make ascorbic acid has been lost by mutation during primate evolution, humans must obtain it from the diet; it is therefore a vitamin.[103] Most other animals are able to produce this compound in their bodies and do not require it in their diets.[104] Ascorbic acid is required for the conversion of the procollagen to collagen by oxidizing proline residues to hydroxyproline. In other cells, it is maintained in its reduced form by reaction with glutathione, which can be catalysed by protein disulfide isomerase and glutaredoxins.[105][106] Ascorbic acid is a redox catalyst which can reduce, and thereby neutralize, reactive oxygen species such as hydrogen peroxide.[107] In addition to its direct antioxidant effects, ascorbic acid is also a substrate for the redox enzyme ascorbate peroxidase, a function that is particularly important in stress resistance in plants.[108] Ascorbic acid is present at high levels in all parts of plants and can reach concentrations of 20 millimolar in chloroplasts.[109]

18.3.3 Glutathione

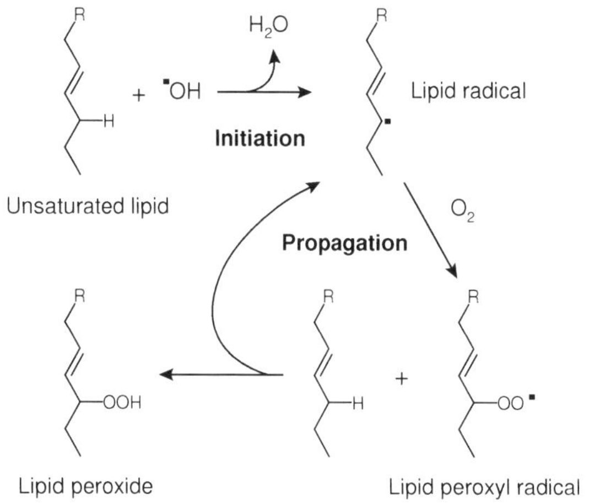

The free radical mechanism of lipid peroxidation.

Glutathione is a cysteine-containing peptide found in most forms of aerobic life.[110] It is not required in the diet and is instead synthesized in cells from its constituent amino acids.[111] Glutathione has antioxidant properties since the thiol group in its cysteine moiety is a reducing agent and can be reversibly oxidized and reduced. In cells, glutathione is maintained in the reduced form by the enzyme glutathione reductase and in turn reduces other metabolites and enzyme systems, such as ascorbate in the glutathione-ascorbate cycle, glutathione peroxidases and glutaredoxins, as well as reacting directly with oxidants.[105] Due to its high concentration and its central role in maintaining the cell's redox state, glutathione is one of the most important cellular antioxidants.[110] In some organisms glutathione is replaced by other thiols, such as by mycothiol in the Actinomycetes, bacillithiol in some Gram-positive bacteria,[112][113] or by trypanothione in the Kinetoplastids.[114][115]

18.3.4 Melatonin

Melatonin is a powerful antioxidant.[116] Melatonin easily crosses cell membranes and the blood–brain barrier.[117] Unlike other antioxidants, melatonin does not undergo redox cycling, which is the ability of a molecule to undergo repeated reduction and oxidation. Redox cycling may allow other antioxidants (such as vitamin C) to act as pro-oxidants and promote free radical formation. Melatonin, once oxidized, cannot be reduced to its former state because it forms several stable end-products upon reacting with free radicals. Therefore, it has been referred to as a terminal (or suicidal) antioxidant.[118]

18.3.5 Vitamin E

Vitamin E is the collective name for a set of eight related tocopherols and tocotrienols, which are fat-soluble vitamins with antioxidant properties.[119][120] Of these, α-tocopherol has been most studied as it has the highest bioavailability, with the body preferentially absorbing and metabolising this form.[121]

It has been claimed that the α-tocopherol form is the most important lipid-soluble antioxidant, and that it protects membranes from oxidation by reacting with lipid radicals produced in the lipid peroxidation chain reaction.[119][122] This removes the free radical intermediates and prevents the propagation reaction from continuing. This reaction produces oxidised α-tocopheroxyl radicals that can be recycled back to the active reduced form through reduction by other antioxidants, such as ascorbate, retinol or ubiquinol.[123] This is in line with findings showing that α-tocopherol, but not water-soluble antioxidants, efficiently protects glutathione peroxidase 4 (GPX4)-deficient cells from cell death.[124] GPx4 is the only known enzyme that efficiently reduces lipid-hydroperoxides within biological membranes.

However, the roles and importance of the various forms of vitamin E are presently unclear,[125][126] and it has even been suggested that the most important function of α-tocopherol is as a signaling molecule, with this molecule having no significant role in antioxidant metabolism.[127][128] The functions of the other forms of vitamin E are even less well-understood, although γ-tocopherol is a nucleophile that may react with electrophilic mutagens,[121] and tocotrienols may be important in protecting neurons from damage.[129]

18.4 Pro-oxidant activities

Further information: Pro-oxidant

Antioxidants that are reducing agents can also act as pro-oxidants. For example, vitamin C has antioxidant activity when it reduces oxidizing substances such as hydrogen peroxide,[130] however, it will also reduce metal ions that generate free radicals through the Fenton reaction.[61][131]

$2\ Fe^{3+}$ + Ascorbate → $2\ Fe^{2+}$ + Dehydroascorbate

$2\ Fe^{2+} + 2\ H_2O_2 \rightarrow 2\ Fe^{3+} + 2\ OH\cdot + 2\ OH^-$

The relative importance of the antioxidant and pro-oxidant activities of antioxidants is an area of current research, but vitamin C, which exerts its effects as a vitamin by oxidizing polypeptides, appears to have a mostly antioxidant action in the human body.[131] However, less data is available for other dietary antioxidants, such as vitamin E,[132] or the polyphenols.[133][134] Likewise, the pathogenesis of diseases involving hyperuricemia likely involve uric acid's direct and indirect pro-oxidant properties.

That is, paradoxically, agents which are normally considered antioxidants can act as conditional pro-oxidants and actually increase oxidative stress. Besides ascorbate, medically important conditional pro-oxidants include uric acid and sulfhydryl amino acids such as homocysteine. Typically, this involves some transition-series metal such as copper or iron as catalyst. The potential role of the pro-oxidant role of uric acid in (e.g.) atherosclerosis and ischemic stroke is considered above. Another example is the postulated role of homocysteine in atherosclerosis.

18.4.1 Negative health effects

Some antioxidant supplements may promote disease and increase mortality in humans under certain conditions.[45][134] Hypothetically, free radicals induce an endogenous response that protects against exogenous radicals (and possibly other toxic compounds).[135] Free radicals may increase life span.[134] This increase may be prevented by antioxidants, providing direct evidence that toxic radicals may mitohormetically exert life extending and health promoting effects.[45][134]

18.5 Enzyme systems

Superoxide dismutase; Peroxidases Catalase

O_2 Oxygen ⟶ $\cdot O_2^-$ Superoxide ⟶ H_2O_2 Hydrogen peroxide ⟶ H_2O Water

Enzymatic pathway for detoxification of reactive oxygen species.

As with the chemical antioxidants, cells are protected against oxidative stress by an interacting network of antioxidant enzymes.[56][57] Here, the superoxide released by processes such as oxidative phosphorylation is first converted to hydrogen peroxide and then further reduced to give water. This detoxification pathway is the result of multiple enzymes, with superoxide dismutases catalysing the first step and then catalases and various peroxidases removing hydrogen peroxide. As with antioxidant metabolites, the contributions of these enzymes to antioxidant defenses can be hard to separate from one another, but the generation of transgenic mice lacking just one antioxidant enzyme can be informative.[136]

18.5.1 Superoxide dismutase, catalase, and peroxiredoxins

Superoxide dismutases (SODs) are a class of closely related enzymes that catalyze the breakdown of the superoxide anion into oxygen and hydrogen peroxide.[137][138] SOD enzymes are present in almost all aerobic cells and in extracellular fluids.[139] Superoxide dismutase enzymes contain metal ion cofactors that, depending on the isozyme, can be copper, zinc, manganese or iron. In humans, the copper/zinc SOD is present in the cytosol, while manganese SOD is present in the mitochondrion.[138] There also exists a third form of SOD in extracellular fluids, which contains copper and zinc in its active sites.[140] The mitochondrial isozyme seems to be the most biologically important of these three, since mice lacking this enzyme die soon after birth.[141] In contrast, the mice lacking copper/zinc SOD (Sod1) are viable but have numerous pathologies and a reduced lifespan (see article on superoxide), while mice without the extracellular SOD have minimal defects (sensitive to hyperoxia).[136][142] In plants, SOD isozymes are present in the cytosol and mitochondria, with an iron SOD found in chloroplasts that is absent from vertebrates and yeast.[143]

Catalases are enzymes that catalyse the conversion of hydrogen peroxide to water and oxygen, using either an iron or manganese cofactor.[144][145] This protein is localized to peroxisomes in most eukaryotic cells.[146] Catalase is an unusual enzyme since, although hydrogen peroxide is its only substrate, it follows a ping-pong mechanism. Here, its cofactor is oxidised by one molecule of hydrogen peroxide and then regenerated by transferring the bound oxygen to a second molecule of substrate.[147] Despite its apparent importance in hydrogen peroxide removal, humans with genetic deficiency of catalase — "acatalasemia" — or mice genetically engineered to lack catalase completely, suffer few ill effects.[148][149]

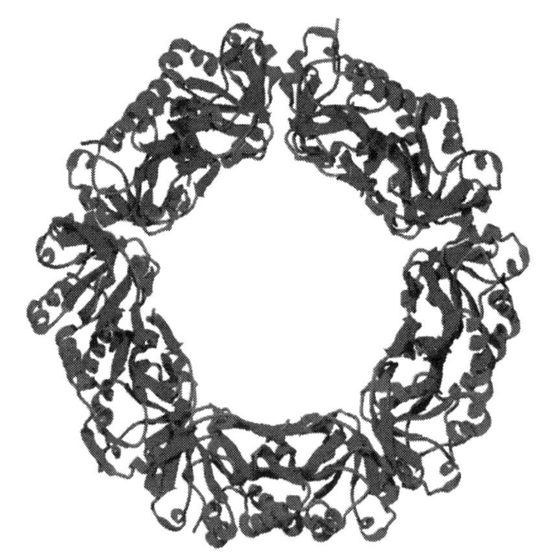

Decameric structure of AhpC, a bacterial 2-cysteine peroxiredoxin from Salmonella typhimurium.[150]

Peroxiredoxins are peroxidases that catalyze the reduction of hydrogen peroxide, organic hydroperoxides, as well as peroxynitrite.[151] They are divided into three classes: typical 2-cysteine peroxiredoxins; atypical 2-cysteine peroxiredoxins; and 1-cysteine peroxiredoxins.[152] These enzymes share the same basic catalytic mechanism, in which a redox-active cysteine (the peroxidatic cysteine) in the active site is oxidized to a sulfenic acid by the peroxide substrate.[153] Over-oxidation of this cysteine residue in peroxiredoxins inactivates these enzymes, but this can be reversed by the action of sulfiredoxin.[154] Peroxiredoxins seem to be important in antioxidant metabolism, as mice lacking peroxiredoxin 1 or 2 have shortened lifespan and suffer from hemolytic anaemia, while plants use peroxiredoxins to remove hydrogen peroxide generated in chloroplasts.[155][156][157]

18.5.2 Thioredoxin and glutathione systems

The thioredoxin system contains the 12-kDa protein thioredoxin and its companion thioredoxin reductase.[158] Proteins related to thioredoxin are present in all sequenced organisms. Plants, such as *Arabidopsis thaliana,* have a particularly great diversity of isoforms.[159] The active site of thioredoxin consists of two neighboring cysteines, as part of a highly conserved CXXC motif, that can cycle between an active dithiol form (reduced) and an oxidized disulfide form. In its active state, thioredoxin acts as an efficient reducing agent, scavenging reactive oxygen species and maintaining other proteins in their reduced state.[160] After being oxidized, the active thioredoxin is regenerated by the action of thioredoxin reductase, using NADPH as an electron donor.[161]

The glutathione system includes glutathione, glutathione reductase, glutathione peroxidases, and glutathione *S*-transferases.[110] This system is found in animals, plants and microorganisms.[110][162] Glutathione peroxidase is an enzyme containing four selenium-cofactors that catalyzes the breakdown of hydrogen peroxide and organic hydroperoxides. There are at least four different glutathione peroxidase isozymes in animals.[163] Glutathione peroxidase 1 is the most abundant and is a very efficient scavenger of hydrogen peroxide, while glutathione peroxidase 4 is most active with lipid hydroperoxides. Surprisingly, glutathione peroxidase 1 is dispensable, as mice lacking this enzyme have normal lifespans,[164] but they are hypersensitive to induced oxidative stress.[165] In addition, the glutathione *S*-transferases show high activity with lipid peroxides.[166] These enzymes are at particularly high levels in the liver and also serve in detoxification metabolism.[167]

18.6 Oxidative stress in disease

Further information: Pathology, Free-radical theory, and Oxidative stress

Oxidative stress is thought to contribute to the development of a wide range of diseases including Alzheimer's disease,[168][169] Parkinson's disease,[170] the pathologies caused by diabetes,[171][172] rheumatoid arthritis,[173] and neurodegeneration in motor neuron diseases.[174] In many of these cases, it is unclear if oxidants trigger the disease, or if they are produced as a secondary consequence of the disease and from general tissue damage;[60] One case in which this link is particularly well-understood is the role of oxidative stress in cardiovascular disease. Here, low density lipoprotein (LDL) oxidation appears to trigger the process of atherogenesis, which results in atherosclerosis, and finally cardiovascular disease.[175][176]

Oxidative damage in DNA can cause cancer. Several antioxidant enzymes such as superoxide dismutase, catalase, glutathione peroxidase, glutathione reductase, glutathione S-transferase etc. protect DNA from oxidative stress. It has been proposed that polymorphisms in these enzymes are associated with DNA damage and subsequently the individual's risk of cancer susceptibility.[177]

A low calorie diet extends median and maximum lifespan in many animals. This effect may involve a reduction in oxidative stress.[178] While there is some evidence to support the role of oxidative stress in aging in model organisms such as *Drosophila melanogaster* and *Caenorhabditis elegans*,[179][180] the evidence in mammals is less clear.[181][182][183] Indeed, a 2009 review of experiments in mice concluded that almost all manipulations of antioxidant systems had no effect on aging.[184]

Diets high in fruit and vegetables, and so possibly being rich in antioxidant vitamins, have no established effect on status of health or aging,[185][186] yet may have more subtle physiological effects, such as modifying cell-to-cell communication.[16][127]

18.7 Uses in technology

18.7.1 Food preservatives

Antioxidants are used as food additives to help guard against food deterioration. Exposure to oxygen and sunlight are the two main factors in the oxidation of food, so food is preserved by keeping in the dark and sealing it in containers or even coating it in wax, as with cucumbers. However, as oxygen is also important for plant respiration, storing plant materials in anaerobic conditions produces unpleasant flavors and unappealing colors.[187] Consequently, packaging of fresh fruits and vegetables contains an ~8% oxygen atmosphere. Antioxidants are an especially important class of preservatives as, unlike bacterial or fungal spoilage, oxidation reactions still occur relatively rapidly in frozen or refrigerated food.[188] These preservatives include natural antioxidants such as ascorbic acid (AA, E300) and tocopherols (E306), as well as synthetic antioxidants such as propyl gallate (PG, E310), tertiary butylhydroquinone (TBHQ), butylated hydroxyanisole (BHA, E320) and butylated hydroxytoluene (BHT, E321).[189][190]

The most common molecules attacked by oxidation are unsaturated fats; oxidation causes them to turn rancid.[191] Since oxidized lipids are often discolored and usually have unpleasant tastes such as metallic or sulfurous flavors, it is important to avoid oxidation in fat-rich foods. Thus, these foods are rarely preserved by drying; instead, they are preserved by smoking, salting or fermenting. Even less fatty foods such as fruits are sprayed with sulfurous antioxidants prior to air drying. Oxidation is often catalyzed by metals, which is why fats such as butter should never be wrapped in aluminium foil or kept in metal containers. Some fatty foods such as olive oil are partially protected from oxidation by their natural content of antioxidants, but remain sensitive to photooxidation.[192] Antioxidant preservatives are also added to fat based cosmetics such as lipstick and moisturizers to prevent rancidity.

18.7.2 Industrial uses

Substituted phenols and derivatives of phenylenediamine are common antioxidants used to inhibit gum formation in gasoline (petrol).

Antioxidants are frequently added to industrial products. A common use is as stabilizers in fuels and lubricants to prevent oxidation, and in gasolines to prevent the polymerization that leads to the formation of engine-fouling residues.[193] In 2014, the worldwide market for natural and synthetic antioxidants was US $2.25 billion with a forecast of growth to $3.25 billion by 2020.[194]

They are widely used to prevent the oxidative degradation of polymers such as rubbers, plastics and adhesives that causes a loss of strength and flexibility in these materials.[195] Polymers containing double bonds in their main chains, such as natural rubber and polybutadiene, are especially susceptible to oxidation and ozonolysis. They can be protected by antiozonants. Solid polymer products start to crack on exposed surfaces as the material degrades and the chains break. The mode of cracking varies between oxygen and ozone attack, the former causing a "crazy paving" effect, while ozone attack produces deeper cracks aligned at right angles to the tensile strain in the product. Oxidation and UV degradation are also frequently linked, mainly because UV radiation creates free radicals by bond breakage. The free radicals then react with oxygen to produce peroxy radicals which cause yet further damage, often in a chain reaction. Other polymers susceptible to oxidation include polypropylene and polyethylene. The former is more sensitive owing to the presence of secondary carbon atoms present in every repeat unit. Attack occurs at this point because the free radical formed is more stable than one formed on a primary carbon atom. Oxidation of polyethylene tends to occur at weak links in the chain, such as branch points in low-density polyethylene.

18.8 Levels in food

Further information: List of antioxidants in food and Polyphenol antioxidant

Antioxidant vitamins are found in vegetables, fruits, eggs, legumes and nuts. Vitamins A, C, and E can be destroyed by long-term storage or prolonged cooking.[197] The effects of cooking and food processing are complex, as these processes can also increase the bioavailability of antioxidants, such as some carotenoids in vegetables.[198] Processed food contains fewer antioxidant vitamins than fresh and uncooked foods, as preparation exposes food to heat and oxygen.[199]

Fruits and vegetables are good sources of antioxidant vitamins A, C and E

Other antioxidants are not obtained from the diet, but instead are made in the body. For example, ubiquinol (coenzyme Q) is poorly absorbed from the gut and is made through the mevalonate pathway.[90] Another example is glutathione, which is made from amino acids. As any glutathione in the gut is broken down to free cysteine, glycine and glutamic acid before being absorbed, even large oral intake has little effect on the concentration of glutathione in the body.[202][203] Although large amounts of sulfur-containing amino acids such as acetylcysteine can increase glutathione,[204] no evidence exists that eating high levels of these glutathione precursors is beneficial for healthy adults.[205]

18.8.1 Measurement and invalidation of ORAC

Measurement of antioxidant content in food is not a straightforward process, as antioxidants collectively are a diverse group of compounds with different reactivities to various reactive oxygen species. In food science, the oxygen radical absorbance capacity (ORAC) was once an industry standard for antioxidant strength of whole foods, juices and food additives.[206][207] However, the United States Department of Agriculture withdrew these ratings in 2012 as biologically invalid, stating that no physiological proof *in vivo* existed to support the free-radical theory or roles for ingested phytochemicals, especially for polyphenols.[208] Consequently, the ORAC method, derived only from *in vitro* experiments, is no longer considered relevant to human diets or biology.

Alternative *in vitro* measurements of antioxidant content in foods include the Folin-Ciocalteu reagent, and the Trolox equivalent antioxidant capacity assay.[209]

18.9 History

As part of their adaptation from marine life, terrestrial plants began producing non-marine antioxidants such as ascorbic acid (vitamin C), polyphenols and tocopherols. The evolution of angiosperm plants between 50 and 200 million years ago resulted in the development of many antioxidant pigments – particularly during the Jurassic period – as chemical defences against reactive oxygen species that are byproducts of photosynthesis.[210] Originally, the term antioxidant specifically referred to a chemical that prevented the consumption of oxygen. In the late 19th and early 20th centuries, extensive study concentrated on the use of antioxidants in important industrial processes, such as the prevention of metal corrosion, the vulcanization of rubber, and the polymerization of fuels in the fouling of internal combustion engines.[211]

Early research on the role of antioxidants in biology focused on their use in preventing the oxidation of unsaturated fats, which is the cause of rancidity.[212] Antioxidant activity could be measured simply by placing the fat in a closed container with oxygen and measuring the rate of oxygen consumption. However, it was the identification of vitamins A, C, and E as antioxidants that revolutionized the field and led to the realization of the importance of antioxidants in the biochemistry of living organisms.[213][214] The possible mechanisms of action of antioxidants were first explored when it was recognized that a substance with anti-oxidative activity is likely to be one that is itself readily oxidized.[215] Research into how vitamin E prevents the process of lipid peroxidation led to the identification of antioxidants as reducing agents that prevent oxidative reactions, often by scavenging reactive oxygen species before they can damage cells.[216]

18.10 See also

- Forensic engineering
- Mitohormesis – Hormesis
- Nootropic
- Polymer degradation

18.11 References

[1] Bjelakovic G, Nikolova D, Gluud C (2013). "Meta-regression analyses, meta-analyses, and trial sequential analyses of the effects of supplementation with beta-carotene, vitamin A, and vitamin E singly or in different combinations on all-cause mortality: do we have evidence for lack of harm?". *PLoS ONE*. **8** (9): e74558. Bibcode:2013PLoSO...874558B. PMC 3765487. PMID 24040282. doi:10.1371/journal.pone.0074558.

[2] Abner EL, Schmitt FA, Mendiondo MS, Marcum JL, Kryscio RJ (Jul 2011). "Vitamin E and all-cause mortality: a meta-analysis". *Current Aging Science*. **4** (2): 158–70. PMC 4030744. PMID 21235492. doi:10.2174/1874609811104020158.

[3] Cortés-Jofré M, Rueda JR, Corsini-Muñoz G, Fonseca-Cortés C, Caraballoso M, Bonfill Cosp X (2012). "Drugs for preventing lung cancer in healthy people". *The Cochrane Database of Systematic Reviews*. **10**: CD002141. PMID 23076895. doi:10.1002/14651858.CD002141.pub2.

[4] Jiang L, Yang KH, Tian JH, Guan QL, Yao N, Cao N, Mi DH, Wu J, Ma B, Yang SH (2010). "Efficacy of antioxidant

vitamins and selenium supplement in prostate cancer prevention: a meta-analysis of randomized controlled trials". *Nutrition and Cancer*. **62** (6): 719–27. PMID 20661819. doi:10.1080/01635581.2010.494335.

[5] Rees K, Hartley L, Day C, Flowers N, Clarke A, Stranges S (2013). "Selenium supplementation for the primary prevention of cardiovascular disease". *The Cochrane Database of Systematic Reviews*. **1** (1): CD009671. PMID 23440843. doi:10.1002/14651858.CD009671.pub2.

[6] Shekelle PG, Morton SC, Jungvig LK, Udani J, Spar M, Tu W, J Suttorp M, Coulter I, Newberry SJ, Hardy M (Apr 2004). "Effect of supplemental vitamin E for the prevention and treatment of cardiovascular disease". *Journal of General Internal Medicine*. **19** (4): 380–9. PMC 1492195. PMID 15061748. doi:10.1111/j.1525-1497.2004.30090.x.

[7] Dabelstein W, Reglitzky A, Schütze A, Reders K (2007). "Automotive Fuels". *Ullmann's Encyclopedia of Industrial Chemistry*. ISBN 3-527-30673-0. doi:10.1002/14356007.a16_719.pub2.

[8] Stanner SA, Hughes J, Kelly CN, Buttriss J (May 2004). "A review of the epidemiological evidence for the 'antioxidant hypothesis'". *Public Health Nutrition*. **7** (3): 407–22. PMID 15153272. doi:10.1079/PHN2003543.

[9] Shenkin A (Feb 2006). "The key role of micronutrients". *Clinical Nutrition*. **25** (1): 1–13. PMID 16376462. doi:10.1016/j.clnu.2005.11.006.

[10] Woodside JV, McCall D, McGartland C, Young IS (Nov 2005). "Micronutrients: dietary intake v. supplement use". *The Proceedings of the Nutrition Society*. **64** (4): 543–53. PMID 16313697. doi:10.1079/PNS2005464.

[11] *Food, Nutrition, Physical Activity, and the Prevention of Cancer: a Global Perspective*. World Cancer Research Fund (2007). ISBN 978-0-9722522-2-5.

[12] Hail N, Cortes M, Drake EN, Spallholz JE (Jul 2008). "Cancer chemoprevention: a radical perspective". *Free Radical Biology & Medicine*. **45** (2): 97–110. PMID 18454943. doi:10.1016/j.freeradbiomed.2008.04.004.

[13] "Flavonoids". Linus Pauling Institute, Oregon State University, Corvallis. 2016. Retrieved 24 July 2016.

[14] Csepregi, K; Neugart, S; Schreiner, M; Hideg, Éva (2016). "Comparative Evaluation of Total Antioxidant Capacities of Plant Polyphenols". *Molecules*. **21** (2): 208. PMID 26867192. doi:10.3390/molecules21020208.

[15] Williams RJ, Spencer JP, Rice-Evans C (Apr 2004). "Flavonoids: antioxidants or signalling molecules?". *Free Radical Biology & Medicine*. **36** (7): 838–49. PMID 15019969. doi:10.1016/j.freeradbiomed.2004.01.001.

[16] Aggarwal BB, Shishodia S (May 2006). "Molecular targets of dietary agents for prevention and therapy of cancer". *Biochemical Pharmacology*. **71** (10): 1397–421. PMID 16563357. doi:10.1016/j.bcp.2006.02.009.

[17] Virgili F, Marino M (Nov 2008). "Regulation of cellular signals from nutritional molecules: a specific role for phytochemicals, beyond antioxidant activity". *Free Radical Biology & Medicine*. **45** (9): 1205–16. PMID 18762244. doi:10.1016/j.freeradbiomed.2008.08.001.

[18] Di Matteo V, Esposito E (Apr 2003). "Biochemical and therapeutic effects of antioxidants in the treatment of Alzheimer's disease, Parkinson's disease, and amyotrophic lateral sclerosis". *Current Drug Targets. CNS and Neurological Disorders*. **2** (2): 95–107. PMID 12769802. doi:10.2174/1568007033482959.

[19] Rao AV, Balachandran B (Oct 2002). "Role of oxidative stress and antioxidants in neurodegenerative diseases". *Nutritional Neuroscience*. **5** (5): 291–309. PMID 12385592. doi:10.1080/1028415021000033767.

[20] Crichton GE, Bryan J, Murphy KJ (Sep 2013). "Dietary antioxidants, cognitive function and dementia--a systematic review". *Plant Foods for Human Nutrition*. **68** (3): 279–92. PMID 23881465. doi:10.1007/s11130-013-0370-0.

[21] Takeda A, Nyssen OP, Syed A, Jansen E, Bueno-de-Mesquita B, Gallo V (2014). "Vitamin A and carotenoids and the risk of Parkinson's disease: a systematic review and meta-analysis". *Neuroepidemiology*. **42** (1): 25–38. PMID 24356061. doi:10.1159/000355849.

[22] Harrison FE (2012). "A critical review of vitamin C for the prevention of age-related cognitive decline and Alzheimer's disease". *Journal of Alzheimer's Disease*. **29** (4): 711–26. PMC 3727637. PMID 22366772. doi:10.3233/JAD-2012-111853.

[23] Sena E, Wheble P, Sandercock P, Macleod M (Feb 2007). "Systematic review and meta-analysis of the efficacy of tirilazad in experimental stroke". *Stroke; A Journal of Cerebral Circulation*. **38** (2): 388–94. PMID 17204689. doi:10.1161/01.STR.0000254462.75851.22.

[24] Zhang S, Wang L, Liu M, Wu B (2010). "Tirilazad for aneurysmal subarachnoid haemorrhage". *The Cochrane Database of Systematic Reviews* (2): CD006778. PMID 20166088. doi:10.1002/14651858.CD006778.pub2.

[25] Bath PM, Iddenden R, Bath FJ, Orgogozo JM (2001). "Tirilazad for acute ischaemic stroke". *The Cochrane Database of Systematic Reviews* (4): CD002087. PMID 11687138. doi:10.1002/14651858.CD002087.

[26] Bath PM, Gray LJ, Bath AJ, Buchan A, Miyata T, Green AR (Aug 2009). "Effects of NXY-059 in experimental stroke: an individual animal meta-analysis". *British Journal of Pharmacology*. **157** (7): 1157–71. PMC 2743834. PMID 19422398. doi:10.1111/j.1476-5381.2009.00196.x.

[27] Green AR, Ashwood T (Apr 2005). "Free radical trapping as a therapeutic approach to neuroprotection in stroke: experimental and clinical studies with NXY-059 and free radical scavengers". *Current Drug Targets. CNS and Neurological Disorders*. **4** (2): 109–18. PMID 15857295. doi:10.2174/1568007053544156.

[28] Lemmo W (Sep 2014). "Potential interactions of prescription and over-the-counter medications having antioxidant capabilities with radiation and chemotherapy". *International Journal of Cancer. Journal International Du Cancer*. **137** (11): 2525–33. PMID 25220632. doi:10.1002/ijc.29208.

[29] D'Andrea GM (2005). "Use of antioxidants during chemotherapy and radiotherapy should be avoided". *CA: A Cancer Journal for Clinicians*. **55** (5): 319–21. PMID 16166076. doi:10.3322/canjclin.55.5.319.

[30] Dekkers JC, van Doornen LJ, Kemper HC (Mar 1996). "The role of antioxidant vitamins and enzymes in the prevention of exercise-induced muscle damage". *Sports Medicine*. **21** (3): 213–38. PMID 8776010. doi:10.2165/00007256-199621030-00005.

[31] Takanami Y, Iwane H, Kawai Y, Shimomitsu T (Feb 2000). "Vitamin E supplementation and endurance exercise: are there benefits?". *Sports Medicine*. **29** (2): 73–83. PMID 10701711. doi:10.2165/00007256-200029020-00001.

[32] Mastaloudis A, Traber MG, Carstensen K, Widrick JJ (Jan 2006). "Antioxidants did not prevent muscle damage in response to an ultramarathon run". *Medicine and Science in Sports and Exercise*. **38** (1): 72–80. PMID 16394956. doi:10.1249/01.mss.0000188579.36272.f6.

[33] Close GL, Ashton T, Cable T, Doran D, Holloway C, McArdle F, MacLaren DP (May 2006). "Ascorbic acid supplementation does not attenuate post-exercise muscle soreness following muscle-damaging exercise but may delay the recovery process". *The British Journal of Nutrition*. **95** (5): 976–81. PMID 16611389. doi:10.1079/BJN20061732.

[34] Gavura S. "Antioxidants and Exercise: More Harm Than Good?". Science Based Medicine. Retrieved 19 December 2011.

[35] Hurrell RF (Sep 2003). "Influence of vegetable protein sources on trace element and mineral bioavailability". *The Journal of Nutrition*. **133** (9): 2973S–7S. PMID 12949395.

[36] Hunt JR (Sep 2003). "Bioavailability of iron, zinc, and other trace minerals from vegetarian diets". *The American Journal of Clinical Nutrition*. **78** (3 Suppl): 633S–639S. PMID 12936958.

[37] Gibson RS, Perlas L, Hotz C (May 2006). "Improving the bioavailability of nutrients in plant foods at the household level". *The Proceedings of the Nutrition Society*. **65** (2): 160–8. PMID 16672077. doi:10.1079/PNS2006489.

[38] Mosha TC, Gaga HE, Pace RD, Laswai HS, Mtebe K (Jun 1995). "Effect of blanching on the content of antinutritional factors in selected vegetables". *Plant Foods for Human Nutrition*. **47** (4): 361–7. PMID 8577655. doi:10.1007/BF01088275.

[39] Sandberg AS (Dec 2002). "Bioavailability of minerals in legumes". *The British Journal of Nutrition*. 88 Suppl 3 (Suppl 3): S281–5. PMID 12498628. doi:10.1079/BJN/2002718.

[40] Beecher GR (Oct 2003). "Overview of dietary flavonoids: nomenclature, occurrence and intake". *The Journal of Nutrition*. **133** (10): 3248S–3254S. PMID 14519822.

[41] Prashar A, Locke IC, Evans CS (Aug 2006). "Cytotoxicity of clove (Syzygium aromaticum) oil and its major components to human skin cells". *Cell Proliferation*. **39** (4): 241–8. PMID 16872360. doi:10.1111/j.1365-2184.2006.00384.x.

[42] Hornig D, Vuilleumier JP, Hartmann D (1980). "Absorption of large, single, oral intakes of ascorbic acid". *International Journal for Vitamin and Nutrition Research*. **50** (3): 309–14. PMID 7429760.

[43] Omenn GS, Goodman GE, Thornquist MD, Balmes J, Cullen MR, Glass A, Keogh JP, Meyskens FL, Valanis B, Williams JH, Barnhart S, Cherniack MG, Brodkin CA, Hammar S (Nov 1996). "Risk factors for lung cancer and for intervention effects in CARET, the Beta-Carotene and Retinol Efficacy Trial". *Journal of the National Cancer Institute*. **88** (21): 1550–9. PMID 8901853. doi:10.1093/jnci/88.21.1550.

[44] Albanes D (Jun 1999). "Beta-carotene and lung cancer: a case study". *The American Journal of Clinical Nutrition*. **69** (6): 1345S–50S. PMID 10359235.

[45] Bjelakovic G, Nikolova D, Gluud LL, Simonetti RG, Gluud C (Feb 2007). "Mortality in randomized trials of antioxidant supplements for primary and secondary prevention: systematic review and meta-analysis". *JAMA*. **297** (8): 842–57. PMID 17327526. doi:10.1001/jama.297.8.842.

[46] Bjelakovic G, Nikolova D, Gluud LL, Simonetti RG, Gluud C (14 March 2012). "Antioxidant supplements for prevention of mortality in healthy participants and patients with various diseases". *The Cochrane Database of Systematic Reviews*. **3** (3): CD007176. PMID 22419320. doi:10.1002/14651858.CD007176.pub2.

[47] Study Citing Antioxidant Vitamin Risks Based On Flawed Methodology, Experts Argue News release from Oregon State University published on ScienceDaily. Retrieved 19 April 2007

[48] Miller ER, Pastor-Barriuso R, Dalal D, Riemersma RA, Appel LJ, Guallar E (Jan 2005). "Meta-analysis: high-dosage vitamin E supplementation may increase all-cause mortality". *Annals of Internal Medicine*. **142** (1): 37–46. PMID 15537682. doi:10.7326/0003-4819-142-1-200501040-00110.

[49] Bjelakovic G, Nagorni A, Nikolova D, Simonetti RG, Bjelakovic M, Gluud C (Jul 2006). "Meta-analysis: antioxidant supplements for primary and secondary prevention of colorectal adenoma". *Alimentary Pharmacology & Therapeutics*. **24** (2): 281–91. PMID 16842454. doi:10.1111/j.1365-2036.2006.02970.x.

[50] Cortés-Jofré M, Rueda JR, Corsini-Muñoz G, Fonseca-Cortés C, Caraballoso M, Bonfill Cosp X (17 October 2012). "Drugs for preventing lung cancer in

healthy people". *The Cochrane Database of Systematic Reviews*. **10**: CD002141. PMID 23076895. doi:10.1002/14651858.CD002141.pub2.

[51] Schumacker PT (Sep 2006). "Reactive oxygen species in cancer cells: live by the sword, die by the sword". *Cancer Cell*. **10** (3): 175–6. PMID 16959608. doi:10.1016/j.ccr.2006.08.015.

[52] Seifried HE, McDonald SS, Anderson DE, Greenwald P, Milner JA (Aug 2003). "The antioxidant conundrum in cancer". *Cancer Research*. **63** (15): 4295–8. PMID 12907593.

[53] Lawenda BD, Kelly KM, Ladas EJ, Sagar SM, Vickers A, Blumberg JB (Jun 2008). "Should supplemental antioxidant administration be avoided during chemotherapy and radiation therapy?". *Journal of the National Cancer Institute*. **100** (11): 773–83. PMID 18505970. doi:10.1093/jnci/djn148.

[54] Block KI, Koch AC, Mead MN, Tothy PK, Newman RA, Gyllenhaal C (Sep 2008). "Impact of antioxidant supplementation on chemotherapeutic toxicity: a systematic review of the evidence from randomized controlled trials". *International Journal of Cancer. Journal International Du Cancer*. **123** (6): 1227–39. PMID 18623084. doi:10.1002/ijc.23754.

[55] Block KI, Koch AC, Mead MN, Tothy PK, Newman RA, Gyllenhaal C (Aug 2007). "Impact of antioxidant supplementation on chemotherapeutic efficacy: a systematic review of the evidence from randomized controlled trials". *Cancer Treatment Reviews*. **33** (5): 407–18. PMID 17367938. doi:10.1016/j.ctrv.2007.01.005.

[56] Davies KJ (1995). "Oxidative stress: the paradox of aerobic life". *Biochemical Society Symposium*. **61**: 1–31. PMID 8660387. doi:10.1042/bss0610001.

[57] Sies H (Mar 1997). "Oxidative stress: oxidants and antioxidants". *Experimental Physiology*. **82** (2): 291–5. PMID 9129943. doi:10.1113/expphysiol.1997.sp004024.

[58] Vertuani S, Angusti A, Manfredini S (2004). "The antioxidants and pro-antioxidants network: an overview". *Current Pharmaceutical Design*. **10** (14): 1677–94. PMID 15134565. doi:10.2174/1381612043384655.

[59] Rhee SG (Jun 2006). "Cell signaling. H2O2, a necessary evil for cell signaling". *Science*. **312** (5782): 1882–3. PMID 16809515. doi:10.1126/science.1130481.

[60] Valko M, Leibfritz D, Moncol J, Cronin MT, Mazur M, Telser J (2007). "Free radicals and antioxidants in normal physiological functions and human disease". *The International Journal of Biochemistry & Cell Biology*. **39** (1): 44–84. PMID 16978905. doi:10.1016/j.biocel.2006.07.001.

[61] Stohs SJ, Bagchi D (Feb 1995). "Oxidative mechanisms in the toxicity of metal ions". *Free Radical Biology & Medicine*. **18** (2): 321–36. PMID 7744317. doi:10.1016/0891-5849(94)00159-H.

[62] Nakabeppu Y, Sakumi K, Sakamoto K, Tsuchimoto D, Tsuzuki T, Nakatsu Y (Apr 2006). "Mutagenesis and carcinogenesis caused by the oxidation of nucleic acids". *Biological Chemistry*. **387** (4): 373–9. PMID 16606334. doi:10.1515/BC.2006.050.

[63] Valko M, Izakovic M, Mazur M, Rhodes CJ, Telser J (Nov 2004). "Role of oxygen radicals in DNA damage and cancer incidence". *Molecular and Cellular Biochemistry*. **266** (1–2): 37–56. PMID 15646026. doi:10.1023/B:MCBI.0000049134.69131.89.

[64] Stadtman ER (Aug 1992). "Protein oxidation and aging". *Science*. **257** (5074): 1220–4. Bibcode:1992Sci...257.1220S. PMID 1355616. doi:10.1126/science.1355616.

[65] Raha S, Robinson BH (Oct 2000). "Mitochondria, oxygen free radicals, disease and ageing". *Trends in Biochemical Sciences*. **25** (10): 502–8. PMID 11050436. doi:10.1016/S0968-0004(00)01674-1.

[66] Lenaz G (2001). "The mitochondrial production of reactive oxygen species: mechanisms and implications in human pathology". *IUBMB Life*. **52** (3–5): 159–64. PMID 11798028. doi:10.1080/15216540152845957.

[67] Finkel T, Holbrook NJ (Nov 2000). "Oxidants, oxidative stress and the biology of ageing". *Nature*. **408** (6809): 239–47. PMID 11089981. doi:10.1038/35041687.

[68] Hirst J, King MS, Pryde KR (Oct 2008). "The production of reactive oxygen species by complex I". *Biochemical Society Transactions*. **36** (Pt 5): 976–80. PMID 18793173. doi:10.1042/BST0360976.

[69] Seaver LC, Imlay JA (Nov 2004). "Are respiratory enzymes the primary sources of intracellular hydrogen peroxide?". *The Journal of Biological Chemistry*. **279** (47): 48742–50. PMID 15361522. doi:10.1074/jbc.M408754200.

[70] Imlay JA (2003). "Pathways of oxidative damage". *Annual Review of Microbiology*. **57**: 395–418. PMID 14527285. doi:10.1146/annurev.micro.57.030502.090938.

[71] Demmig-Adams B, Adams WW (Dec 2002). "Antioxidants in photosynthesis and human nutrition". *Science*. **298** (5601): 2149–53. Bibcode:2002Sci...298.2149D. PMID 12481128. doi:10.1126/science.1078002.

[72] Krieger-Liszkay A (Jan 2005). "Singlet oxygen production in photosynthesis". *Journal of Experimental Botany*. **56** (411): 337–46. PMID 15310815. doi:10.1093/jxb/erh237.

[73] Küpper FC, Carpenter LJ, McFiggans GB, Palmer CJ, Waite TJ, Boneberg EM, Woitsch S, Weiller M, Abela R, Grolimund D, Potin P, Butler A, Luther GW, Kroneck PM, Meyer-Klaucke W, Feiters MC (May 2008). "Iodide accumulation provides kelp with an inorganic antioxidant impacting atmospheric chemistry". *Proceedings of the National Academy of Sciences of the United States of America*. **105** (19): 6954–8.

Bibcode:2008PNAS..105.6954K. PMC 2383960. PMID 18458346. doi:10.1073/pnas.0709959105.

[74] Szabó I, Bergantino E, Giacometti GM (Jul 2005). "Light and oxygenic photosynthesis: energy dissipation as a protection mechanism against photo-oxidation". *EMBO Reports*. **6** (7): 629–34. PMC 1369118. PMID 15995679. doi:10.1038/sj.embor.7400460.

[75] Kerfeld CA (Oct 2004). "Water-soluble carotenoid proteins of cyanobacteria". *Archives of Biochemistry and Biophysics*. **430** (1): 2–9. PMID 15325905. doi:10.1016/j.abb.2004.03.018.

[76] Miller RA, Britigan BE (Jan 1997). "Role of oxidants in microbial pathophysiology". *Clinical Microbiology Reviews*. **10** (1): 1–18. PMC 172912. PMID 8993856.

[77] Chaudière J, Ferrari-Iliou R (1999). "Intracellular antioxidants: from chemical to biochemical mechanisms". *Food and Chemical Toxicology*. **37** (9–10): 949–62. PMID 10541450. doi:10.1016/S0278-6915(99)00090-3.

[78] Sies H (Jul 1993). "Strategies of antioxidant defense". *European Journal of Biochemistry / FEBS*. **215** (2): 213–9. PMID 7688300. doi:10.1111/j.1432-1033.1993.tb18025.x.

[79] Ames BN, Cathcart R, Schwiers E, Hochstein P (Nov 1981). "Uric acid provides an antioxidant defense in humans against oxidant- and radical-caused aging and cancer: a hypothesis". *Proceedings of the National Academy of Sciences of the United States of America*. **78** (11): 6858–62. Bibcode:1981PNAS...78.6858A. PMC 349151. PMID 6947260. doi:10.1073/pnas.78.11.6858.

[80] Khaw KT, Woodhouse P (Jun 1995). "Interrelation of vitamin C, infection, haemostatic factors, and cardiovascular disease". *BMJ*. **310** (6994): 1559–63. PMC 2549940. PMID 7787643. doi:10.1136/bmj.310.6994.1559.

[81] Evelson P, Travacio M, Repetto M, Escobar J, Llesuy S, Lissi EA (Apr 2001). "Evaluation of total reactive antioxidant potential (TRAP) of tissue homogenates and their cytosols". *Archives of Biochemistry and Biophysics*. **388** (2): 261–6. PMID 11368163. doi:10.1006/abbi.2001.2292.

[82] Morrison JA, Jacobsen DW, Sprecher DL, Robinson K, Khoury P, Daniels SR (Nov 1999). "Serum glutathione in adolescent males predicts parental coronary heart disease". *Circulation*. **100** (22): 2244–7. PMID 10577998. doi:10.1161/01.CIR.100.22.2244.

[83] Teichert J, Preiss R (Nov 1992). "HPLC-methods for determination of lipoic acid and its reduced form in human plasma". *International Journal of Clinical Pharmacology, Therapy, and Toxicology*. **30** (11): 511–2. PMID 1490813.

[84] Akiba S, Matsugo S, Packer L, Konishi T (May 1998). "Assay of protein-bound lipoic acid in tissues by a new enzymatic method". *Analytical Biochemistry*. **258** (2): 299–304. PMID 9570844. doi:10.1006/abio.1998.2615.

[85] Glantzounis GK, Tsimoyiannis EC, Kappas AM, Galaris DA (2005). "Uric acid and oxidative stress". *Current Pharmaceutical Design*. **11** (32): 4145–51. PMID 16375736. doi:10.2174/138161205774913255.

[86] El-Sohemy A, Baylin A, Kabagambe E, Ascherio A, Spiegelman D, Campos H (Jul 2002). "Individual carotenoid concentrations in adipose tissue and plasma as biomarkers of dietary intake". *The American Journal of Clinical Nutrition*. **76** (1): 172–9. PMID 12081831.

[87] Sowell AL, Huff DL, Yeager PR, Caudill SP, Gunter EW (Mar 1994). "Retinol, alpha-tocopherol, lutein/zeaxanthin, beta-cryptoxanthin, lycopene, alpha-carotene, trans-beta-carotene, and four retinyl esters in serum determined simultaneously by reversed-phase HPLC with multiwavelength detection". *Clinical Chemistry*. **40** (3): 411–6. PMID 8131277.

[88] Stahl W, Schwarz W, Sundquist AR, Sies H (Apr 1992). "cis-trans isomers of lycopene and beta-carotene in human serum and tissues". *Archives of Biochemistry and Biophysics*. **294** (1): 173–7. PMID 1550343. doi:10.1016/0003-9861(92)90153-N.

[89] Zita C, Overvad K, Mortensen SA, Sindberg CD, Moesgaard S, Hunter DA (2003). "Serum coenzyme Q10 concentrations in healthy men supplemented with 30 mg or 100 mg coenzyme Q10 for two months in a randomised controlled study". *BioFactors*. **18** (1–4): 185–93. PMID 14695934. doi:10.1002/biof.5520180221.

[90] Turunen M, Olsson J, Dallner G (Jan 2004). "Metabolism and function of coenzyme Q". *Biochimica et Biophysica Acta*. **1660** (1–2): 171–99. PMID 14757233. doi:10.1016/j.bbamem.2003.11.012.

[91] Enomoto A, Endou H (Sep 2005). "Roles of organic anion transporters (OATs) and a urate transporter (URAT1) in the pathophysiology of human disease". *Clinical and Experimental Nephrology*. **9** (3): 195–205. PMID 16189627. doi:10.1007/s10157-005-0368-5.

[92] Wu XW, Lee CC, Muzny DM, Caskey CT (Dec 1989). "Urate oxidase: primary structure and evolutionary implications". *Proceedings of the National Academy of Sciences of the United States of America*. **86** (23): 9412–6. Bibcode:1989PNAS...86.9412W. PMC 298506. PMID 2594778. doi:10.1073/pnas.86.23.9412.

[93] Wu XW, Muzny DM, Lee CC, Caskey CT (Jan 1992). "Two independent mutational events in the loss of urate oxidase during hominoid evolution". *Journal of Molecular Evolution*. **34** (1): 78–84. PMID 1556746. doi:10.1007/BF00163854.

[94] Álvarez-Lario B, Macarrón-Vicente J (Nov 2010). "Uric acid and evolution". *Rheumatology*. **49** (11): 2010–5. PMID 20627967. doi:10.1093/rheumatology/keq204.

[95] Watanabe S, Kang DH, Feng L, Nakagawa T, Kanellis J, Lan H, Mazzali M, Johnson RJ (Sep 2002). "Uric

acid, hominoid evolution, and the pathogenesis of salt-sensitivity". *Hypertension*. **40** (3): 355–60. PMID 12215479. doi:10.1161/01.HYP.0000028589.66335.AA.

[96] Johnson RJ, Andrews P, Benner SA, Oliver W (2010). "Theodore E. Woodward award. The evolution of obesity: insights from the mid-Miocene". *Transactions of the American Clinical and Climatological Association*. **121**: 295–305; discussion 305–8. PMC 2917125. PMID 20697570.

[97] Baillie JK, Bates MG, Thompson AA, Waring WS, Partridge RW, Schnopp MF, Simpson A, Gulliver-Sloan F, Maxwell SR, Webb DJ (May 2007). "Endogenous urate production augments plasma antioxidant capacity in healthy lowland subjects exposed to high altitude". *Chest*. **131** (5): 1473–8. PMID 17494796. doi:10.1378/chest.06-2235.

[98] Becker BF (Jun 1993). "Towards the physiological function of uric acid". *Free Radical Biology & Medicine*. **14** (6): 615–31. PMID 8325534. doi:10.1016/0891-5849(93)90143-I.

[99] Sautin YY, Johnson RJ (Jun 2008). "Uric acid: the oxidant-antioxidant paradox". *Nucleosides, Nucleotides & Nucleic Acids*. **27** (6): 608–19. PMC 2895915. PMID 18600514. doi:10.1080/15257770802138558.

[100] Eggebeen AT (Sep 2007). "Gout: an update". *American Family Physician*. **76** (6): 801–8. PMID 17910294.

[101] Campion EW, Glynn RJ, DeLabry LO (Mar 1987). "Asymptomatic hyperuricemia. Risks and consequences in the Normative Aging Study". *The American Journal of Medicine*. **82** (3): 421–6. PMID 3826098. doi:10.1016/0002-9343(87)90441-4.

[102] Nazarewicz RR, Ziolkowski W, Vaccaro PS, Ghafourifar P (Dec 2007). "Effect of short-term ketogenic diet on redox status of human blood". *Rejuvenation Research*. **10** (4): 435–40. PMID 17663642. doi:10.1089/rej.2007.0540.

[103] Smirnoff N (2001). "L-ascorbic acid biosynthesis". *Vitamins and Hormones*. Vitamins & Hormones. **61**: 241–66. ISBN 978-0-12-709861-6. PMID 11153268. doi:10.1016/S0083-6729(01)61008-2.

[104] Linster CL, Van Schaftingen E (Jan 2007). "Vitamin C. Biosynthesis, recycling and degradation in mammals". *The FEBS Journal*. **274** (1): 1–22. PMID 17222174. doi:10.1111/j.1742-4658.2006.05607.x.

[105] Meister A (Apr 1994). "Glutathione-ascorbic acid antioxidant system in animals". *The Journal of Biological Chemistry*. **269** (13): 9397–400. PMID 8144521.

[106] Wells WW, Xu DP, Yang YF, Rocque PA (Sep 1990). "Mammalian thioltransferase (glutaredoxin) and protein disulfide isomerase have dehydroascorbate reductase activity". *The Journal of Biological Chemistry*. **265** (26): 15361–4. PMID 2394726.

[107] Padayatty SJ, Katz A, Wang Y, Eck P, Kwon O, Lee JH, Chen S, Corpe C, Dutta A, Dutta SK, Levine M (Feb 2003). "Vitamin C as an antioxidant: evaluation of its role in disease prevention". *Journal of the American College of Nutrition*. **22** (1): 18–35. PMID 12569111. doi:10.1080/07315724.2003.10719272.

[108] Shigeoka S, Ishikawa T, Tamoi M, Miyagawa Y, Takeda T, Yabuta Y, Yoshimura K (May 2002). "Regulation and function of ascorbate peroxidase isoenzymes". *Journal of Experimental Botany*. **53** (372): 1305–19. PMID 11997377. doi:10.1093/jexbot/53.372.1305.

[109] Smirnoff N, Wheeler GL (2000). "Ascorbic acid in plants: biosynthesis and function". *Critical Reviews in Biochemistry and Molecular Biology*. **35** (4): 291–314. PMID 11005203. doi:10.1080/10409230008984166.

[110] Meister A, Anderson ME (1983). "Glutathione". *Annual Review of Biochemistry*. **52**: 711–60. PMID 6137189. doi:10.1146/annurev.bi.52.070183.003431.

[111] Meister A (Nov 1988). "Glutathione metabolism and its selective modification". *The Journal of Biological Chemistry*. **263** (33): 17205–8. PMID 3053703.

[112] Gaballa A, Newton GL, Antelmann H, Parsonage D, Upton H, Rawat M, Claiborne A, Fahey RC, Helmann JD (Apr 2010). "Biosynthesis and functions of bacillithiol, a major low-molecular-weight thiol in Bacilli". *Proceedings of the National Academy of Sciences of the United States of America*. **107** (14): 6482–6. Bibcode:2010PNAS..107.6482G. PMC 2851989. PMID 20308541. doi:10.1073/pnas.1000928107.

[113] Newton GL, Rawat M, La Clair JJ, Jothivasan VK, Budiarto T, Hamilton CJ, Claiborne A, Helmann JD, Fahey RC (Sep 2009). "Bacillithiol is an antioxidant thiol produced in Bacilli". *Nature Chemical Biology*. **5** (9): 625–627. PMC 3510479. PMID 19578333. doi:10.1038/nchembio.189.

[114] Fahey RC (2001). "Novel thiols of prokaryotes". *Annual Review of Microbiology*. **55**: 333–56. PMID 11544359. doi:10.1146/annurev.micro.55.1.333.

[115] Fairlamb AH, Cerami A (1992). "Metabolism and functions of trypanothione in the Kinetoplastida". *Annual Review of Microbiology*. **46**: 695–729. PMID 1444271. doi:10.1146/annurev.mi.46.100192.003403.

[116] Tan DX, Manchester LC, Terron MP, Flores LJ, Reiter RJ (Jan 2007). "One molecule, many derivatives: a never-ending interaction of melatonin with reactive oxygen and nitrogen species?". *Journal of Pineal Research*. **42** (1): 28–42. PMID 17198536. doi:10.1111/j.1600-079X.2006.00407.x.

[117] Reiter RJ, Paredes SD, Manchester LC, Tan DX (2009). "Reducing oxidative/nitrosative stress: a newly-discovered genre for melatonin". *Critical Reviews in Biochemistry and Molecular Biology*. **44** (4): 175–200. PMID 19635037. doi:10.1080/10409230903044914.

[118] Tan DX, Manchester LC, Reiter RJ, Qi WB, Karbownik M, Calvo JR (2000). "Significance of melatonin in antioxidative defense system: reactions and products". *Biological Signals and Receptors*. **9** (3–4): 137–59. PMID 10899700. doi:10.1159/000014635.

[119] Herrera E, Barbas C (Mar 2001). "Vitamin E: action, metabolism and perspectives". *Journal of Physiology and Biochemistry*. **57** (2): 43–56. PMID 11579997. doi:10.1007/BF03179812.

[120] Packer L, Weber SU, Rimbach G (Feb 2001). "Molecular aspects of alpha-tocotrienol antioxidant action and cell signalling". *The Journal of Nutrition*. **131** (2): 369S–73S. PMID 11160563.

[121] Brigelius-Flohé R, Traber MG (Jul 1999). "Vitamin E: function and metabolism". *FASEB Journal*. **13** (10): 1145–55. CiteSeerX 10.1.1.337.5276 ô. PMID 10385606.

[122] Traber MG, Atkinson J (Jul 2007). "Vitamin E, antioxidant and nothing more". *Free Radical Biology & Medicine*. **43** (1): 4–15. PMC 2040110 ô. PMID 17561088. doi:10.1016/j.freeradbiomed.2007.03.024.

[123] Wang X, Quinn PJ (Jul 1999). "Vitamin E and its function in membranes". *Progress in Lipid Research*. **38** (4): 309–36. PMID 10793887. doi:10.1016/S0163-7827(99)00008-9.

[124] Seiler A, Schneider M, Förster H, Roth S, Wirth EK, Culmsee C, Plesnila N, Kremmer E, Rådmark O, Wurst W, Bornkamm GW, Schweizer U, Conrad M (Sep 2008). "Glutathione peroxidase 4 senses and translates oxidative stress into 12/15-lipoxygenase dependent- and AIF-mediated cell death". *Cell Metabolism*. **8** (3): 237–48. PMID 18762024. doi:10.1016/j.cmet.2008.07.005.

[125] Brigelius-Flohé R, Davies KJ (Jul 2007). "Is vitamin E an antioxidant, a regulator of signal transduction and gene expression, or a 'junk' food? Comments on the two accompanying papers: "Molecular mechanism of alpha-tocopherol action" by A. Azzi and "Vitamin E, antioxidant and nothing more" by M. Traber and J. Atkinson". *Free Radical Biology & Medicine*. **43** (1): 2–3. PMID 17561087. doi:10.1016/j.freeradbiomed.2007.05.016.

[126] Atkinson J, Epand RF, Epand RM (Mar 2008). "Tocopherols and tocotrienols in membranes: a critical review". *Free Radical Biology & Medicine*. **44** (5): 739–64. PMID 18160049. doi:10.1016/j.freeradbiomed.2007.11.010.

[127] Azzi A (Jul 2007). "Molecular mechanism of alpha-tocopherol action". *Free Radical Biology & Medicine*. **43** (1): 16–21. PMID 17561089. doi:10.1016/j.freeradbiomed.2007.03.013.

[128] Zingg JM, Azzi A (May 2004). "Non-antioxidant activities of vitamin E". *Current Medicinal Chemistry*. **11** (9): 1113–33. PMID 15134510. doi:10.2174/0929867043365332. Archived from the original on 6 October 2011.

[129] Sen CK, Khanna S, Roy S (Mar 2006). "Tocotrienols: Vitamin E beyond tocopherols". *Life Sciences*. **78** (18): 2088–98. PMC 1790869 ô. PMID 16458936. doi:10.1016/j.lfs.2005.12.001.

[130] Duarte TL, Lunec J (Jul 2005). "Review: When is an antioxidant not an antioxidant? A review of novel actions and reactions of vitamin C". *Free Radical Research*. **39** (7): 671–86. PMID 16036346. doi:10.1080/10715760500104025.

[131] Carr A, Frei B (Jun 1999). "Does vitamin C act as a pro-oxidant under physiological conditions?". *FASEB Journal*. **13** (9): 1007–24. PMID 10336883.

[132] Schneider C (Jan 2005). "Chemistry and biology of vitamin E". *Molecular Nutrition & Food Research*. **49** (1): 7–30. PMID 15580660. doi:10.1002/mnfr.200400049.

[133] Halliwell B (Aug 2008). "Are polyphenols antioxidants or pro-oxidants? What do we learn from cell culture and in vivo studies?". *Archives of Biochemistry and Biophysics*. **476** (2): 107–112. PMID 18284912. doi:10.1016/j.abb.2008.01.028.

[134] Ristow M, Zarse K (Jun 2010). "How increased oxidative stress promotes longevity and metabolic health: The concept of mitochondrial hormesis (mitohormesis)". *Experimental Gerontology*. **45** (6): 410–418. PMID 20350594. doi:10.1016/j.exger.2010.03.014.

[135] Tapia PC (2006). "Sublethal mitochondrial stress with an attendant stoichiometric augmentation of reactive oxygen species may precipitate many of the beneficial alterations in cellular physiology produced by caloric restriction, intermittent fasting, exercise and dietary phytonutrients: "Mitohormesis" for health and vitality". *Medical Hypotheses*. **66** (4): 832–43. PMID 16242247. doi:10.1016/j.mehy.2005.09.009.

[136] Ho YS, Magnenat JL, Gargano M, Cao J (Oct 1998). "The nature of antioxidant defense mechanisms: a lesson from transgenic studies". *Environmental Health Perspectives*. 106 Suppl 5 (Suppl 5): 1219–28. JSTOR 3433989. PMC 1533365 ô. PMID 9788901. doi:10.2307/3433989.

[137] Zelko IN, Mariani TJ, Folz RJ (Aug 2002). "Superoxide dismutase multigene family: a comparison of the CuZn-SOD (SOD1), Mn-SOD (SOD2), and EC-SOD (SOD3) gene structures, evolution, and expression". *Free Radical Biology & Medicine*. **33** (3): 337–49. PMID 12126755. doi:10.1016/S0891-5849(02)00905-X.

[138] Bannister JV, Bannister WH, Rotilio G (1987). "Aspects of the structure, function, and applications of superoxide dismutase". *CRC Critical Reviews in Biochemistry*. **22** (2): 111–80. PMID 3315461. doi:10.3109/10409238709083738.

[139] Johnson F, Giulivi C (2005). "Superoxide dismutases and their impact upon human health". *Molecular Aspects of Medicine*. **26** (4–5): 340–52. PMID 16099495. doi:10.1016/j.mam.2005.07.006.

[140] Nozik-Grayck E, Suliman HB, Piantadosi CA (Dec 2005). "Extracellular superoxide dismutase". *The International Journal of Biochemistry & Cell Biology*. **37** (12): 2466–71. PMID 16087389. doi:10.1016/j.biocel.2005.06.012.

[141] Melov S, Schneider JA, Day BJ, Hinerfeld D, Coskun P, Mirra SS, Crapo JD, Wallace DC (Feb 1998). "A novel neurological phenotype in mice lacking mitochondrial manganese superoxide dismutase". *Nature Genetics*. **18** (2): 159–63. PMID 9462746. doi:10.1038/ng0298-159.

[142] Reaume AG, Elliott JL, Hoffman EK, Kowall NW, Ferrante RJ, Siwek DF, Wilcox HM, Flood DG, Beal MF, Brown RH, Scott RW, Snider WD (May 1996). "Motor neurons in Cu/Zn superoxide dismutase-deficient mice develop normally but exhibit enhanced cell death after axonal injury". *Nature Genetics*. **13** (1): 43–7. PMID 8673102. doi:10.1038/ng0596-43.

[143] Van Camp W, Inzé D, Van Montagu M (1997). "The regulation and function of tobacco superoxide dismutases". *Free Radical Biology & Medicine*. **23** (3): 515–20. PMID 9214590. doi:10.1016/S0891-5849(97)00112-3.

[144] Chelikani P, Fita I, Loewen PC (Jan 2004). "Diversity of structures and properties among catalases". *Cellular and Molecular Life Sciences*. **61** (2): 192–208. PMID 14745498. doi:10.1007/s00018-003-3206-5.

[145] Zámocký M, Koller F (1999). "Understanding the structure and function of catalases: clues from molecular evolution and in vitro mutagenesis". *Progress in Biophysics and Molecular Biology*. **72** (1): 19–66. PMID 10446501. doi:10.1016/S0079-6107(98)00058-3.

[146] del Río LA, Sandalio LM, Palma JM, Bueno P, Corpas FJ (Nov 1992). "Metabolism of oxygen radicals in peroxisomes and cellular implications". *Free Radical Biology & Medicine*. **13** (5): 557–80. PMID 1334030. doi:10.1016/0891-5849(92)90150-F.

[147] Hiner AN, Raven EL, Thorneley RN, García-Cánovas F, Rodríguez-López JN (Jul 2002). "Mechanisms of compound I formation in heme peroxidases". *Journal of Inorganic Biochemistry*. **91** (1): 27–34. PMID 12121759. doi:10.1016/S0162-0134(02)00390-2.

[148] Mueller S, Riedel HD, Stremmel W (Dec 1997). "Direct evidence for catalase as the predominant H2O2 -removing enzyme in human erythrocytes". *Blood*. **90** (12): 4973–8. PMID 9389716.

[149] Ogata M (Feb 1991). "Acatalasemia". *Human Genetics*. **86** (4): 331–40. PMID 1999334. doi:10.1007/BF00201829.

[150] Parsonage D, Youngblood D, Sarma G, Wood Z, Karplus P, Poole L (2005). "Analysis of the link between enzymatic activity and oligomeric state in AhpC, a bacterial peroxiredoxin". *Biochemistry*. **44** (31): 10583–92. PMC 3832347 ©. PMID 16060667. doi:10.1021/bi050448i. PDB 1YEX

[151] Rhee SG, Chae HZ, Kim K (Jun 2005). "Peroxiredoxins: a historical overview and speculative preview of novel mechanisms and emerging concepts in cell signaling". *Free Radical Biology & Medicine*. **38** (12): 1543–52. PMID 15917183. doi:10.1016/j.freeradbiomed.2005.02.026.

[152] Wood ZA, Schröder E, Robin Harris J, Poole LB (Jan 2003). "Structure, mechanism and regulation of peroxiredoxins". *Trends in Biochemical Sciences*. **28** (1): 32–40. PMID 12517450. doi:10.1016/S0968-0004(02)00003-8.

[153] Claiborne A, Yeh JI, Mallett TC, Luba J, Crane EJ, Charrier V, Parsonage D (Nov 1999). "Protein-sulfenic acids: diverse roles for an unlikely player in enzyme catalysis and redox regulation". *Biochemistry*. **38** (47): 15407–16. PMID 10569923. doi:10.1021/bi992025k.

[154] Jönsson TJ, Lowther WT (2007). "The peroxiredoxin repair proteins". *Sub-Cellular Biochemistry*. Subcellular Biochemistry. **44**: 115–41. ISBN 978-1-4020-6050-2. PMC 2391273 ©. PMID 18084892. doi:10.1007/978-1-4020-6051-9_6.

[155] Neumann CA, Krause DS, Carman CV, Das S, Dubey DP, Abraham JL, Bronson RT, Fujiwara Y, Orkin SH, Van Etten RA (Jul 2003). "Essential role for the peroxiredoxin Prdx1 in erythrocyte antioxidant defence and tumour suppression". *Nature*. **424** (6948): 561–5. Bibcode:2003Natur.424..561N. PMID 12891360. doi:10.1038/nature01819.

[156] Lee TH, Kim SU, Yu SL, Kim SH, Park DS, Moon HB, Dho SH, Kwon KS, Kwon HJ, Han YH, Jeong S, Kang SW, Shin HS, Lee KK, Rhee SG, Yu DY (Jun 2003). "Peroxiredoxin II is essential for sustaining life span of erythrocytes in mice". *Blood*. **101** (12): 5033–8. PMID 12586629. doi:10.1182/blood-2002-08-2548.

[157] Dietz KJ, Jacob S, Oelze ML, Laxa M, Tognetti V, de Miranda SM, Baier M, Finkemeier I (2006). "The function of peroxiredoxins in plant organelle redox metabolism". *Journal of Experimental Botany*. **57** (8): 1697–709. PMID 16606633. doi:10.1093/jxb/erj160.

[158] Nordberg J, Arnér ES (Dec 2001). "Reactive oxygen species, antioxidants, and the mammalian thioredoxin system". *Free Radical Biology & Medicine*. **31** (11): 1287–312. PMID 11728801. doi:10.1016/S0891-5849(01)00724-9.

[159] Vieira Dos Santos C, Rey P (Jul 2006). "Plant thioredoxins are key actors in the oxidative stress response". *Trends in Plant Science*. **11** (7): 329–34. PMID 16782394. doi:10.1016/j.tplants.2006.05.005.

[160] Arnér ES, Holmgren A (Oct 2000). "Physiological functions of thioredoxin and thioredoxin reductase". *European Journal of Biochemistry / FEBS*. **267** (20): 6102–9. PMID 11012661. doi:10.1046/j.1432-1327.2000.01701.x.

[161] Mustacich D, Powis G (Feb 2000). "Thioredoxin reductase". *The Biochemical Journal*. **346** (1): 1–8. PMC 1220815 ©. PMID 10657232. doi:10.1042/0264-6021:3460001.

[162] Creissen G, Broadbent P, Stevens R, Wellburn AR, Mullineaux P (May 1996). "Manipulation of glutathione metabolism in transgenic plants". *Biochemical Society Transactions*. **24** (2): 465–9. PMID 8736785. doi:10.1042/bst0240465.

[163] Brigelius-Flohé R (Nov 1999). "Tissue-specific functions of individual glutathione peroxidases". *Free Radical Biology & Medicine*. **27** (9–10): 951–65. PMID 10569628. doi:10.1016/S0891-5849(99)00173-2.

[164] Ho YS, Magnenat JL, Bronson RT, Cao J, Gargano M, Sugawara M, Funk CD (Jun 1997). "Mice deficient in cellu lar glutathione peroxidase develop normally and show no increased sensitivity to hyperoxia". *The Journal of Biological Chemistry*. **272** (26): 16644–51. PMID 9195979. doi:10.1074/jbc.272.26.16644.

[165] de Haan JB, Bladier C, Griffiths P, Kelner M, O'Shea RD, Cheung NS, Bronson RT, Silvestro MJ, Wild S, Zheng SS, Beart PM, Hertzog PJ, Kola I (Aug 1998). "Mice with a homozygous null mutation for the most abundant glutathione peroxidase, Gpx1, show increased susceptibility to the oxidative stress-inducing agents paraquat and hydrogen peroxide". *The Journal of Biological Chemistry*. **273** (35): 22528–36. PMID 9712879. doi:10.1074/jbc.273.35.22528.

[166] Sharma R, Yang Y, Sharma A, Awasthi S, Awasthi YC (Apr 2004). "Antioxidant role of glutathione S-transferases: protection against oxidant toxicity and regulation of stress-mediated apoptosis". *Antioxidants & Redox Signaling*. **6** (2): 289–300. PMID 15025930. doi:10.1089/152308604322899350.

[167] Hayes JD, Flanagan JU, Jowsey IR (2005). "Glutathione transferases". *Annual Review of Pharmacology and Toxicology*. **45**: 51–88. PMID 15822171. doi:10.1146/annurev.pharmtox.45.120403.095857.

[168] Christen Y (Feb 2000). "Oxidative stress and Alzheimer disease". *The American Journal of Clinical Nutrition*. **71** (2): 621S–629S. PMID 10681270.

[169] Nunomura A, Castellani RJ, Zhu X, Moreira PI, Perry G, Smith MA (Jul 2006). "Involvement of oxidative stress in Alzheimer disease". *Journal of Neuropathology and Experimental Neurology*. **65** (7): 631–41. PMID 16825950. doi:10.1097/01.jnen.0000228136.58062.bf.

[170] Wood-Kaczmar A, Gandhi S, Wood NW (Nov 2006). "Understanding the molecular causes of Parkinson's disease". *Trends in Molecular Medicine*. **12** (11): 521–8. PMID 17027339. doi:10.1016/j.molmed.2006.09.007.

[171] Davì G, Falco A, Patrono C (2005). "Lipid peroxidation in diabetes mellitus". *Antioxidants & Redox Signaling*. **7** (1–2): 256–68. PMID 15650413. doi:10.1089/ars.2005.7.256.

[172] Giugliano D, Ceriello A, Paolisso G (Mar 1996). "Oxidative stress and diabetic vascular complications". *Diabetes Care*. **19** (3): 257–67. PMID 8742574. doi:10.2337/diacare.19.3.257.

[173] Hitchon CA, El-Gabalawy HS (2004). "Oxidation in rheumatoid arthritis". *Arthritis Research & Therapy*. **6** (6): 265–78. PMC 1064874. PMID 15535839. doi:10.1186/ar1447.

[174] Cookson MR, Shaw PJ (Jan 1999). "Oxidative stress and motor neurone disease". *Brain Pathology*. **9** (1): 165–86. PMID 9989458. doi:10.1111/j.1750-3639.1999.tb00217.x.

[175] Van Gaal LF, Mertens IL, De Block CE (Dec 2006). "Mechanisms linking obesity with cardiovascular disease". *Nature*. **444** (7121): 875–80. Bibcode:2006Natur.444..875V. PMID 17167476. doi:10.1038/nature05487.

[176] Aviram M (Nov 2000). "Review of human studies on oxidative damage and antioxidant protection related to cardiovascular diseases". *Free Radical Research*. 33 Suppl: S85–97. PMID 11191279.

[177] Khan MA, Tania M, Zhang D, Chen H (2010). "Antioxidant enzymes and cancer". *Chin J Cancer Res*. **22** (2): 87–92. doi:10.1007/s11670-010-0087-7.

[178] López-Lluch G, Hunt N, Jones B, Zhu M, Jamieson H, Hilmer S, Cascajo MV, Allard J, Ingram DK, Navas P, de Cabo R (Feb 2006). "Calorie restriction induces mitochondrial biogenesis and bioenergetic efficiency". *Proceedings of the National Academy of Sciences of the United States of America*. **103** (6): 1768–1773. Bibcode:2006PNAS..103.1768L. PMC 1413655. PMID 16446459. doi:10.1073/pnas.0510452103.

[179] Larsen PL (Oct 1993). "Aging and resistance to oxidative damage in Caenorhabditis elegans". *Proceedings of the National Academy of Sciences of the United States of America*. **90** (19): 8905–9. Bibcode:1993PNAS...90.8905L. PMC 47469. PMID 8415630. doi:10.1073/pnas.90.19.8905.

[180] Helfand SL, Rogina B (2003). "Genetics of aging in the fruit fly, Drosophila melanogaster". *Annual Review of Genetics*. **37**: 329–48. PMID 14616064. doi:10.1146/annurev.genet.37.040103.095211.

[181] Sohal RS, Mockett RJ, Orr WC (Sep 2002). "Mechanisms of aging: an appraisal of the oxidative stress hypothesis". *Free Radical Biology & Medicine*. **33** (5): 575–86. PMID 12208343. doi:10.1016/S0891-5849(02)00886-9.

[182] Sohal RS (Jul 2002). "Role of oxidative stress and protein oxidation in the aging process". *Free Radical Biology & Medicine*. **33** (1): 37–44. PMID 12086680. doi:10.1016/S0891-5849(02)00856-0.

[183] Rattan SI (Dec 2006). "Theories of biological aging: genes, proteins, and free radicals". *Free Radical Research*. **40** (12): 1230–8. PMID 17090411. doi:10.1080/10715760600911303.

[184] Pérez VI, Bokov A, Van Remmen H, Mele J, Ran Q, Ikeno Y, Richardson A (Oct 2009). "Is the oxidative stress theory of aging dead?". *Biochimica et Biophysica Acta*. **1790**

(10): 1005–1014. PMC 2789432 . PMID 19524016. doi:10.1016/j.bbagen.2009.06.003.

[185] Thomas DR (May 2004). "Vitamins in health and aging". *Clinics in Geriatric Medicine*. **20** (2): 259–74. PMID 15182881. doi:10.1016/j.cger.2004.02.001.

[186] Ward JA (Mar 1998). "Should antioxidant vitamins be routinely recommended for older people?". *Drugs & Aging*. **12** (3): 169–75. PMID 9534018. doi:10.2165/00002512-199812030-00001.

[187] Kader AA, Zagory D, Kerbel EL (1989). "Modified atmosphere packaging of fruits and vegetables". *Critical Reviews in Food Science and Nutrition*. **28** (1): 1–30. PMID 2647417. doi:10.1080/10408398909527490.

[188] Zallen EM, Hitchcock MJ, Goertz GE (Dec 1975). "Chilled food systems. Effects of chilled holding on quality of beef loaves". *Journal of the American Dietetic Association*. **67** (6): 552–7. PMID 1184900.

[189] Iverson F (Jun 1995). "Phenolic antioxidants: Health Protection Branch studies on butylated hydroxyanisole". *Cancer Letters*. **93** (1): 49–54. PMID 7600543. doi:10.1016/0304-3835(95)03787-W.

[190] "E number index". UK food guide. Archived from the original on 4 March 2007. Retrieved 5 March 2007.

[191] Robards K, Kerr AF, Patsalides E (Feb 1988). "Rancidity and its measurement in edible oils and snack foods. A review". *The Analyst*. **113** (2): 213–24. Bibcode:1988Ana...113..213R. PMID 3288002. doi:10.1039/an9881300213.

[192] Del Carlo M, Sacchetti G, Di Mattia C, Compagnone D, Mastrocola D, Liberatore L, Cichelli A (Jun 2004). "Contribution of the phenolic fraction to the antioxidant activity and oxidative stability of olive oil". *Journal of Agricultural and Food Chemistry*. **52** (13): 4072–9. PMID 15212450. doi:10.1021/jf049806z.

[193] Boozer CE, Hammond GS, Hamilton CE, Sen JN (1955). "Air Oxidation of Hydrocarbons.1II. The Stoichiometry and Fate of Inhibitors in Benzene and Chlorobenzene". *Journal of the American Chemical Society*. **77** (12): 3233–7. Bibcode:1955JAChS..77.1678G. doi:10.1021/ja01617a026.

[194] "Global Antioxidants (Natural and Synthetic) Market Poised to Surge From USD 2.25 Billion in 2014 to USD 3.25 Billion by 2020, Growing at 5.5% CAGR". GlobalNewswire, El Segundo, CA. 19 January 2016. Retrieved 30 January 2017.

[195] "Why use Antioxidants?". SpecialChem Adhesives. Archived from the original on 11 February 2007. Retrieved 27 February 2007.

[196] "Fuel antioxidants". Innospec Chemicals. Archived from the original on 15 October 2006. Retrieved 27 February 2007.

[197] Rodriguez-Amaya DB (2003). "Food carotenoids: analysis, composition and alterations during storage and processing of foods". *Forum of Nutrition*. **56**: 35–7. PMID 15806788.

[198] Maiani G, Castón MJ, Catasta G, Toti E, Cambrodón IG, Bysted A, Granado-Lorencio F, Olmedilla-Alonso B, Knuthsen P, Valoti M, Böhm V, Mayer-Miebach E, Behsnilian D, Schlemmer U (Sep 2009). "Carotenoids: actual knowledge on food sources, intakes, stability and bioavailability and their protective role in humans". *Molecular Nutrition & Food Research*. 53 Suppl 2: S194–218. PMID 19035552. doi:10.1002/mnfr.200800053.

[199] Henry CJ, Heppell N (Feb 2002). "Nutritional losses and gains during processing: future problems and issues". *The Proceedings of the Nutrition Society*. **61** (1): 145–8. PMID 12002789. doi:10.1079/PNS2001142.

[200] "Antioxidants and Cancer Prevention: Fact Sheet". National Cancer Institute. Archived from the original on 4 March 2007. Retrieved 27 February 2007.

[201] Ortega R (Dec 2006). "Importance of functional foods in the Mediterranean diet". *Public Health Nutrition*. **9** (8A): 1136–40. PMID 17378953. doi:10.1017/S1368980007668530.

[202] Witschi A, Reddy S, Stofer B, Lauterburg BH (1992). "The systemic availability of oral glutathione". *European Journal of Clinical Pharmacology*. **43** (6): 667–9. PMID 1362956. doi:10.1007/BF02284971.

[203] Flagg EW, Coates RJ, Eley JW, Jones DP, Gunter EW, Byers TE, Block GS, Greenberg RS (1994). "Dietary glutathione intake in humans and the relationship between intake and plasma total glutathione level". *Nutrition and Cancer*. **21** (1): 33–46. PMID 8183721. doi:10.1080/01635589409514302.

[204] Dodd S, Dean O, Copolov DL, Malhi GS, Berk M (Dec 2008). "N-acetylcysteine for antioxidant therapy: pharmacology and clinical utility". *Expert Opinion on Biological Therapy*. **8** (12): 1955–62. PMID 18990082. doi:10.1517/14728220802517901.

[205] van de Poll MC, Dejong CH, Soeters PB (Jun 2006). "Adequate range for sulfur-containing amino acids and biomarkers for their excess: lessons from enteral and parenteral nutrition". *The Journal of Nutrition*. **136** (6 Suppl): 1694S–1700S. PMID 16702341.

[206] Cao G, Alessio HM, Cutler RG (Mar 1993). "Oxygen-radical absorbance capacity assay for antioxidants". *Free Radical Biology & Medicine*. **14** (3): 303–11. PMID 8458588. doi:10.1016/0891-5849(93)90027-R.

[207] Ou B, Hampsch-Woodill M, Prior RL (Oct 2001). "Development and validation of an improved oxygen radical absorbance capacity assay using fluorescein as the fluorescent probe". *Journal of Agricultural and Food Chemistry*. **49** (10): 4619–26. PMID 11599998. doi:10.1021/jf010586o.

[208] "Withdrawn: Oxygen Radical Absorbance Capacity (ORAC) of Selected Foods, Release 2 (2010)". United States Department of Agriculture, Agricultural Research Service. 16 May 2012. Retrieved 13 June 2012.

[209] Prior RL, Wu X, Schaich K (May 2005). "Standardized methods for the determination of antioxidant capacity and phenolics in foods and dietary supplements". *Journal of Agricultural and Food Chemistry*. **53** (10): 4290–302. PMID 15884874. doi:10.1021/jf0502698.

[210] Benzie IF (Sep 2003). "Evolution of dietary antioxidants". *Comparative Biochemistry and Physiology A*. **136** (1): 113–26. PMID 14527634. doi:10.1016/S1095-6433(02)00368-9.

[211] Mattill HA (1947). "Antioxidants". *Annual Review of Biochemistry*. **16**: 177–92. PMID 20259061. doi:10.1146/annurev.bi.16.070147.001141.

[212] German JB (1999). "Food processing and lipid oxidation". *Advances in Experimental Medicine and Biology*. Advances in Experimental Medicine and Biology. **459**: 23–50. ISBN 978-0-306-46051-7. PMID 10335367. doi:10.1007/978-1-4615-4853-9_3.

[213] Jacob RA (1996). "Three eras of vitamin C discovery". *Sub-Cellular Biochemistry*. Subcellular Biochemistry. **25**: 1–16. ISBN 978-1-4613-7998-0. PMID 8821966. doi:10.1007/978-1-4613-0325-1_1.

[214] Knight JA (1998). "Free radicals: their history and current status in aging and disease". *Annals of Clinical and Laboratory Science*. **28** (6): 331–46. PMID 9846200.

[215] Moureu C, Dufraisse C (1922). "Sur l'autoxydation: Les antioxygènes". *Comptes Rendus des Séances et Mémoires de la Société de Biologie* (in French). **86**: 321–322.

[216] Wolf G (Mar 2005). "The discovery of the antioxidant function of vitamin E: the contribution of Henry A. Mattill". *The Journal of Nutrition*. **135** (3): 363–6. PMID 15735064.

18.12 Further reading

- Nick Lane *Oxygen: The Molecule That Made the World* (Oxford University Press, 2003) ISBN 0-19-860783-0
- Barry Halliwell and John M.C. Gutteridge *Free Radicals in Biology and Medicine* (Oxford University Press, 2007) ISBN 0-19-856869-X
- Jan Pokorny, Nelly Yanishlieva and Michael H. Gordon *Antioxidants in Food: Practical Applications* (CRC Press Inc, 2001) ISBN 0-8493-1222-1

18.13 External links

Chapter 19

Biodegradation

Yellow slime mold growing on a bin of wet paper

IUPAC definition

Degradation caused by enzymatic process resulting from the action of cells.

Note: Modified to exclude *abiotic enzymatic* processes.[1]

Biodegradation is the disintegration of materials by bacteria, fungi, or other biological means.

The term is often used in relation to: biomedicine, waste management, ecology, and the bioremediation of the natural environment. It is now commonly associated with environmentally-friendly products, capable of decomposing back into natural elements.

Although often conflated, biodegradable is distinct in meaning from: compostable. While biodegradable simply means *can be consumed by microorganisms*, compostable makes the specific demand that the object break down under composting conditions.

Organic material can be degraded: aerobically (with oxygen) or anaerobically (without oxygen). Decomposition of biodegradable substances may include both biological and abiotic steps.

Biodegradable matter is generally organic material that provides a nutrient for microorganisms. These are so numerous and diverse that a huge range of compounds can be biodegraded, including: hydrocarbons (oils), polycyclic aromatic hydrocarbons (PAHs), polychlorinated biphenyls (PCBs) and pharmaceutical substances. Microorganisms secrete biosurfactant, an extracellular surfactant, to enhance this process.

19.1 Factors affecting rate

In practice, almost all chemical compounds and materials are subject to biodegradation, the key is the relative rates of such processes - minutes, days, years, centuries... A number of factors determine the degradation rate of organic compounds.[2] Salient factors include light, water and oxygen. Temperature is also important because chemical reactions proceed more quickly at higher temperatures. The degradation rate of many organic compounds is limited by their bioavailability. Compounds must be released into solution before organisms can degrade them.[3]

Biodegradability can be measured in a number of ways. Respirometry tests can be used for aerobic microbes. First one places a solid waste sample in a container with microorganisms and soil, and then aerate the mixture. Over the course of several days, microorganisms digest the sample bit by bit and produce carbon dioxide – the resulting amount of CO_2 serves as an indicator of degradation. Biodegradability can also be measured by anaerobic microbes and the amount of methane or alloy that they are able to produce. In formal scientific literature, the process is termed bio-remediation.[4]

19.2 Detergents

In advanced societies, laundry detergents are based on *linear* alkylbenzenesulfonates. Branched alkybenzenesulfonates (below right), used in former times, were abandoned because they biodegrade too slowly.[6]

- 4-(5-Dodecyl) benzenesulfonate, a linear dodecylbenzenesulfonate
- A branched dodecylbenzenesulfonate, which has been phased out in developed countries.

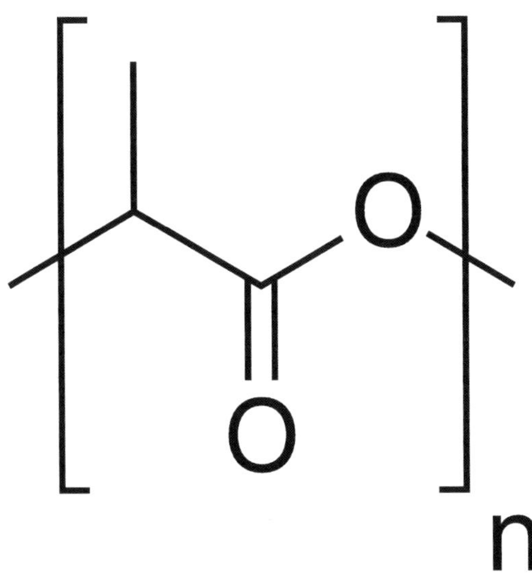

Polylactic acid is an example of a plastic that biodegrades quickly.

19.3 Plastics

Main article: Biodegradable plastic § Examples of biodegradable plastics

Plastics biodegrade at highly variable rates. PVC-based plumbing is specifically selected for handing sewage because PVC biodegrades very slowly. Some packaging materials on the other hand are being developed that would degrade readily upon exposure to the environment.[7] Illustrative synthetic polymers that are biodegrade quickly include polycaprolactone, others are polyesters and aromatic-aliphatic esters, due to their ester bonds being susceptible to attack by water. A prominent example is poly-3-hydroxybutyrate, the renewably derived polylactic acid, and the synthetic polycaprolactone. Others are the cellulose-based cellulose acetate and celluloid (cellulose nitrate).

Under low oxygen conditions biodegradable plastics break down slower and with the production of methane, like other organic materials do. The breakdown process is accelerated in a dedicated compost heap. Starch-based plastics will degrade within two to four months in a home compost bin, while polylactic acid is largely undecomposed, requiring higher temperatures.[8] Polycaprolactone and polycaprolactone-starch composites decompose slower, but the starch content accelerates decomposition by leaving behind a porous, high surface area polycaprolactone. Nevertheless, it takes many months.[9] In 2016, a bacterium named Ideonella sakaiensis was found to biodegrade PET.

Many plastic producers have gone so far even to say that their plastics are compostable, typically listing corn starch as an ingredient. However, these claims are questionable because the plastics industry operates under its own definition of compostable:

> "that which is capable of undergoing biological decomposition in a compost site such that the material is not visually distinguishable and breaks down into carbon dioxide, water, inorganic compounds and biomass at a rate consistent with known compostable materials." (Ref: ASTM D 6002)[10]

The term "composting" is often used informally to describe the biodegradation of packaging materials. Legal definitions exist for compostability, the process that leads to compost. Four criteria are offered by the European Union:[11][12]

- Biodegradability, the conversion of >90% material material into CO2 and water by the action of microorganisms within 6 months.
- Disintegrability, the fragmentation of 90% of the original mass to particles that then pass through a 2 mm sieve.
- Absence of toxic substances and other substances that impede composting.

19.4 Biodegradable technology

In 1973 it was proven for the first time that polyester degrades when disposed in bioactive material such as soil. Polyesters are water resistant and can be melted and shaped into sheets, bottles, and other products, making certain plastics now available as a biodegradable product. Following, Polyhydroxylalkanoates (PHAs) were produced directly from renewable resources by microbes. They are approximately 95% cellular bacteria and can be manipulated by genetic strategies. The composition and biodegradability of PHAs can be regulated by blending it with other natural polymers. In the 1980s the company ICI Zenecca commercialized PHAs under the name Biopol. It was used for the production of shampoo bottles and other cosmetic products. Consumer response was unusual. Consumers were willing to pay more for this product because it was natural and biodegradable, which had not occurred before.[13]

Now biodegradable technology is a highly developed market with applications in product packaging, production and medicine. Biodegradable technology is concerned with the manufacturing science of biodegradable materials. It imposes science-based mechanisms of plant genetics into the processes of today. Scientists and manufacturing corporations can help impact climate change by developing a use of plant genetics that would mimic some technologies. By looking to plants, such as biodegradable material harvested through photosynthesis, waste and toxins can be minimized.[14]

Oxo-biodegradable technology, which has further developed biodegradable plastics, has also emerged. Oxo-biodegradation is defined by CEN (the European Standards Organisation) as "degradation resulting from oxidative and cell-mediated phenomena, either simultaneously or successively." Whilst sometimes described as "oxo-fragmentable," and "oxo-degradable" this describes only the first or oxidative phase. These descriptions should not be used for material which degrades by the process of oxo-biodegradation defined by CEN, and the correct description is "oxo-biodegradable."

By combining plastic products with very large polymer molecules, which contain only carbon and hydrogen, with oxygen in the air, the product is rendered capable of decomposing in anywhere from a week to one to two years. This reaction occurs even without prodegradant additives but at a very slow rate. That is why conventional plastics, when discarded, persist for a long time in the environment. Oxo-biodegradable formulations catalyze and accelerate the biodegradation process but it takes considerable skill and experience to balance the ingredients within the formulations so as to provide the product with a useful life for a set period, followed by degradation and biodegradation.[15]

Biodegradable technology is especially utilized by the biomedical community. Biodegradable polymers are classified into three groups: medical, ecological, and dual application, while in terms of origin they are divided into two groups: natural and synthetic.[16] The Clean Technology Group is exploiting the use of supercritical carbon dioxide, which under high pressure at room temperature is a solvent that can use biodegradable plastics to make polymer drug coatings. The polymer (meaning a material composed of molecules with repeating structural units that form a long chain) is used to encapsulate a drug prior to injection in the body and is based on lactic acid, a compound normally produced in the body, and is thus able to be excreted naturally. The coating is designed for controlled release over a period of time, reducing the number of injections required and maximizing the therapeutic benefit. Professor Steve Howdle states that biodegradable polymers are particularly attractive for use in drug delivery, as once introduced into the body they require no retrieval or further manipulation and are degraded into soluble, non-toxic by-products. Different polymers degrade at different rates within the body and therefore polymer selection can be tailored to achieve desired release rates.[17]

Other biomedical applications include the use of biodegradable, elastic shape-memory polymers. Biodegradable implant materials can now be used for minimally invasive surgical procedures through degradable thermoplastic polymers. These polymers are now able to change their shape with increase of temperature, causing shape memory capabilities as well as easily degradable sutures. As a result, implants can now fit through small incisions, doctors can easily perform complex deformations, and sutures and other material aides can naturally biodegrade after a completed surgery.[18]

19.5 Etymology of "biodegradable"

The first known use of the word in biological text was in 1961 when employed to describe the breakdown of material into the base components of carbon, hydrogen, and oxygen by microorganisms. Now biodegradable is commonly associated with environmentally friendly products that are part of the earth's innate cycle and capable of decomposing back into natural elements.

19.6 See also

- Anaerobic digestion
- Biodegradability prediction
- Biodegradable electronics

- Biodegradable polythene film
- Biodegradation (journal)
- Bioplastic – biodegradable, bio-based plastics
- Bioremediation
- Decomposition – reduction of the body of a formerly living organism into simpler forms of matter
- Landfill gas monitoring
- List of environment topics
- Microbial biodegradation
- Photodegradation

19.7 References

[1] "Terminology for biorelated polymers and applications (IUPAC Recommendations 2012)" (PDF). *Pure and Applied Chemistry*. **84** (2): 377–410. 2012. doi:10.1351/PAC-REC-10-12-04.

[2] Sims, G. K. and A.M. Cupples. 1999. Factors controlling degradation of pesticides in soil. Pesticide Science 55:598–601.

[3] Sims, G.K. (1991). *The effects of sorption on the bioavailability of pesticides*. London: Springer Verlag. pp. 119–137.

[4] "Measuring Biodegradability", *The University of Waikato*, June 19, 2008

[5] "Marine Debris Biodegradation Time Line". C-MORE, citing Mote Marine Laboratory, 1993.

[6] Kurt Kosswig,"Surfactants" in Ullmann's Encyclopedia of Industrial Chemistry, Wiley-VCH, 2005, Weinheim. doi:10.1002/14356007.a25_747

[7] Kyrikou, Ioanna; Briassoulis, Demetres (12 Apr 2007). "Biodegradation of Agricultural Plastic Films: A Critical Review". *Journal of Polymers and the Environment*. SpringerLink . **15** (2): 125–150. doi:10.1007/s10924-007-0053-8. Retrieved 30 May 2015.

[8] "Microsoft Word - SECTION 6 BIODEGRADABILITY OF PACKAGING WASTE.doc" (PDF). Www3.imperial.ac.uk. Retrieved 2014-03-02.

[9] Fig.9

[10] "Compostable.info". Compostable.info. Retrieved 2014-03-02.

[11] http://greenplastics.com/wiki/EN_13432

[12] M. Breulmann et al. "Polymers, Biodegradable" in Ullmann's Encyclopedia of Industrial Chemistry 2012 Wiley-VCH, Weinheim.doi:10.1002/14356007.n21_n01

[13] Gross,Richard. "Biodegradable Polymers for the Environment", American Association of Advanced Science, August 2, 2002, p. 804.

[14] Luzier, W. D. "Materials Derived from Biomass/Biodegradable Materials." Proceedings of the National Academy of Sciences 89.3 (1992): 839–42. Print.

[15] Agamuthu, P."Biodegradability and Degradability of Plastic Waste", "International Solid Waste Association" November 9, 2004

[16] Yoshito, Ikada. "Biodegradable Polyesters for Medical and Ecological Applications", "Massachusetts Institute of Technology", 2000. p117

[17] "Using Green Chemistry to Deliver Cutting Edge Drugs". The University of Nottingham. September 13, 2007.

[18] Lendlein, Andreas. "Biodegradable, Elastic Shape-Memory Polymers for Potential Biomedical Applications". American Association of Advancement of Science, 2002, p 1673.

19.7.1 Standards by ASTM International

- D5210- Standard Test Method for Determining the Anaerobic Biodegradation of Plastic Materials in the Presence of Municipal Sewage Sludge
- D5526- Standard Test Method for Determining Anaerobic Biodegradation of Plastic Materials Under Accelerated Landfill Conditions
- D5338- Standard Test Method for Determining Aerobic Biodegradation of Plastic Materials Under Controlled Composting Conditions, Incorporating Thermophilic Temperatures
- D5511- Standard Test Method for Determining Anaerobic Biodegradation of Plastic Materials Under High-Solids Anaerobic-Digestion Conditions
- D5864- Standard Test Method for Determining Aerobic Aquatic Biodegradation of Lubricants or Their Components
- D5988- Standard Test Method for Determining Aerobic Biodegradation of Plastic Materials in Soil
- D6139- Standard Test Method for Determining the Aerobic Aquatic Biodegradation of Lubricants or Their Components Using the Gledhill Shake Flask
- D6006- Standard Guide for Assessing Biodegradability of Hydraulic Fluids
- D6340- Standard Test Methods for Determining Aerobic Biodegradation of Radiolabeled Plastic Materials in an Aqueous or Compost Environment

- D6691- Standard Test Method for Determining Aerobic Biodegradation of Plastic Materials in the Marine Environment by a Defined Microbial Consortium or Natural Sea Water Inoculum
- D6731-Standard Test Method for Determining the Aerobic, Aquatic Biodegradability of Lubricants or Lubricant Components in a Closed Respirometer
- D6954- Standard Guide for Exposing and Testing Plastics that Degrade in the Environment by a Combination of Oxidation and Biodegradation
- D7044- Standard Specification for Biodegradable Fire Resistant Hydraulic Fluids
- D7373-Standard Test Method for Predicting Biodegradability of Lubricants Using a Bio-kinetic Model
- D7475- Standard Test Method for Determining the Aerobic Degradation and Anaerobic Biodegradation of Plastic Materials under Accelerated Bioreactor Landfill Conditions
- D7665- Standard Guide for Evaluation of Biodegradable Heat Transfer Fluids

19.8 External links

- European Bioplastics Association
- The Science of Biodegradable Plastics: The Reality Behind Biodegradable Plastic Packaging Material
- Biodegradable Polyesters for Medical and Ecological Applications
- Biodegradable Plastic Definition

Chapter 20

Microbial biodegradation

Microbial biodegradation is the use of bioremediation and biotransformation methods to harness the naturally occurring ability of microbial xenobiotic metabolism to degrade, transform or accumulate environmental pollutants, including hydrocarbons (e.g. oil), polychlorinated biphenyls (PCBs), polyaromatic hydrocarbons (PAHs), heterocyclic compounds (such as pyridine or quinoline), pharmaceutical substances, radionuclides and metals.

Interest in the microbial biodegradation of pollutants has intensified in recent years,[1][2] and recent major methodological breakthroughs have enabled detailed genomic, metagenomic, proteomic, bioinformatic and other high-throughput analyses of environmentally relevant microorganisms, providing new insights into biodegradative pathways and the ability of organisms to adapt to changing environmental conditions.

Biological processes play a major role in the removal of contaminants and take advantage of the catabolic versatility of microorganisms to degrade or convert such compounds. In environmental microbiology, genome-based global studies are increasing the understanding of metabolic and regulatory networks, as well as providing new information on the evolution of degradation pathways and molecular adaptation strategies to changing environmental conditions.

20.1 Aerobic biodegradation of pollutants

The increasing amount of bacterial genomic data provides new opportunities for understanding the genetic and molecular bases of the degradation of organic pollutants. Aromatic compounds are among the most persistent of these pollutants and lessons can be learned from the recent genomic studies of *Burkholderia xenovorans* LB400 and *Rhodococcus* sp. strain RHA1, two of the largest bacterial genomes completely sequenced to date. These studies have helped expand our understanding of bacterial catabolism, non-catabolic physiological adaptation to organic compounds, and the evolution of large bacterial genomes. First, the metabolic pathways from phylogenetically diverse isolates are very similar with respect to overall organization. Thus, as originally noted in pseudomonads, a large number of "peripheral aromatic" pathways funnel a range of natural and xenobiotic compounds into a restricted number of "central aromatic" pathways. Nevertheless, these pathways are genetically organized in genus-specific fashions, as exemplified by the b-ketoadipate and Paa pathways. Comparative genomic studies further reveal that some pathways are more widespread than initially thought. Thus, the Box and Paa pathways illustrate the prevalence of non-oxygenolytic ring-cleavage strategies in aerobic aromatic degradation processes. Functional genomic studies have been useful in establishing that even organisms harboring high numbers of homologous enzymes seem to contain few examples of true redundancy. For example, the multiplicity of ring-cleaving dioxygenases in certain rhodococcal isolates may be attributed to the cryptic aromatic catabolism of different terpenoids and steroids. Finally, analyses have indicated that recent genetic flux appears to have played a more significant role in the evolution of some large genomes, such as LB400's, than others. However, the emerging trend is that the large gene repertoires of potent pollutant degraders such as LB400 and RHA1 have evolved principally through more ancient processes. That this is true in such phylogenetically diverse species is remarkable and further suggests the ancient origin of this catabolic capacity.[3]

20.2 Anaerobic biodegradation of pollutants

Anaerobic microbial mineralization of recalcitrant organic pollutants is of great environmental significance and involves intriguing novel biochemical reactions. In particular, hydrocarbons and halogenated compounds have long been doubted to be degradable in the absence of oxygen, but the isolation of hitherto unknown anaerobic hydrocarbon-degrading and reductively dehalogenating bacteria during the last decades provided ultimate proof for these processes in nature. While such research involved mostly chlorinated compounds initially, recent studies have revealed reductive dehalogenation of bromine and iodine moieties in aromatic pesticides.[4] Other reactions, such as biologically induced abiotic reduction by soil minerals,[5] has been shown to deactivate relatively persistent aniline-based herbicides far more rapidly than observed in aerobic environments. Many novel biochemical reactions were discovered enabling the respective metabolic pathways, but progress in the molecular understanding of these bacteria was rather slow, since genetic systems are not readily applicable for most of them. However, with the increasing application of genomics in the field of environmental microbiology, a new and promising perspective is now at hand to obtain molecular insights into these new metabolic properties. Several complete genome sequences were determined during the last few years from bacteria capable of anaerobic organic pollutant degradation. The ~4.7 Mb genome of the facultative denitrifying *Aromatoleum aromaticum* strain EbN1 was the first to be determined for an anaerobic hydrocarbon degrader (using toluene or ethylbenzene as substrates). The genome sequence revealed about two dozen gene clusters (including several paralogs) coding for a complex catabolic network for anaerobic and aerobic degradation of aromatic compounds. The genome sequence forms the basis for current detailed studies on regulation of pathways and enzyme structures. Further genomes of anaerobic hydrocarbon degrading bacteria were recently completed for the iron-reducing species *Geobacter metallireducens* (accession nr. NC_007517) and the perchlorate-reducing *Dechloromonas aromatica* (accession nr. NC_007298), but these are not yet evaluated in formal publications. Complete genomes were also determined for bacteria capable of anaerobic degradation of halogenated hydrocarbons by halorespiration: the ~1.4 Mb genomes of *Dehalococcoides ethenogenes* strain 195 and *Dehalococcoides* sp. strain CBDB1 and the ~5.7 Mb genome of *Desulfitobacterium hafniense* strain Y51. Characteristic for all these bacteria is the presence of multiple paralogous genes for reductive dehalogenases, implicating a wider dehalogenating spectrum of the organisms than previously known. Moreover, genome sequences provided unprecedented insights into the evolution of reductive dehalogenation and differing strategies for niche adaptation.[6]

Recently, it has become apparent that some organisms, including *Desulfitobacterium chlororespirans*, originally evaluated for halorespiration on chlorophenols, can also use certain brominated compounds, such as the herbicide bromoxynil and its major metabolite as electron acceptors for growth. Iodinated compounds may be dehalogenated as well, though the process may not satisfy the need for an electron acceptor.[4]

20.3 Bioavailability, chemotaxis, and transport of pollutants

Bioavailability, or the amount of a substance that is physiochemically accessible to microorganisms is a key factor in the efficient biodegradation of pollutants. O'Loughlin *et al.* (2000)[7] showed that, with the exception of kaolinite clay, most soil clays and cation exchange resins attenuated biodegradation of 2-picoline by *Arthrobacter* sp. strain R1, as a result of adsorption of the substrate to the clays. Chemotaxis, or the directed movement of motile organisms towards or away from chemicals in the environment is an important physiological response that may contribute to effective catabolism of molecules in the environment. In addition, mechanisms for the intracellular accumulation of aromatic molecules via various transport mechanisms are also important.[8]

20.4 Oil biodegradation

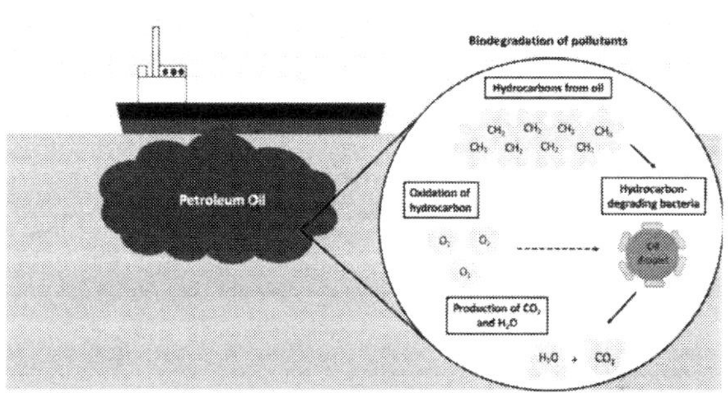

General overview of microbial biodegradation of petroleum oil by microbial communities. Some microorganisms, such as A. borkumensis, *are able to use hydrocarbons as their source for carbon in metabolism. They are able to oxidize the environmentally harmful hydrocarbons while producing harmless products, following the general equation* $CnHn + O_2 \rightarrow H_2O + CO_2$. *In the figure, carbon is represented as yellow circles, oxygen as pink circles, and hydrogen as blue circles. This type of special metabolism allows these microbes to thrive in areas affected by oil spills and are important in the elimination of environmental pollutants.*

Petroleum oil contains aromatic compounds that are toxic to most life forms. Episodic and chronic pollution of the environment by oil causes major disruption to the local ecological environment. Marine environments in particular are especially vulnerable, as oil spills near coastal regions and in the open sea are difficult to contain and make mitigation efforts more complicated. In addition to pollution through human activities, approximately 250 million litres of petroleum enter the marine environment every year from natural seepages.[9] Despite its toxicity, a considerable fraction of petroleum oil entering marine systems is eliminated by the hydrocarbon-degrading activities of microbial communities, in particular by a recently discovered group of specialists, the hydrocarbonoclastic bacteria (HCB).[10] *Alcanivorax borkumensis* was the first HCB to have its genome sequenced.[11] In addition to hydrocarbons, crude oil often contains various heterocyclic compounds, such as pyridine, which appear to be degraded by similar mechanisms to hydrocarbons.[12]

20.5 Cholesterol biodegradation

Many synthetic steroidic compounds like some sexual hormones frequently appear in municipal and industrial wastewaters, acting as environmental pollutants with strong metabolic activities negatively affecting the ecosystems. Since these compounds are common carbon sources for many different microorganisms their aerobic and anaerobic mineralization has been extensively studied. The interest of these studies lies on the biotechnological applications of sterol transforming enzymes for the industrial synthesis of sexual hormones and corticoids. Very recently, the catabolism of cholesterol has acquired a high relevance because it is involved in the infectivity of the pathogen *Mycobacterium tuberculosis* (*Mtb*).[11][13] *Mtb* causes tuberculosis disease, and it has been demonstrated that novel enzyme architectures have evolved to bind and modify steroid compounds like cholesterol in this organism and other steroid-utilizing bacteria as well.[14][15] These new enzymes might be of interest for their potential in the chemical modification of steroid substrates.

20.6 Analysis of waste biotreatment

Sustainable development requires the promotion of environmental management and a constant search for new technologies to treat vast quantities of wastes generated by increasing anthropogenic activities. Biotreatment, the processing of wastes using living organisms, is an environmentally friendly, relatively simple and cost-effective alternative to physico-chemical clean-up options. Confined environments, such as bioreactors, have been engineered to overcome the physical, chemical and biological limiting factors of biotreatment processes in highly controlled systems. The great versatility in the design of confined environments allows the treatment of a wide range of wastes under optimized conditions. To perform a correct assessment, it is necessary to consider various microorganisms having a variety of genomes and expressed transcripts and proteins. A great number of analyses are often required. Using traditional genomic techniques, such assessments are limited and time-consuming. However, several high-throughput techniques originally developed for medical studies can be applied to assess biotreatment in confined environments.[16]

20.7 Metabolic engineering and biocatalytic applications

The study of the fate of persistent organic chemicals in the environment has revealed a large reservoir of enzymatic reactions with a large potential in preparative organic synthesis, which has already been exploited for a number of oxygenases on pilot and even on industrial scale. Novel catalysts can be obtained from metagenomic libraries and DNA sequence based approaches. Our increasing capabilities in adapting the catalysts to specific reactions and process requirements by rational and random mutagenesis broadens the scope for application in the fine chemical industry, but also in the field of biodegradation. In many cases, these catalysts need to be exploited in whole cell bioconversions or in fermentations, calling for system-wide approaches to understanding strain physiology and metabolism and rational approaches to the engineering of whole cells as they are increasingly put forward in the area of systems biotechnology and synthetic biology.[17]

20.8 Fungal biodegradation

In the ecosystem, different substrates are attacked at different rates by consortia of organisms from different kingdoms. *Aspergillus* and other moulds play an important role in these consortia because they are adept at recycling starches, hemicelluloses, celluloses, pectins and other sugar polymers. Some aspergilli are capable of degrading more refractory compounds such as fats, oils, chitin, and keratin. Maximum decomposition occurs when there is sufficient nitrogen, phosphorus and other essential inorganic nutrients. Fungi also provide food for many soil organisms.[18]

For *Aspergillus* the process of degradation is the means of obtaining nutrients. When these moulds degrade human-made substrates, the process usually is called biodeterioration. Both paper and textiles (cotton, jute, and linen) are particularly vulnerable to *Aspergillus* degradation. Our artistic heritage is also subject to *Aspergillus* assault. To give but one example, after Florence in Italy flooded in 1969, 74% of the isolates from a damaged Ghirlandaio fresco in the Ognissanti church were *Aspergillus versicolor*.[19]

20.9 See also

- Biodegradation
- Bioremediation
- Biotransformation
- Bioavailability
- Chemotaxis
- Microbiology
- Environmental microbiology
- Industrial microbiology

20.10 References

[1] Koukkou, Anna-Irini, ed. (2011). *Microbial Bioremediation of Non-metals: Current Research*. Caister Academic Press. ISBN 978-1-904455-83-7.

[2] Díaz, Eduardo, ed. (2008). *Microbial Biodegradation: Genomics and Molecular Biology* (1st ed.). Caister Academic Press. ISBN 978-1-904455-17-2.

[3] McLeod MP & Eltis LD (2008). "Genomic Insights Into the Aerobic Pathways for Degradation of Organic Pollutants". *Microbial Biodegradation: Genomics and Molecular Biology*. Caister Academic Press. ISBN 978-1-904455-17-2.

[4] Cupples, A. M., R. A. Sanford, and G. K. Sims. 2005. Dehalogenation of Bromoxynil (3,5-Dibromo-4-Hydroxybenzonitrile) and Ioxynil (3,5-Diiodino-4-Hydroxybenzonitrile) by Desulfitobacterium chlororespirans. Appl. Env. Micro. 71(7):3741-3746.

[5] Tor, J., C. Xu, J. M. Stucki, M. Wander, G. K. Sims. 2000. Trifluralin degradation under micro-biologically induced nitrate and Fe(III) reducing conditions. Env. Sci. Tech. 34:3148-3152.

[6] Heider J & Rabus R (2008). "Genomic Insights in the Anaerobic Biodegradation of Organic Pollutants". *Microbial Biodegradation: Genomics and Molecular Biology*. Caister Academic Press. ISBN 978-1-904455-17-2.

[7] O'Loughlin, E. J; Traina, S. J.; Sims, G. K. (2000). "Effects of sorption on the biodegradation of 2-methylpyridine in aqueous suspensions of reference clay minerals". *Environ. Toxicol. and Chem.* **19**: 2168–2174.

[8] Parales RE, et al. (2008). "Bioavailability, Chemotaxis, and Transport of Organic Pollutants". *Microbial Biodegradation: Genomics and Molecular Biology*. Caister Academic Press. ISBN 978-1-904455-17-2.

[9] I. R. MacDonald (2002). "Transfer of hydrocarbons from natural seeps to the water column and atmosphere." (PDF). *Geofluids*.

[10] Yakimov MM, Timmis KN, Golyshin PN (June 2007). "Obligate oil-degrading marine bacteria". *Curr. Opin. Biotechnol.* **18** (3): 257–66. PMID 17493798. doi:10.1016/j.copbio.2007.04.006.

[11] Martins dos Santos VA, et al. (2008). "Genomic Insights into Oil Biodegradation in Marine Systems". In Díaz E. *Microbial Biodegradation: Genomics and Molecular Biology*. Caister Academic Press. ISBN 978-1-904455-17-2.

[12] Sims, G. K. and E.J. O'Loughlin. 1989. Degradation of pyridines in the environment. CRC Critical Reviews in Environmental Control. 19(4): 309-340.

[13] Wipperman, Matthew, F.; Sampson, Nicole, S.; Thomas, Suzanne, T. (2014). "Pathogen roid rage: Cholesterol utilization by Mycobacterium tuberculosis". *Crit Rev Biochem Mol Biol.* **49** (4): 269–93. PMID 24611808. doi:10.3109/10409238.2014.895700.

[14] Thomas, S.T.; Sampson, N.S. (2013). "Mycobacterium tuberculosis utilizes a unique heterotetrameric structure for dehydrogenation of the cholesterol side chain". *Biochemistry*. **52** (17): 2895–2904. PMC 3726044. PMID 23560677. doi:10.1021/bi4002979.

[15] Wipperman, M.F.; Yang, M.; Thomas, S.T.; Sampson, N.S. (2013). "Shrinking the FadE Proteome of *Mycobacterium tuberculosis*: Insights into Cholesterol Metabolism through Identification of an $\alpha_2\beta_2$ Heterotetrameric Acyl Coenzyme A Dehydrogenase Family". *J. Bacteriol.* **195** (19): 4331–4341. PMC 3807453. PMID 23836861. doi:10.1128/JB.00502-13.

[16] Watanabe K & Kasai Y (2008). "Emerging Technologies to Analyze Natural Attenuation and Bioremediation". *Microbial Biodegradation: Genomics and Molecular Biology*. Caister Academic Press. ISBN 978-1-904455-17-2.

[17] Meyer A & Panke S (2008). "Genomics in Metabolic Engineering and Biocatalytic Applications of the Pollutant Degradation Machinery". *Microbial Biodegradation: Genomics and Molecular Biology*. Caister Academic Press. ISBN 978-1-904455-17-2.

[18] Machida, Masayuki; Gomi, Katsuya, eds. (2010). Aspergillus*: Molecular Biology and Genomics*. Caister Academic Press. ISBN 978-1-904455-53-0.

[19] Bennett JW (2010). "An Overview of the Genus *Aspergillus*" (PDF). Aspergillus*: Molecular Biology and Genomics*. Caister Academic Press. ISBN 978-1-904455-53-0.

Chapter 21

Genotyping

Genotyping is the process of determining differences in the genetic make-up (genotype) of an individual by examining the individual's DNA sequence using biological assays and comparing it to another individual's sequence or a reference sequence. It reveals the alleles an individual has inherited from their parents.[1] Traditionally genotyping is the use of DNA sequences to define biological populations by use of molecular tools. It does not usually involve defining the genes of an individual.

Current methods of genotyping include restriction fragment length polymorphism identification (RFLPI) of genomic DNA, random amplified polymorphic detection (RAPD) of genomic DNA, amplified fragment length polymorphism detection (AFLPD), polymerase chain reaction (PCR), DNA sequencing, allele specific oligonucleotide (ASO) probes, and hybridization to DNA microarrays or beads. Genotyping is important in research of genes and gene variants associated with disease. Due to current technological limitations, almost all genotyping is partial. That is, only a small fraction of an individual's genotype is determined, such as with (epi)GBS (Genotyping by sequencing) or RADseq. New [2] mass-sequencing technologies promise to provide whole-genome genotyping (or whole genome sequencing) in the future.

Genotyping applies to a broad range of individuals, including microorganisms. For example, viruses and bacteria can be genotyped. Genotyping in this context may help in controlling the spreading of pathogens, by tracing the origin of outbreaks. This area is often referred to as molecular epidemiology or forensic microbiology.

Humans can also be genotyped. For example, when testing fatherhood or motherhood, scientists typically only need to examine 10 or 20 genomic regions (like single-nucleotide polymorphism (SNPs)). That is a tiny fraction of the human genome, which consists of three billion or so nucleotides.

When genotyping transgenic organisms, a single genomic region may be all that needs to be examined to determine the genotype. A single PCR assay is typically enough to genotype a transgenic mouse; the mouse is the mammalian model of choice for much of medical research today.

21.1 See also

- Mendelian error
- Quantitative trait locus
- SNP genotyping

21.2 References

[1] "Genotyping definition". NIH. 2011-09-21. Retrieved 2011-09-21.

[2] "Genotyping at Illumina, Inc". Illumina.com. Archived from the original on 2011-04-16. Retrieved 2010-12-04.

'Genotyping Services' from Source Bioscience

21.3 External links

- International HapMap Project
- UCLA Genotyping Core
- resources for genotyping microorganisms
- Custom snp genotyping

Chapter 22

Chemogenomics

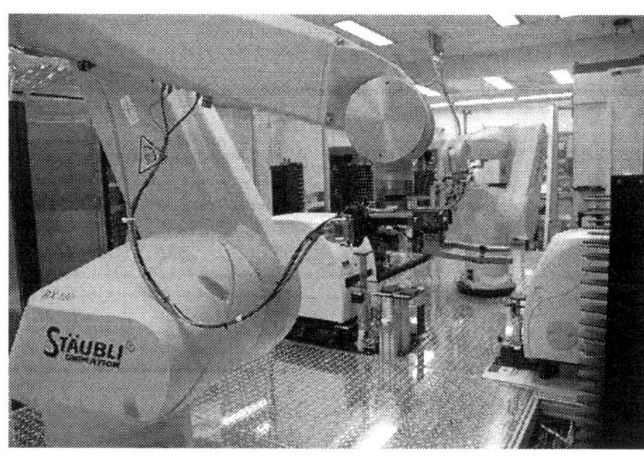

Chemogenomics robot retrieves assay plates from incubators

Chemogenomics, or **chemical genomics**, is the systematic screening of targeted chemical libraries of small molecules against individual drug target families (e.g., GPCRs, nuclear receptors, kinases, proteases, etc.) with the ultimate goal of identification of novel drugs and drug targets.[1] Typically some members of a target library have been well characterized where both the function has been determined and compounds that modulate the function of those targets (ligands in the case of receptors, inhibitors of enzymes, or blockers of ion channels) have been identified. Other members of the target family may have unknown function with no known ligands and hence are classified as orphan receptors. By identifying screening hits that modulate the activity of the less well characterized members of the target family, the function of these novel targets can be elucidated. Furthermore, the hits for these targets can be used as a starting point for drug discovery. The completion of the human genome project has provided an abundance of potential targets for therapeutic intervention. Chemogenomics strives to study the intersection of all possible drugs on all of these potential targets.[2]

A common method to construct a targeted chemical library is to include known ligands of at least one and preferably several members of the target family. Since a portion of ligands that were designed and synthesized to bind to one family member will also bind to additional family members, the compounds contained in a targeted chemical library should collectively bind to a high percentage of the target family.[3]

22.1 Strategy

Chemogenomics integrates target and drug discovery by using active compounds, which function as ligands, as probes to characterize proteome functions. The interaction between a small compound and a protein induces a phenotype. Once the phenotype is characterized, we could associate a protein to a molecular event. Compared with genetics, chemogenomics techniques are able to modify the function of a protein rather than the gene. Also, chemogenomics is able to observe the interaction as well as reversibility in real-time. For example, the modification of a phenotype can be observed only after addition of a specific compound and can be interrupted after its withdrawal from the medium.

Currently, there are two experimental chemogenomic approaches: forward (classical) chemogenomics and reverse chemogenomics. Forward chemogenomics attempt to identify drug targets by searching for molecules which give a certain phenotype on cells or animals, while reverse chemogenomics aim to validate phenotypes by searching for molecules that interact specifically with a given protein.[4] Both of these approaches require a suitable collection of compounds and an appropriate model system for screening the compounds and looking for the parallel identification of biological targets and biologically active compounds. The biologically active compounds that are discovered through forward or reverse chemogenomics approaches are known as modulators because they bind to and modulate specific molecular targets, thus they could be used as 'targeted therapeutics'.[1]

22.1.1 Forward chemogenomics

In forward chemogenomics, which is also known as classical chemogenomics, a particular phenotype is studied and small compound interacting with this function are identified. The molecular basis of this desired phenotype is unknown. Once the modulators have been identified, they will be used as tools to look for the protein responsible for the phenotype. For example, a loss-of-function phenotype could be an arrest of tumor growth. Once compounds that lead to a target phenotype have been identified, identifying the gene and protein targets should be the next step.[5] The main challenge of forward chemogenomics strategy lies in designing phenotypic assays that lead immediately from screening to target identification.

22.1.2 Reverse chemogenomics

In reverse chemogenomics, small compounds that perturb the function of an enzyme in the context of an in vitro enzymatic test will be identified. Once the modulators have been identified, the phenotype induced by the molecule is analyzed in a test on cells or on whole organisms. This method will identify or confirm the role of the enzyme in the biological response.[5] Reverse chemogenomics used to be virtually identical to the target-based approaches that have been applied in drug discovery and molecular pharmacology over the past decade. This strategy is now enhanced by parallel screening and by the ability to perform lead optimization on many targets that belong to one target family.

22.2 Applications

22.2.1 Determining mode of action

Chemogenomics has been used to identify mode of action (MOA) for traditional Chinese medicine (TCM) and Ayurveda. Compounds contained in traditional medicines are usually more soluble than synthetic compounds, have "privileged structures" (chemical structures that are more frequently found to bind in different living organisms), and have more comprehensively known safety and tolerance factors. Therefore, this makes them especially attractive as a resource for lead structures in when developing new molecular entities. Databases containing chemical structures of compounds used in alternative medicine along with their phenotypic effects, in silico analysis may be of use to assist in determining MOA for example, by predicting ligand targets that were relevant to known phenotypes for traditional medicines.[6] In a case study for TCM, the therapeutic class of 'toning and replenishing medicine" was evaluated. Therapeutic actions (or phenotypes) for that class include anti-inflammatory, antioxidant, neuroprotective, hypoglycemic activity, immunomodulatory, antimetastatic, and hypotensive. Sodium-glucose transport proteins and PTP1B (an insulin signaling regulator) were identified as targets which link to the hypoglycemic phenotype suggested. The case study for Ayurveda involved anti-cancer formulations. In this case, the target prediction program enriched for targets directly connected to cancer progression such as steroid-5-alpha-reductase and synergistic targets like the efflux pump P-gp. These target-phenotype links can help identify novel MOAs.

Beyond TCM and Ayurveda, chemogenomics can be applied early in drug discovery to determine a compound's mechanism of action and take advantage of genomic biomarkers of toxicity and efficacy for application to Phase I and II clinical trials.[7]

22.2.2 Identifying new drug targets

Chemogenomics profiling can be used to identify totally new therapeutic targets, for example new antibacterial agents.[8] The study capitalized on the availability of an existing ligand library for an enzyme called murD that is used in the peptidoglycan synthesis pathway. Relying on the chemogenomics similarity principle, the researchers mapped the murD ligand library to other members of the mur ligase family (murC, murE, murF, murA, and murG) to identify new targets for the known ligands. Ligands identified would be expected to be broad-spectrum Gram-negative inhibitors in experimental assays since peptidoglycan synthesis is exclusive to bacteria. Structural and molecular docking studies revealed candidate ligands for murC and murE ligases.

22.2.3 Identifying genes in biological pathway

Thirty years after the posttranslationally modified histidine derivative diphthamide was determined, chemogenomics was used to discover the enzyme responsible for the final step in its synthesis.[9] Dipthamide is a posttranslationally modified histidine residue found on the translation elongation factor 2 (eEF-2). The first two steps of the biosynthesis pathway leading to dipthine have been known, but the enzyme responsible for the amidation of dipthine to diphthamide remained a mystery. The researchers capitalized on Saccharomyces cerevisiae cofitness data. Cofitness data is data representing the similarity of growth fitness under various conditions between any two different deletion strains. Under the assumption that strains lacking the diphthamide synthetase gene should have high cofitness with strain lacking other diphthamide biosynthesis genes, they identified ylr143w as the strain with the highest cofitness to the all other strains lacking known diphthamide biosynthesis genes. Subsequent experimental assays confirmed that YLR143W was required for diphthamide synthesis and was the missing diphthamide synthetase.

22.3 See also

- Chemical biology
- Chemical genetics
- Drug discovery
- High-throughput screening
- Personalized medicine
- Phenotypic screening

22.4 References

[1] Bredel M, Jacoby E (Apr 2004). "Chemogenomics: an emerging strategy for rapid target and drug discovery". *Nature Reviews Genetics*. **5** (4): 262–75. PMID 15131650. doi:10.1038/nrg1317.

[2] Namchuk M (2002). "Finding the molecules to fuel chemogenomics". *Targets*. **1** (4): 125–129. doi:10.1016/S1477-3627(02)02206-7.

[3] Caron PR, Mullican MD, Mashal RD, Wilson KP, Su MS, Murcko MA (Aug 2001). "Chemogenomic approaches to drug discovery". *Current Opinion in Chemical Biology*. **5** (4): 464–70. PMID 11470611. doi:10.1016/S1367-5931(00)00229-5.

[4] Ambroise Y. "Chemogenomic techniques". Retrieved 28 July 2013.

[5] Wuster A, Madan Babu M (May 2008). "Chemogenomics and biotechnology". *Trends in Biotechnology*. **26** (5): 252–8. PMID 18346803. doi:10.1016/j.tibtech.2008.01.004.

[6] Mohd Fauzi F, Koutsoukas A, Lowe R, Joshi K, Fan TP, Glen RC, Bender A (Mar 2013). "Chemogenomics approaches to rationalizing the mode-of-action of traditional Chinese and Ayurvedic medicines". *Journal of Chemical Information and Modeling*. **53** (3): 661–73. PMID 23351136. doi:10.1021/ci3005513.

[7] Engelberg A (Sep 2004). "Iconix Pharmaceuticals, Inc.--removing barriers to efficient drug discovery through chemogenomics". *Pharmacogenomics*. **5** (6): 741–4. PMID 15335294. doi:10.1517/14622416.5.6.741.

[8] Bhattacharjee B, Simon RM, Gangadharaiah C, Karunakar P (Jun 2013). "Chemogenomics profiling of drug targets of peptidoglycan biosynthesis pathway in Leptospira interrogans by virtual screening approaches". *Journal of Microbiology and Biotechnology*. **23** (6): 779–84. PMID 23676922. doi:10.4014/jmb.1206.06050.

[9] Cheung-Ong K, Song KT, Ma Z, Shabtai D, Lee AY, Gallo D, Heisler LE, Brown GW, Bierbach U, Giaever G, Nislow C (Nov 2012). "Comparative chemogenomics to examine the mechanism of action of dna-targeted platinum-acridine anticancer agents". *ACS Chemical Biology*. **7** (11): 1892–901. PMC 3500413 🔓. PMID 22928710. doi:10.1021/cb300320d.

22.5 Further reading

- Folkers G, Kubinyi H, Müller G, Mannhold R (2004). *Chemogenomics in drug discovery: a medicinal chemistry perspective*. Weinheim: Wiley-VCH. ISBN 3-527-30987-X.
- Jacoby E (2009). *Chemogenomics: methods and applications*. Totowa, NJ: Humana Press. ISBN 1-60761-273-9.
- Weill N (2011). "Chemogenomic approaches for the exploration of GPCR space". *Current Topics in Medicinal Chemistry*. **11** (15): 1944–55. PMID 21470168. doi:10.2174/156802611796391212.

22.6 External links

- GLASS: A comprehensive database for experimentally-validated GPCR-ligand associations
- Kubinyi's slides

Chapter 23

Personalized medicine

Personalized medicine, also termed precision medicine, is a medical procedure that separates patients into different groups—with medical decisions, practices, interventions and/or products being tailored to the individual patient based on their predicted response or risk of disease.[1] The terms personalized medicine, precision medicine, **stratified medicine** and P4 medicine are used interchangeably to describe this concept[1][2] though some authors and organisations use these expressions separately to indicate particular nuances.[2]

While the tailoring of treatment to patients dates back at least to the time of Hippocrates,[3] the term has risen in usage in recent years given the growth of new diagnostic and informatics approaches that provide understanding of the molecular basis of disease, particularly genomics. This provides a clear evidence base on which to stratify (group) related patients.[1][4][5]

23.1 Development of concept

In personalised medicine, diagnostic testing is often employed for selecting appropriate and optimal therapies based on the context of a patient's genetic content or other molecular or cellular analysis.[6] The use of genetic information has played a major role in certain aspects of personalized medicine (e.g. pharmacogenomics), and the term was first coined in the context of genetics, though it has since broadened to encompass all sorts of personalization measures.[6]

23.2 Background

23.2.1 Basics

Every person has a unique variation of the human genome.[7] Although most of the variation between individuals has no effect on health, an individual's health stems from genetic variation with behaviors and influences from the environment.[8][9]

Modern advances in personalized medicine rely on technology that confirms a patient's fundamental biology, DNA, RNA, or protein, which ultimately leads to confirming disease. For example, personalised techniques such as genome sequencing can reveal mutations in DNA that influence diseases ranging from cystic fibrosis to cancer. Another method, called RNA-seq, can show which RNA molecules are involved with specific diseases. Unlike DNA, levels of RNA can change in response to the environment. Therefore, sequencing RNA can provide a broader understanding of a person's state of health. Recent studies have linked genetic differences between individuals to RNA expression,[10] translation,[11] and protein levels.[12]

The concepts of personalised medicine can be applied to new and transformative approaches to health care. Personalised health care is based on the dynamics of systems biology and uses predictive tools to evaluate health risks and to design personalised health plans to help patients mitigate risks, prevent disease and to treat it with precision when it occurs. The concepts of personalised health care are receiving increasing acceptance with the Veterans Administration committing to personalised, proactive patient driven care for all veterans.[13]

23.2.2 Method

In order for physicians to know if a mutation is connected to a certain disease, researchers often do a study called a "genome-wide association study" (GWAS). A GWAS study will look at one disease, and then sequence the genome of many patients with that particular disease to look for shared mutations in the genome. Mutations that are determined to be related to a disease by a GWAS study can then be used to diagnose that disease in future patients, by looking at their genome sequence to find that same mutation. The first GWAS, conducted in 2005, studied patients with age-related macular degeneration (ARMD).[14] It found two different mutations, each containing only a variation in only one nucleotide (called single nucleotide polymorphisms, or SNPs), which were associated with ARMD. GWAS studies like this have been very successful in identifying common genetic variations associated with diseases. As of early 2014, over 1,300 GWAS studies have been completed.[15]

23.2.3 Disease risk assessment

Multiple genes collectively influence the likelihood of developing many common and complex diseases.[8] Personalised medicine can also be used to predict a person's risk for a particular disease, based on one or even several genes. This approach uses the same sequencing technology to focus on the evaluation of disease risk, allowing the physician to initiate preventative treatment before the disease presents itself in their patient. For example, if it is found that a DNA mutation increases a person's risk of developing Type 2 Diabetes, this individual can begin lifestyle changes that will lessen their chances of developing Type 2 Diabetes later in life.

23.3 Applications

Advances in personalised medicine will create a more unified treatment approach specific to the individual and their genome. Personalised medicine may provide better diagnoses with earlier intervention, and more efficient drug development and therapies.[16]

23.3.1 Diagnosis and intervention

Having the ability to look at a patient on an individual basis will allow for a more accurate diagnosis and specific treatment plan. Genotyping is the process of obtaining an individual's DNA sequence by using biological assays.[17] By having a detailed account of an individual's DNA sequence, their genome can then be compared to a reference genome, like that of the Human Genome Project, to assess the existing genetic variations that can account for possible diseases. A number of private companies, such as 23andMe, Navigenics, and Illumina, have created Direct-to-Consumer genome sequencing accessible to the public.[7] Having this information from individuals can then be applied to effectively treat them. An individual's genetic make-up also plays a large role in how well they respond to a certain treatment, and therefore, knowing their genetic content can change the type of treatment they receive.

An aspect of this is pharmacogenomics, which uses an individual's genome to provide a more informed and tailored drug prescription.[18] Often, drugs are prescribed with the idea that it will work relatively the same for everyone, but in the application of drugs, there are a number of factors that must be considered. The detailed account of genetic information from the individual will help prevent adverse events, allow for appropriate dosages, and create maximum efficacy with drug prescriptions.[7] The pharmacogenomic process for discovery of genetic variants that predict adverse events to a specific drug has been termed toxgnostics.[19]

Another aspect is theranostics or therapeutic diagnostics in medicine, which is the use of diagnostic tests to guide therapy. The term «theranostics» is derived as a combination of therapy and diagnostics. The test may involve medical imaging such as MRI contrast agents (T1 and T2 agents), fluorescent markers (organic dyes and inorganic quantum dots), and nuclear imaging agents (PET radiotracers or SPECT agents).[20][21] or in vitro lab test[22] including DNA sequencing[23] and often involve deep learning algorithms that weigh the result of testing for several biomarkers.[24]

In addition to specific treatment, personalised medicine can greatly aid the advancements of preventive care. For instance, many women are already being genotyped for certain mutations in the BRCA1 and BRCA2 gene if they are predisposed because of a family history of breast cancer or ovarian cancer.[25] As more causes of diseases are mapped out according to mutations that exist within a genome, the easier they can be identified in an individual. Measures can then be taken to prevent a disease from developing. Even if mutations were found within a genome, having the details of their DNA can reduce the impact or delay the onset of certain diseases.[16] Having the genetic content of an individual will allow better guided decisions in determining the source of the disease and thus treating it or preventing its progression. This will be extremely useful for diseases like Alzheimer's or cancers that are thought to be linked to certain mutations in our DNA.[16]

A tool that is being used now to test efficacy and safety of a drug specific to a targeted patient group/sub-group is companion diagnostics. This technology is an assay that is developed during or after a drug is made available on the market and is helpful in enhancing the therapeutic treatment available based on the individual.[26] These companion diagnostics have incorporated the pharmacogenomic information related to the drug into their prescription label in an effort to assist in making the most optimal treatment decision possible for the patient.[26]

23.3.2 Drug development and usage

Having an individual's genomic information can be significant in the process of developing drugs as they await approval from the FDA for public use. Having a detailed account of an individual's genetic make-up can be a major asset in deciding if a patient can be chosen for inclusion or exclusion in the final stages of a clinical trial.[16] Being able to identify patients who will benefit most from a clinical trial will increase the safety of patients from adverse outcomes caused by the product in testing, and will allow smaller and faster trials that lead to lower overall costs.[27] In addition, drugs that are deemed ineffective for the larger population can gain approval by the FDA by using personal genomes to qualify the effectiveness and need for that specific drug or therapy even though it may only be needed by a small percentage of the population.,[16][28]

Today in medicine, it is common that physicians often use a trial and error strategy until they find the treatment therapy that is most effective for their patient.[16] With personalised medicine, these treatments can be more specifically tailored to an individual and give insight into how their body will respond to the drug and if that drug will work based on their genome.[7] The personal genotype can allow physicians to have more detailed information that will guide them in their decision in treatment prescriptions, which will be more cost-effective and accurate.[16] As quoted from the article *Pharmacogenomics: The Promise of Personalised Medicine*, “therapy with the right drug at the right dose in the right patient” is a description of how personalized medicine will affect the future of treatment.[29] For instance, Tamoxifen used to be a drug commonly prescribed to women with ER+ breast cancer, but 65% of women initially taking it developed resistance. After some research by people such as David Flockhart, it was discovered that women with certain mutation in their CYP2D6 gene, a gene that encodes the metabolizing enzyme, were not able to efficiently break down Tamoxifen, making it an ineffective treatment for their cancer.[30] Since then, women are now genotyped for those specific mutations, so that immediately these women can have the most effective treatment therapy.

Screening for these mutations is carried out via high-throughput screening or phenotypic screening. Several drug discovery and pharmaceutical companies are currently utilizing these technologies to not only advance the study of personalised medicine, but also to amplify genetic research; these companies include Alacris Theranostics, Persomics, Flatiron Health, Novartis, OncoDNA and Foundation Medicine, among others. Alternative multi-target approaches to the traditional approach of “forward” transfection library screening can entail reverse transfection or chemogenomics.

Pharmacy compounding is yet another application of personalised medicine. Though not necessarily utilizing genetic information, the customized production of a drug whose various properties (e.g. dose level, ingredient selection, route of administration, etc.) are selected and crafted for an individual patient is accepted as an area of personalised medicine (in contrast to mass-produced unit doses or fixed-dose combinations).

23.3.3 Cancer genomics

Over recent decades cancer research has discovered a great deal about the genetic variety of types of cancer that appear the same in traditional pathology. There has also been increasing awareness of tumour heterogeneity, or genetic diversity within a single tumour. Among other prospects, these discoveries raise the possibility of finding that drugs that have not given good results applied to a general population of cases may yet be successful for a proportion of cases with particular genetic profiles.

Cancer Genomics, or “Oncogenomics,” is the application of genomics and personalized medicine to cancer research and treatment. High-throughput sequencing methods are used to characterize genes associated with cancer to better understand disease pathology and improve drug development. Oncogenomics is one of the most promising branches of genomics, particularly because of its implications in drug therapy. Examples of this include:

- Trastuzumab (trade names Herclon, Herceptin) is a monoclonal antibody drug that interferes with the HER2/neu receptor. Its main use is to treat certain breast cancers. This drug is only used if a patient's cancer is tested for over-expression of the HER2/neu receptor. Two tissue-typing tests are used to screen patients for possible benefit from Herceptin treatment. The tissue tests are immunohistochemistry(IHC) and Fluorescence In Situ Hybridization(FISH)[31] Only Her2+ patients will be treated with Herceptin therapy (trastuzumab)[32]
- Tyrosine kinase inhibitors such as imatinib (marketed as Gleevec) have been developed to treat chronic myeloid leukemia (CML), in which the BCR-ABL fusion gene (the product of a reciprocal translocation between chromosome 9 and chromosome 22) is present in >95% of cases and produces hyperactivated abl-driven protein signaling. These medications specifically inhibit the Ableson tyrosine kinase (ABL) protein and are thus a prime example of "**rational drug design**" based on knowledge of disease pathophysiology.[33]

23.4 Challenges

As personalised medicine is practiced more widely, a number of challenges arise. The current approaches to intellectual property rights, reimbursement policies, patient privacy and confidentiality as well as regulatory oversight will have to be redefined and restructured to accommodate the changes personalised medicine will bring to healthcare.[34] Furthermore, the analysis of acquired diagnostic data is a recent challenge of personalized medicine and its adoption.[35] For example, genetic data obtained from next-generation sequencing requires computer-intensive data processing prior to its analysis.[36] In the future, adequate tools will be required to accelerate the adoption of personalised medicine to further fields of medicine, which requires the interdisciplinary cooperation of experts from specific fields of research, such as medicine, clinical oncology, biology, and artificial intelligence.

23.4.1 Regulatory oversight

The FDA has already started to take initiatives to integrate personalised medicine into their regulatory policies. They developed a report in October 2013 entitled, "*Paving the Way for Personalized Medicine: FDA's role in a New Era of Medical Product Development*," in which they outlined steps they would have to take to integrate genetic and biomarker information for clinical use and drug development.[37] They determined that they would have to develop specific regulatory science standards, research methods, reference material and other tools in order to incorporate personalised medicine into their current regulatory practices. For example, they are working on a "genomic reference library" for regulatory agencies to compare and test the validity of different sequencing platforms in an effort to uphold reliability.[37]

23.4.2 Intellectual property rights

As with any innovation in medicine, investment and interest in personalised medicine is influenced by intellectual property rights.[34] There has been a lot of controversy regarding patent protection for diagnostic tools, genes, and biomarkers.[38] In June 2013, the U.S Supreme Court ruled that natural occurring genes cannot be patented, while "synthetic DNA" that is edited or artificially- created can still be patented. The Patent Office is currently reviewing a number of issues related to patent laws for personalised medicine, such as whether "confirmatory" secondary genetic tests post initial diagnosis, can have full immunity from patent laws. Those who oppose patents argue that patents on DNA sequences are an impediment to ongoing research while proponents point to research exemption and stress that patents are necessary to entice and protect the financial investments required for commercial research and the development and advancement of services offered.[38]

23.4.3 Reimbursement policies

Reimbursement policies will have to be redefined to fit the changes that personalised medicine will bring to the healthcare system. Some of the factors that should be considered are the level of efficacy of various genetic tests in the general population, cost-effectiveness relative to benefits, how to deal with payment systems for extremely rare conditions, and how to redefine the insurance concept of "shared risk" to incorporate the effect of the newer concept of "individual risk factors".[34]

23.4.4 Patient privacy and confidentiality

Perhaps the most critical issue with the commercialization of personalised medicine is the protection of patients. One of the largest issues is the fear and potential consequences for patients who are predisposed after genetic testing or found to be non-responsive towards certain treatments. This includes the psychological effects on patients due to genetic testing results. The right of family members who do not directly consent is another issue, considering that genetic predispositions and risks are inheritable. The implications for certain ethnic groups and presence of a common allele would also have to be considered.[34] In 2008, the Genetic Information Nondiscrimination Act (GINA) was passed in an effort to minimize the fear of patients participating in genetic research by ensuring that their genetic information will not be misused by employers or insurers.[34] On February 19, 2015 FDA issued a press release titled: "FDA permits marketing of first direct-to-consumer genetic carrier test for Bloom syndrome.[6]

23.5 See also

- Chemogenomics
- Drug discovery
- Foundation Medicine
- Genetics
- Next-generation sequencing
- Pharmacogenetics
- Pharmacogenomics
- Phenotypic screening
- Theranostics

23.6 References

[1] *Stratified, personalised or P4 medicine: a new direction for placing the patient at the centre of healthcare and health education* (Technical report). Academy of Medical Sciences. May 2015. Retrieved 6 Jan 2016.

[2] "Many names for one concept or many concepts in one name?". PHG Foundation. Retrieved 6 Jan 2015.

[3] Egnew, Thomas (1 Mar 2009). "Suffering, Meaning, and Healing: Challenges of Contemporary Medicine". *Annals of Family Medicine*. **7** (2): 170–175. doi:10.1370/afm.943. Retrieved 6 January 2016.

[4] "The Case for Personalized Medicine" (PDF). Personalized Medicine Coalition. 2014. Retrieved 6 Jan 2016.

[5] Smith, Richard (15 Oct 2012). "Stratified, personalised, or precision medicine". *British Medical Journal*. Retrieved 6 January 2016.

[6] "Personalized Medicine 101". Personalized Medicine Coalition. Retrieved 26 April 2014.

[7] Dudley, J; Karczewski, K. (2014). *Exploring Personal Genomics*. Oxford: Oxford University Press.

[8] "Personalized Medicine 101: The Science". Personalized Medicine Coalition. Retrieved 26 April 2014.

[9] Lu, YF; Goldstein, DB; Angrist, M; Cavalleri, G (24 July 2014). "Personalized medicine and human genetic diversity". *Cold Spring Harbor perspectives in medicine*. **4** (9): a008581. PMC 4143101. PMID 25059740. doi:10.1101/cshperspect.a008581.

[10] Alexis Battle, Sara Mostafavi, Xiaowei Zhu, James B. Potash, Myrna M. Weissman, Courtney McCormick, Christian D. Haudenschild, Kenneth B. Beckman, Jianxin Shi, Rui Mei, Alexander E. Urban, Stephen B. Montgomery, Douglas F. Levinson & Daphne Koller (2014). "Characterizing the genetic basis of transcriptome diversity through RNA-sequencing of 922 individuals". *Genome research*. **24** (1): 14–24. PMC 3875855. PMID 24092820. doi:10.1101/gr.155192.113.

[11] Cenik, Can; Cenik, Elif Sarinay; Byeon, Gun W; Candille, Sophie P.; Spacek, Damek; Araya, Carlos L; Tang, Hua; Ricci, Emiliano; Snyder, Michael P. (Nov 2015). "Integrative analysis of RNA, translation, and protein levels reveals distinct regulatory variation across humans.". *Genome Research*. **25**: 1610–21. PMC 4617958. PMID 26297486. doi:10.1101/gr.193342.115.

[12] Linfeng Wu; Sophie I. Candille; Yoonha Choi; Dan Xie; Lihua Jiang; Jennifer Li-Pook-Than; Hua Tang; Michael Snyder (2013). "Variation and genetic control of protein abundance in humans". *Nature*. **499** (7456): 79–82. PMC 3789121. PMID 23676674. doi:10.1038/nature12223.

[13] Snyderman, R. Personalized Health Care from Theory to Practice, Biotechnology J. 2012, 7

[14] Haines, J.L. (Apr 15, 2005). "Complement Factor H Variant Increases the Risk of Age-Related Macular Degeneration". *Science*. **308** (5720): 419–21. PMID 15761120. doi:10.1126/science.1110359.

[15] "A Catalog of Published Genome-Wide Association Studies". Retrieved 28 June 2015.

[16] "Personalized Medicine 101: The Promise". Personalized Medicine Coalition. Retrieved April 26, 2014.

[17] "Research Portfolio Online Reporting Tools: Human Genome Project". National Institutes of Health (NIH). Retrieved April 28, 2014.

[18] "Genetics Home Reference: What is pharmacogenomics?". National Institutes of Health (NIH). Retrieved April 28, 2014.

[19] Church D, Kerr R, Domingo E, Rosmarin D, Palles C, Maskell K, Tomlinson I, Kerr D (2014). "'Toxgnostics': an unmet need in cancer medicine". Nat Rev Cancer (6):440-5

[20] Xie, Jin; Lee, Seulki; Chen, Xiaoyuan (2010-08-30). "Nanoparticle-based theranostic agents". *Advanced Drug Delivery Reviews*. Development of Theranostic Agents that Co-Deliver Therapeutic and Imaging Agents. **62** (11): 1064–1079. PMC 2988080. PMID 20691229. doi:10.1016/j.addr.2010.07.009.

[21] Kelkar, Sneha S.; Reineke, Theresa M. (2011-10-19). "Theranostics: Combining Imaging and Therapy". *Bioconjugate Chemistry*. **22** (10): 1879–1903. ISSN 1043-1802. doi:10.1021/bc200151q.

[22] Perkovic, MN; Erjavec, GN; Strac, DS; Uzun, S; Kozumplik, O; Pivac, N (30 March 2017). "Theranostic Biomarkers for Schizophrenia.". *International journal of molecular sciences*. **18** (4). PMC 5412319. PMID 28358316.

[23] Kamps, R; Brandão, RD; Bosch, BJ; Paulussen, AD; Xanthoulea, S; Blok, MJ; Romano, A (31 January 2017). "Next-Generation Sequencing in Oncology: Genetic Diagnosis, Risk Prediction and Cancer Classification.". *International journal of molecular sciences*. **18** (2). PMC 5343844. PMID 28146134.

[24] Yahata, N; Kasai, K; Kawato, M (April 2017). "Computational neuroscience approach to biomarkers and treatments for mental disorders.". *Psychiatry and clinical neurosciences*. **71** (4): 215–237. PMID 28032396. doi:10.1111/pcn.12502.

[25] "Fact Sheet: BRCA1 and BRCA2: Cancer and Genetic Testing". National Cancer Institute (NCI). Retrieved April 28, 2014.

[26] "BIOMARKER TOOLKIT: Companion Diagnostics" (PDF). Amgen. Archived from the original (PDF) on August 1, 2014. Retrieved May 2, 2014.

[27] "Paving the Way for Personalized Medicine: FDA's Role in a New Era of Medical Product Development" (PDF). Federal Drug Administration (FDA). Retrieved April 28, 2014.

[28] Hamburg MA, Collins FS (July 22, 2010). "The Path to Personalized Medicine". *New England Journal of Medicine (NEJM)*. **363**: 301–304. PMID 20551152. doi:10.1056/nejmp1006304. Retrieved April 28, 2014.

[29] Mancinelli L, Cronin M, Sadée W (2000). "Pharmacogenomics. The Promise of Personalized Medicine". *AAPS PharmSci*. **2** (1): E4. PMC 2750999. PMID 11741220.

[30] Ellsworth RE, Decewicz DJ, Shriver CD, Ellsworth DL. "Breast Cancer in the Personal Genomics Era". *Current Genomics: Bentham Science*. **11**: 146–61. PMC 2878980. PMID 21037853. doi:10.2174/138920210791110951.

[31] Carney, Walter (2006). "HER2/neu Status is an Important Biomarker in Guiding Personalized HER2/neu Therapy" (PDF). *Connection*. **9**: 25–27.

[32] Telli, M. L.; Hunt, S. A.; Carlson, R. W.; Guardino, A. E. (2007). "Trastuzumab-Related Cardiotoxicity: Calling Into Question the Concept of Reversibility". *Journal of Clinical Oncology*. **25** (23): 3525–3533. ISSN 0732-183X. PMID 17687157. doi:10.1200/JCO.2007.11.0106.

[33] Saglio G; Morotti A; Mattioli G; et al. (December 2004). "Rational approaches to the design of therapeutics targeting molecular markers: the case of chronic myelogenous leukemia". *Ann. N. Y. Acad. Sci*. **1028** (1): 423–31. Bibcode:2004NYASA1028..423S. PMID 15650267. doi:10.1196/annals.1322.050.

[34] "Personalized Medicine 101: The Challenges". Personalized Medicine Coalition. Retrieved April 26, 2014.

[35] Huser, V; Sincan, M; Cimino, J. J. (2014). "Developing genomic knowledge bases and databases to support clinical management: Current perspectives". *Pharmacogenomics and Personalized Medicine*. **7**: 275–83. PMC 4175027. PMID 25276091. doi:10.2147/PGPM.S49904.

[36] "Analyze Genomes: Motivation". Schapranow, Matthieu-P. Retrieved July 20, 2014.

[37] "Paving the Way for Personalized Medicine: FDA's Role in a New Era of Medical Product Development" (PDF). U.S Food and Drug Administration. Retrieved April 26, 2014.

[38] "Intellectual Property Issues Impacting the Future of Personalized Medicine". American Intellectual Property Law Association. Retrieved April 26, 2014.

Chapter 24

Pharmacovigilance

Pharmacovigilance (**PV** or **PhV**), also known as **drug safety**, is the pharmacological science relating to the *collection, detection, assessment, monitoring, and prevention* of adverse effects with pharmaceutical products.[1] The etymological roots for the word "pharmacovigilance" are: *pharmakon* (Greek for drug) and *vigilare* (Latin for to keep watch). As such, pharmacovigilance heavily focuses on adverse drug reactions, or ADRs, which are defined as any response to a drug which is noxious and unintended, including lack of efficacy (the condition that this definition only applies with the doses normally used for the prophylaxis, diagnosis or therapy of disease, or for the modification of physiological disorder function was excluded with the latest amendment of the applicable legislation[2]). Medication errors such as overdose, and misuse and abuse of a drug as well as drug exposure during pregnancy and breastfeeding, are also of interest, even without an adverse event, because they may result in an adverse drug reaction.[3]

Information received from patients and healthcare providers via pharmacovigilance agreements (PVAs), as well as other sources such as the medical literature, plays a critical role in providing the data necessary for pharmacovigilance to take place. In fact, in order to market or to test a pharmaceutical product in most countries, adverse event data received by the license holder (usually a pharmaceutical company) must be submitted to the local drug regulatory authority. (See Adverse Event Reporting below.)

Ultimately, pharmacovigilance is concerned with identifying the hazards associated with pharmaceutical products and with minimizing the risk of any harm that may come to patients. Companies must conduct a comprehensive drug safety and pharmacovigilance audit to assess their compliance with worldwide laws, regulations, and guidance.[4]

24.1 Terms commonly used in drug safety

Pharmacovigilance has its own unique terminology that is important to understand. Most of the following terms are used within this article and are peculiar to drug safety, although some are used by other disciplines within the pharmaceutical sciences as well.

- *Adverse drug reaction* is a side effect (non intended reaction to the drug) occurring with a drug where a positive (direct) causal relationship between the event and the drug is thought, or has been proven, to exist.
- *Adverse event (AE)* is a side effect occurring with a drug. By definition, the causal relationship between the AE and the drug is unknown.
- *Benefits* are commonly expressed as the proven therapeutic good of a product but should also include the patient's subjective assessment of its effects.
- *Causal relationship* is said to exist when a drug is thought to have caused or contributed to the occurrence of an adverse drug reaction.
- *Clinical trial* (or study) refers to an organised program to determine the safety and/or efficacy of a drug (or drugs) in patients. The design of a clinical trial will depend on the drug and the phase of its development.
- *Control group* is a group (or cohort) of individual patients that is used as a standard of comparison within a clinical trial. The control group may be taking a placebo (where no active drug is given) or where a different active drug is given as a comparator.
- *Dechallenge* and *rechallenge* refer to a drug being stopped and restarted in a patient, respectively. A positive dechallenge has occurred, for example, when an adverse event abates or resolves completely following the drug's discontinuation. A positive rechallenge has

occurred when the adverse event re-occurs after the drug is restarted. Dechallenge and rechallenge play an important role in determining whether a causal relationship between an event and a drug exists.

- *Effectiveness* is the extent to which a drug works under real world circumstances, i.e., clinical practice.
- *Efficacy* is the extent to which a drug works under ideal circumstances, i.e., in clinical trials.
- *Event* refers to an adverse event (AE).
- *Harm* is the nature and extent of the actual damage that could be or has been caused.
- *Implied causality* refers to spontaneously reported AE cases where the causality is always presumed to be positive unless the reporter states otherwise.
- *Individual Case Study Report (ICSR)* is an adverse event report for an individual patient.
- *Life-threatening* refers to an adverse event that places a patient at the *immediate* risk of death.
- *Phase* refers to the four phases of clinical research and development: I – small safety trials early on in a drug's development; II – medium-sized trials for both safety and efficacy; III – large trials, which includes key (or so-called "pivotal") trials; IV – large, post-marketing trials, typically for safety reasons. There are also intermediate phases designated by an "a" or "b", e.g. Phase IIb.
- *Risk* is the probability of harm being caused, usually expressed as a percent or ratio of the treated population.
- *Risk factor* is an attribute of a patient that may predispose, or increase the risk, of that patient developing an event that may or may not be drug-related. For instance, obesity is considered a risk factor for a number of different diseases and, potentially, ADRs. Others would be high blood pressure, diabetes, possessing a specific mutated gene, for example, mutations in the BRCA1 and BRCA2 genes increase propensity to develop breast cancer.
- *Signal* is a new safety finding within safety data that requires further investigation. There are three categories of signals: *confirmed signals* where the data indicate that there is a causal relationship between the drug and the AE; *refuted* (or false) signals where after investigation the data indicate that no causal relationship exists; and *unconfirmed* signals which require further investigation (more data) such as the conducting of a post-marketing trial to study the issue.
- *Temporal relationship* is said to exist when an adverse event occurs when a patient is taking a given drug. Although a temporal relationship is absolutely necessary in order to establish a causal relationship between the drug and the AE, a temporal relationship does not necessarily in and of itself prove that the event was caused by the drug.
- *Triage* refers to the process of placing a potential adverse event report into one of three categories: 1) non-serious case; 2) serious case; or 3) no case (minimum criteria for an AE case are not fulfilled).

24.2 Adverse event reporting

The activity that is most commonly associated with pharmacovigilance (PV), and which consumes a significant amount of resources for drug regulatory authorities (or similar government agencies) and drug safety departments in pharmaceutical companies, is that of adverse event reporting. Adverse event (AE) reporting involves the receipt, triage, data entering, assessment, distribution, reporting (if appropriate), and archiving of AE data and documentation. The source of AE reports may include: spontaneous reports from healthcare professionals or patients (or other intermediaries); solicited reports from patient support programs; reports from clinical or post-marketing studies; reports from literature sources; reports from the media (including social media and websites); and reports reported to drug regulatory authorities themselves. For pharmaceutical companies, AE reporting is a regulatory requirement in most countries. AE reporting also provides data to these companies and drug regulatory authorities that play a key role in assessing the risk-benefit profile of a given drug. The following are several facets of AE reporting:

24.2.1 The "4 Elements" of an Individual Case Safety Report

One of the fundamental principles of adverse event reporting is the determination of what constitutes an Individual Case Safety Report (ICSR). During the triage phase of a potential adverse event report, it is important to determine if the "four elements" of a valid ICSR are present: (1) an identifiable patient, (2) an identifiable reporter, (3) a suspect drug, and (4) an adverse event.

If one or more of these four elements is missing, the case is not a valid ICSR. Although there are no exceptions to this rule there may be circumstances that may require a judgment call. For example, the term "identifiable" may not always be clear-cut. If a physician reports that he/she has a patient X taking drug Y who experienced Z (an AE), but refuses to provide any specifics about patient X, the report is still a valid case even though the patient is not specifically identified. This is because the reporter has first-hand information about the patient and is *identifiable* (i.e. a real person) to the physician. Identifiability is important so as not only to prevent duplicate reporting of the same case, but also to permit follow-up for additional information.

The concept of identifiability also applies to the other three elements. Although uncommon, it is not unheard of for fictitious adverse event "cases" to be reported to a company by an anonymous individual (or on behalf of an anonymous patient, disgruntled employee, or former employee) trying to damage the company's reputation or a company's product. In these and all other situations, the source of the report should be ascertained (if possible). But anonymous reporting is also important, as whistle blower protection is not granted in all countries. In general, the drug must also be specifically named. Note that in different countries and regions of the world, drugs are sold under various tradenames. In addition, there are a large number of generics which may be mistaken for the trade product. Finally, there is the problem of counterfeit drugs producing adverse events. If at all possible, it is best to try to obtain the sample which induced the adverse event, and send it to either the EMA, FDA or other government agency responsible for investigating AE reports.

If a reporter can't recall the name of the drug they were taking when they experienced an adverse event, this would not be a valid case. This concept also applies to adverse events. If a patient states that they experienced "symptoms", but cannot be more specific, such a report might technically be considered valid, but will be of very limited value to the pharmacovigilance department of the company or to drug regulatory authorities.[5]

24.2.2 Coding of adverse events

Adverse event coding is the process by which information from an AE reporter, called the "verbatim", is coded using standardized terminology from a medical coding dictionary, such as MedDRA (the most commonly used medical coding dictionary). The purpose of medical coding is to convert adverse event information into terminology that can be readily identified and analyzed. For instance, Patient 1 may report that they had experienced "a very bad headache that felt like their head was being hit by a hammer" [Verbatim 1] when taking Drug X. Or, Patient 2 may report that they had experienced a "slight, throbbing headache that occurred daily at about two in the afternoon" [Verbatim 2] while taking Drug Y. Neither Verbatim 1 nor Verbatim 2 will exactly match a code in the MedDRA coding dictionary. However, both quotes describe different manifestations of a headache. As a result, in this example both quotes would be coded as PT Headache (PT = Preferred Term in MedDRA).

24.2.3 Seriousness determination

Although somewhat intuitive, there are a set of criteria within pharmacovigilance that are used to distinguish a serious adverse event from a non-serious one. An adverse event is considered serious if it meets one or more of the following criteria:

1. results in death, or is life-threatening;
2. requires inpatient hospitalization or prolongation of existing hospitalization;
3. results in persistent or significant disability or incapacity;
4. results in a congenital anomaly (birth defect); or
5. is otherwise "medically significant" —i.e., that it does not meet preceding criteria, but is considered serious because treatment/intervention would be required to prevent one of the preceding criteria.[5]

Aside from death, each of these categories is subject to some interpretation. Life-threatening, as it used in the drug safety world, specifically refers to an adverse event that places the patient at an *immediate risk of death*, such as cardiac or respiratory arrest. By this definition, events such as myocardial infarction, which would be hypothetically life-threatening, would not be considered life-threatening unless the patient went into cardiac arrest following the MI. Defining what constitutes hospitalization can be problematic as well. Although typically straightforward, it's possible for a hospitalization to occur even if the events being treated are not serious. By the same token, serious events may be treated without hospitalization, such as the treatment of anaphylaxis may be successfully performed with epinephrine. Significant disability and incapacity, as a concept, is also subject to debate. While permanent disability following a stroke would no doubt be serious, would "complete blindness for 30 seconds" be considered "significant disability"? For birth defects, the seriousness of the event is usually not in dispute so much as the attribution of the event to the drug. Finally, "medically significant events" is a category that includes events that may be always serious, or sometimes serious, but will not fulfill any of the other criteria. Events such as cancer might always be considered serious, whereas liver disease, depending on its CTCAE (Common Terminology Criteria for Adverse Events) grade—Grades 1 or 2 are generally considered non-serious and Grades 3-5 serious—may be considered non-serious.[6]

24.2.4 Expedited reporting

This refers to ICSRs (individual case safety reports) that involve a serious and unlisted event (an event not described in the drug's labeling) that is considered related to the use of the drug. (Spontaneous reports are typically considered to have a positive causality, whereas a clinical trial case will typically be assessed for causality by the clinical trial investigator and/or the license holder.) In most countries, the timeframe for reporting expedited cases is 7/15 calendar days from the time a drug company receives notification (referred to as "Day 0") of such a case. Within clinical trials such a case is referred to as a SUSAR (a Suspected Unexpected Serious Adverse Reaction). If the SUSAR involves an event that is life-threatening or fatal, it may be subject to a 7-day "clock". Cases that do not involve a serious, unlisted event may be subject to non-expedited or periodic reporting.

24.2.5 Clinical trial reporting

Also known as SAE (serious adverse event) reporting from clinical trials, safety information from clinical studies is used to establish a drug's safety profile in humans and is a key component that drug regulatory authorities consider in the decision-making as to whether to grant or deny market authorization (market approval) for a drug. SAE reporting occurs as a result of study patients (subjects) who experience serious adverse events during the conducting of clinical trials. (Non-serious adverse events are also captured separately.) SAE information, which may also include relevant information from the patient's medical background, are reviewed and assessed for causality by the study investigator. This information is forwarded to a sponsoring entity (typically a pharmaceutical company) that is responsible for the reporting of this information, as appropriate, to drug regulatory authorities.

24.2.6 Spontaneous reporting

Spontaneous reports are termed spontaneous as they take place during the clinician's normal diagnostic appraisal of a patient, when the clinician is drawing the conclusion that the drug may be implicated in the causality of the event. Spontaneous reporting system relies on vigilant physicians and other healthcare professionals who not only generate a suspicion of an ADR, but also report it. It is an important source of regulatory actions such as taking a drug off the market or a label change due to safety problems. Spontaneous reporting is the core data-generating system of international pharmacovigilance, relying on healthcare professionals (and in some countries consumers) to identify and report any adverse events to their national pharmacovigilance center, health authority (such as EMA or FDA), or to the drug manufacturer itself.[7] Spontaneous reports are, by definition, submitted voluntarily although under certain circumstances these reports may be encouraged, or "stimulated", by media reports or articles published in medical or scientific publications, or by product lawsuits. In many parts of the world adverse event reports are submitted electronically using a defined message standard.[8][9]

One of the major weaknesses of spontaneous reporting is that of under-reporting, where, unlike in clinical trials, less than 100% of those adverse events occurring are reported. Further complicating the assessment of adverse events, AE reporting behavior varies greatly between countries and in relation to the seriousness of the events, but in general probably less than 10% (some studies suggest less than 5%) of all adverse events that occur are actually reported. The rule-of-thumb is that on a scale of 0 to 10, with 0 being least likely to be reported and 10 being the most likely to be reported, an uncomplicated non-serious event such as a mild headache will be closer to a "0" on this scale, whereas a life-threatening or fatal event will be closer to a "10" in terms of its likelihood of being reported. In view of this, medical personnel may not always see AE reporting as a priority, especially if the symptoms are not serious. And even if the symptoms are serious, the symptoms may not be recognized as a possible side effect of a particular drug or combination thereof. In addition, medical personnel may not feel compelled to report events that are viewed as expected. This is why reports from patients themselves are of high value. The confirmation of these events by a healthcare professional is typically considered to increase the value of these reports. Hence it is important not only for the patient to report the AE to his health care provider (who may neglect to report the AE), but also report the AE to both the biopharmaceutical company and the FDA, EMA, ... This is especially important when one has obtained one's pharmaceutical from a compounding pharmacy.

As such, spontaneous reports are a crucial element in the worldwide enterprise of pharmacovigilance and form the core of the World Health Organization Database, which includes around 4.6 million reports (January 2009),[10] growing annually by about 250,000.[11]

24.2.7 Aggregate reporting

Aggregate reporting, also known as periodic reporting, plays a key role in the safety assessment of drugs. Aggregate reporting involves the compilation of safety data for a drug over a prolonged period of time (months or years), as opposed to single-case reporting which, by definition, involves only individual AE reports. The advantage of aggregate reporting is that it provides a broader view of the safety profile of a drug. Worldwide, the most important aggregate report is the Periodic Safety Update Report (PSUR) and Development safety updated report (DSUR). This is a document that is submitted to drug regulatory agencies in Europe, the US and Japan (ICH countries), as well as other countries around the world. The PSUR was updated in 2012 and is now referred to in many countries as the Periodic Benefit Risk Evaluation report (PBRER). As the title suggests, the PBRER's focus is on the benefit-risk profile of the drug, which includes a review of relevant safety data compiled for a drug product since its development.

24.2.8 Other reporting methods

Some countries legally oblige spontaneous reporting by physicians. In most countries, manufacturers are required to submit, through its Qualified Person for Pharmacovigilance (QPPV), all of the reports they receive from healthcare providers to the national authority. Others have intensive, focused programmes concentrating on new drugs, or on controversial drugs, or on the prescribing habits of groups of doctors, or involving pharmacists in reporting. All of these generate potentially useful information. Such intensive schemes, however, tend to be the exception.

24.3 Risk management

Risk management is the discipline within pharmacovigilance that is responsible for signal detection and the monitoring of the risk-benefit profile of drugs. Other key activities within the area of risk management are that of the compilation of risk management plans (RMPs) and aggregate reports such as the Periodic Safety Update Report (PSUR), Periodic Benefit-Risk Evaluation Report (PBRER), and the Development Safety Update Report (DSUR).

24.3.1 Causality assessment

One of the most important, and challenging, problems in pharmacovigilance is that of the determination of causality. Causality refers to the relationship of a given adverse event to a specific drug. Causality determination (or assessment) is often difficult because of the lack of clear-cut or reliable data. While one may assume that a positive temporal relationship might "prove" a positive causal relationship, this is not always the case. Indeed, a "bee sting" AE—where the AE can clearly be attributed to a specific cause—is by far the exception rather than the rule. This is due to the complexity of human physiology as well as that of disease and illnesses. By this reckoning, in order to determine causality between an adverse event and a drug, one must first exclude the possibility that there were other possible causes or contributing factors. If the patient is on a number of medications, it may be the combination of these drugs which causes the AE, and not any one individually. There have been a number of recent high-profile cases where the AE led to the death of an individual. The individual(s) were not overdosed with any one of the many medications they were taking, but the combination there appeared to cause the AE. Hence it is important to include in your/one's AE report, not only the drug being reported, but also all other drugs the patient was also taking.

For instance, if a patient were to start Drug X and then three days later were to develop an AE, one might be tempted to attribute blame Drug X. However, before that can be done, the patient's medical history would need to be reviewed to look for possible risk factors for the AE. In other words, did the AE occur with the drug or *because* of the drug? This is because a patient on any drug may develop or be diagnosed with a condition that could not have possibly been caused by the drug. This is especially true for diseases, such as cancer, which develop over an extended period of time, being diagnosed in a patient who has been taken a drug for a relatively short period of time. On the other hand, certain adverse events, such as blood clots (thrombosis), can occur with certain drugs with only short-term exposure. Nevertheless, the determination of risk factors is an important step of confirming or ruling-out a causal relationship between an event and a drug.

Often the only way to confirm the existence of a causal relationship of an event to a drug is to conduct an observational study where the incidence of the event in a patient population taking the drug is compared to a control group. This may be necessary to determine if the background incidence of an event is less than that found in a group taking a drug. If the incidence of an event is statistically significantly higher in the "active" group versus the placebo group (or other control group), it is possible that a causal relationship may exist to a drug, unless other confounding factors may exist.

24.3.2 Signal detection

Signal detection (SD) involves a range of techniques (CIOMS VIII). The WHO defines a safety signal as: "Reported information on a possible causal relationship between an adverse event and a drug, the relationship being unknown or incompletely documented previously". Usually more than a single report is required to generate a signal, depending upon the event and quality of the information available.

Data mining pharmacovigilance databases is one approach

that has become increasingly popular with the availability of extensive data sources and inexpensive computing resources. The data sources (databases) may be owned by a pharmaceutical company, a drug regulatory authority, or a large healthcare provider. Individual Case Safety Reports (ICSRs) in these databases are retrieved and converted into structured format, and statistical methods (usually a mathematical algorithm) are applied to calculate statistical measures of association. If the statistical measure crosses an arbitrarily set threshold, a signal is declared for a given drug associated with a given adverse event. All signals deemed worthy of investigation, require further analysis using all available data in an attempt to confirm or refute the signal. If the analysis is inconclusive, additional data may be needed such as a post-marketing observational trial.

SD is an essential part of drug use and safety surveillance. Ideally, the goal of SD is to identify ADRs that were previously considered unexpected and to be able to provide guidance in the product's labeling as to how to minimize the risk of using the drug in a given patient population.

24.3.3 Risk management plans

A risk management plan (RMP) is a documented plan that describes the risks (adverse drug reactions and potential adverse reactions) associated with the use of a drug and how they are being handled (warning on drug label or on packet inserts of possible side effects which if observed should cause the patient to inform/see his physician and/or pharmacist and/or the manufacturer of the drug and/or the FDA, EMA)). The overall goal of an RMP is to assure a positive risk-benefit profile once the drug is (has been) marketed. The document is required to be submitted, in a specified format, with all new market authorization requests within the European Union (EU). Although not necessarily required, RMPs may also be submitted in countries outside the EU. The risks described in an RMP fall into one of three categories: identified risks, potential risks, and unknown risks. Also described within an RMP are the measures that the Market Authorization Holder, usually a pharmaceutical company, will undertake to minimize the risks associated with the use of the drug. These measures are usually focused on the product's labeling and healthcare professionals. Indeed, the risks that are documented in a pre-authorization RMP will inevitably become part of the product's post-marketing labeling. Since a drug, once authorized, may be used in ways not originally studied in clinical trials, this potential "off-label use", and its associated risks, is also described within the RMP. RMPs can be very lengthy documents, running in some cases hundreds of pages and, in rare instances, up to a thousand pages long.

In the US, under certain circumstances, the FDA may require a company to submit a document called a Risk Evaluation and Mitigation Strategies (REMS) for a drug that has a specific risk that FDA believes requires mitigation. While not as comprehensive as an RMP, a REMS can require a sponsor to perform certain activities or to follow a protocol, referred to as Elements to Assure Safe Use (ETASU),[12] to assure that a positive risk-benefit profile for the drug is maintained for the circumstances under which the product is marketed.

24.3.4 Risk/benefit profile of drugs

Pharmaceutical companies are required by law in most countries to perform clinical trials, testing new drugs on people before they are made generally available. This occurs after a drug has been pre-screened for toxicity, sometimes using animals for testing. The manufacturers or their agents usually select a representative sample of patients for whom the drug is designed – at most a few thousand – along with a comparable control group. The control group may receive a placebo and/or another drug, often a so-called "gold standard" that is "best" drug marketed for the disease.

The purpose of clinical trials is to determine:

- if a drug works and how well it works
- if it has any harmful effects, and
- if it does more good than harm, and how much more? If it has a potential for harm, how probable and how serious is the harm?

Clinical trials do, in general, tell a good deal about how well a drug works. They provide information that should be reliable for larger populations with the same characteristics as the trial group – age, gender, state of health, ethnic origin, and so on though target clinical populations are typically very different from trial populations with respect to such characteristics.

The variables in a clinical trial are specified and controlled, but a clinical trial can never tell you the whole story of the effects of a drug in all situations. In fact, nothing could tell you the whole story, but a clinical trial must tell you enough; "enough" being determined by legislation and by contemporary judgements about the acceptable balance of benefit and harm. Ultimately, when a drug is marketed it may be used in patient populations that were not studied during clinical trials (children, the elderly, pregnant women, patients with co-morbidities not found in the clinical trial population, etc.) and a different set of warnings, precautions or contraindications (where the drug should not be used at all) for the product's labeling may be necessary in order to maintain a positive risk/benefit profile in all known populations using the drug.

24.3.5 Pharmacoepidemiology

Pharmacoepidemiology is the study of the incidence of adverse drug reactions in patient populations using drug agents.[13]

24.3.6 Pharmacogenetics and pharmacogenomics

Although often used interchangeably, there are subtle differences between the two disciplines. Pharmacogenetics is generally regarded as the study or clinical testing of genetic variation that gives rise to differing responses to drugs, including adverse drug reactions. It is hoped that pharmacogenetics will eventually provide information as to which genetic profiles in patients will place those patients at greatest risk, or provide the greatest benefit, for using a particular drug or drugs. Pharmacogenomics, on the other hand, is the broader application of genomic technologies to new drug discovery and further characterization of older drugs.

24.4 International collaboration

The following organizations play a key collaborative role in the global oversight of pharmacovigilance.

24.4.1 The World Health Organization (WHO)

The principle of international collaboration in the field of pharmacovigilance is the basis for the WHO International Drug Monitoring Programme, through which over 100 member nations have systems in place that encourage healthcare personnel to record and report adverse effects of drugs in their patients, reports that are assessed locally and may lead to action within the country. Member countries send their reports to the Uppsala Monitoring Centre where they are processed, evaluated and entered into the WHO International Database; through membership in the WHO Programme one country can know if similar reports are being made elsewhere. When there are several reports of adverse reactions to a particular drug, this process may lead to the detection of a signal, and an alert about a possible hazard communicated to members countries after detailed evaluation and expert review.

24.4.2 The International Council for Harmonisation (ICH)

The ICH is a global organization with members from the European Union, the United States and Japan; its goal is to recommend global standards for drug companies and drug regulatory authorities around the world, with the ICH Steering Committee (SC) overseeing harmonization activities. Established in 1990, each of its six co-sponsors—the EU, the European Federation of Pharmaceutical Industries and Associations (EFPIA), Japan's Ministry of Health, Labor and Welfare (MHLW), the Japanese Pharmaceutical Manufacturers Association (JPMA), the U.S. Food and Drug Administration (FDA), and the Pharmaceutical Research and Manufacturers of America (PhRMA)—have two seats on the SC. Other parties have a significant interest in ICH and have been invited to nominate Observers to the SC; three current observers are the WHO, Health Canada, and the European Free Trade Association (EFTA), with the International Federation of Pharmaceutical Manufacturers Association (IFPMA) participating as a non-voting member of the SC.[14][15]

24.4.3 The Council for International Organizations of Medical Science (CIOMS)

The CIOMS, a part of the WHO, is a globally oriented think tank that provides guidance on drug safety related topics through its Working Groups. The CIOMS prepares reports that are used as a reference for developing future drug regulatory policy and procedures, and over the years, many of CIOMS' proposed policies have been adopted. Examples of topics these reports have covered include: Current Challenges in Pharmacovigilance: Pragmatic Approaches (CIOMS V); Management of Safety Information from Clinical Trials (CIOMS VI); the Development Safety Update Report (DSUR): Harmonizing the Format and Content for Periodic Safety Reporting During Clinical Trials (CIOMS VII); and Practical Aspects of Signal Detection in Pharmacovigilance: Report of CIOMS Working Group (CIOMS VIII).

24.4.4 The International Society of Pharmacovigilance (ISoP)

The ISoP is an international non-profit scientific organization, which aims to foster pharmacovigilance both scientifically and educationally, and enhance all aspects of the safe and proper use of medicines, in all countries.[16] It was established in 1992 as the European Society of Pharmacovigilance.[17]

24.5 Regulatory authorities

Drug regulatory authorities play a key role in national or regional oversight of pharmacovigilance. Some of the agencies involved are listed below (in order of 2011 spending on pharmaceuticals, from the IMS Institute for Healthcare Informatics).[18]

24.5.1 United States

See also: Regulation of therapeutic goods in the United States

In the U.S., with about a third of all global 2011 pharmaceutical expenditures,[18] the drug industry is regulated by the FDA, the largest national drug regulatory authority in the world. FDA authority is exercised through enforcement of regulations derived from legislation, as published in the U.S. Code of Federal Regulations (CFR); the principal drug safety regulations are found in 21 CFR Part 312 (IND regulations) and 21 CFR Part 314 (NDA regulations). While those regulatory efforts address pre-marketing concerns, pharmaceutical manufacturers and academic/non-profit organizations such as RADAR and Public Citizen do play a role in pharmacovigilance in the US. The post-legislative rule-making process of the U.S. federal government provides for significant input from both the legislative and executive branches, which also play specific, distinct roles in determining FDA policy.

24.5.2 Emerging economies (including Latin America)

The "pharmerging", or emerging pharmaceutical market economies, which include Brazil, India, Russia, Argentina, Egypt, Indonesia, Mexico, Pakistan, Poland, Romania, South Africa, Thailand, Turkey, Ukraine, Venezuela, and Vietnam, accrued one fifth of global 2011 pharmaceutical expenditures; in future, aggregated data for this set will include China as well.[18]

China's economy is anticipated to pass Japan to become second in the ranking of individual countries' in pharmaceutical purchases by 2015, and so its PV regulation will become increasing important; China's regulation of PV is through its National Center for Adverse Drug Reaction (ADR) Monitoring, under China's Ministry of Health.[19]

As JE Sackman notes, as of April 2013 "there is no Latin American equivalent of the European Medicines Agency—no common body with the power to facilitate greater consistency across countries".[20] For simplicity, and per sources, 17 smaller economies are discussed alongside the 4 pharmemerging large economies of Argentina, Brazil, Mexico and Venzuala—Bolivia, Chile, Colombia, Costa Rica, Cuba, Dominican Republic, Ecuador, El Salvador, Guatemala, Haiti, Honduras, Nicaragua, Panama, Paraguay, Peru, Suriname, and Uruguay.[21] As of June 2012, 16 of this total of 21 countries have systems for immediate reporting and 9 have systems for periodic reporting of adverse events for on-market agents, while 10 and 8, respectively, have systems for immediate and periodic reporting of adverse events during clinical trials; most of these have PV requirements that rank as "high or medium...in line with international standards" (*ibid.*). The WHO's Pan American Network for Drug Regulatory Harmonization[22] seeks to assist Latin American countries in develop harmonized PV regulations.[21]

Some further PV regulatory examples from the pharmerging nations are as follows. In India, the PV regulatory authority is the Indian Pharmacopoeia Commission, with a National Coordination Centre under the Pharmacovigilance Program of India, in the Ministry of Health and Family Welfare.[23][24] Scientists working on pharmacovigilance share their experiences, findings, innovative ideas and researches during the annual meeting of Society of Pharmacovigilance, India. In Egypt, PV is regulated by the Egyptian Pharmacovigilance Center of the Egyptian Ministry of Health.

24.5.3 European Union

The EU5 (France, Germany, Italy, Spain, United Kingdom) accrued ~17% of global 2011 pharmaceutical expenditures.[18] PV efforts in the EU are coordinated by the European Medicines Agency (EMA) and are conducted by the national competent authorities (NCAs). The main responsibility of the EMA is to maintain and develop the pharmacovigilance database consisting of all suspected serious adverse reactions to medicines observed in the European Community; the data processing network and management system is called EudraVigilance and contains separate but similar databases of human and veterinary reactions.[25] The EMA requires the individual marketing authorization holders to submit all received adverse reactions in electronic form, except in exceptional circumstances; the reporting obligations of the various stakeholders are defined by EEC legislation, namely Regulation (EC) No 726/2004, and for human medicines, European Union Directive 2001/83/EC as amended and Directive 2001/20/EC. In 2002, Heads of Medicines Agencies[26] agreed on a mandate for an ad hoc Working Group on establishing a European risk management strategy; the Working Group considered the conduct of a high level survey of EU pharmacovigilance resources to promote the utilization of expertise and encourage collaborative working.

In conjunction with this oversight, individual countries maintain their distinct regulatory agencies with PV responsibility.[27] For instance, in Spain, PV is regulated by the Agencia Española de Medicamentos y Productos Sanitarios (AEMPS), a legal entity that retains the right to suspend or withdraw the authorization of pharmaceuticals already on-market if the evidence shows that safety (or quality or efficacy) of an agent are unsatisfactory.[28]

24.5.4 Rest of Europe, including non-EU

The remaining EU and non-EU countries outside the EU5 accrued ~7% of global 2011 pharmaceutical expenditures.[18] Regulation of those outside the EU being managed by specific governmental agencies. For instance, in Switzerland, PV "inspections" for clinical trials of medicinal products are conducted by the Swiss Agency for Therapeutic Products.[29]

24.5.5 Japan

In Japan, with ~12% of all global 2011 pharmaceutical expenditures,[18] PV matters are regulated by the Pharmaceuticals and Medical Devices Agency (PMDA) and the Ministry of Health, Labour, and Welfare MHLW.

24.5.6 Canada

In Canada, with ~2% of all global 2011 pharmaceutical expenditures,[18] PV is regulated by the Marketed Health Products Directorate of the Health Products and Food Branch (Health Canada).

24.5.7 South Korea

The Republic of Korea, with ~1% of all global 2011 pharmaceutical expenditures,[18] PV matters are regulated in South Korea by the Ministry Of Food And Drug Safety (MFDS) and the Ministry of Health, Labour, and Welfare MHLW.

24.5.8 Africa

Kenya

In Kenya, PV is regulated by the Pharmacy and Poisons Board.The Pharmacy and Poisons Board provides a Pharmacovigilance Electronic Reporting System which allows for the online reporting of suspected adverse drug reactions as well as suspected poor quality of medicinal products.[30]

The Pharmacovigilance activities in Kenya are supported by the School of Pharmacy, University of Nairobi through its Master of Pharmacy in Pharmacoepidemiology & Pharmacovigilance (M.Pharm. EpiVigil) program offered by the Department of Pharmacology and Pharmacognosy.[31]

In Uganda, PV is regulated by the National Drug Authority.

24.5.9 Rest of world (ROW)

ROW accrued ~7% of global 2011 pharmaceutical expenditures.[18] Some examples of PV regulatory agencies in ROW are as follows. In Iraq, PV is regulated by the Iraqi Pharmacovigilance Center of the Iraqi Ministry of Health.

24.6 Ecopharmacovigilance (EPV; pharmacoenvironmentology)

Despite attention from the FDA and regulatory agencies of the European Union, procedures for monitoring drug concentrations and adverse effects in the environment are lacking. Pharmaceuticals, their metabolites, and related substances may enter the environment after patient excretion, after direct release to waste streams during manufacturing or administration, or via terrestrial deposits (e.g., from waste sludges or leachates).[32] A concept combining pharmacovigilance and environmental pharmacology, intended to focus attention on this area, was introduced first as *ecopharmacology* and later as *ecopharmacovigilance* (EPV) by Velo et al., with further concurrent and later terms for the same concept (pharmacoenvironmentology, environmental pharmacology, ecopharmacostewardship).[32][33][34][35]

The first of these routes to the environment, elimination through living organisms subsequent to pharmacotherapy, is suggested as the principal source of environmental contamination (apart from cases where norms for treatment of manufacturing and other wastes are violated), and EPV is intended to deal specifically with this impact of pharmacological agents on the environment.[32][36]

Activities of EPV have been suggested to include:

- Increasing, generally, the availability of environmental data on medicinal products;
- Tracking emerging data on environmental exposure, effects and risks after product launch;
- Using Environmental Risk Management Plans (ERMPs) to manage risk throughout a drug's life cycle;

- Following risk identification, promoting further research and environmental monitoring, and
- In general, promoting a global perspective on EPV issues.[32]

24.7 Related to medical devices

A medical device is an instrument, apparatus, implant, *in vitro* reagent, or similar or related article that is used to diagnose, prevent, or treat disease or other conditions, and does not achieve its purposes through chemical action within or on the body (which would make it a drug). Whereas medicinal products (also called pharmaceuticals) achieve their principal action by pharmacological, metabolic or immunological means, medical devices act by physical, mechanical, or thermal means. Medical devices vary greatly in complexity and application. Examples range from simple devices such as tongue depressors, medical thermometers, and disposable gloves to advanced devices such as medical robots, cardiac pacemakers, and neuroprosthetics.

Given the inherent difference between medicinal products and medical products, the vigilance of medical devices is also different from that of medicinal products. To reflect this difference, a classification system has been adopted in some countries to stratify the risk of failure with the different classes of devices. The classes of devices typically run on a 1-3 or 1-4 scale, with Class 1 being the least likely to cause significant harm with device failure versus Classes 3 or 4 being the most likely to cause significant harm with device failure. An example of a device in the "low risk" category would be contact lenses. An example of a device in the "high risk" category would be cardiac pacemakers.

Medical device reporting (MDR), which is the reporting of adverse events with medical devices, is similar to that with medicinal products, although there are differences. For instance, in the US user-facilities such as hospitals and nursing homes are legally required to report suspected medical device-related deaths to both FDA and the manufacturer, if known, and serious injuries to the manufacturer or to FDA, if the manufacturer is unknown.[37] This is in contrast to the voluntary reporting of AEs with medicinal products.

24.8 For herbal medicines

The safety of herbal medicines has become a major concern to both national health authorities and the general public.[38] The use of herbs as traditional medicines continues to expand rapidly across the world; many people now take herbal medicines or herbal products for their health care in different national health-care settings. However, mass media reports of adverse events with herbal medicines can be incomplete and therefore misleading. Moreover, it can be difficult to identify the causes of herbal medicine-associated adverse events since the amount of data on each event is generally less than for pharmaceuticals formally regulated as drugs (since the requirements for adverse event reporting are either non-existent or are less stringent for herbal supplements and medications).[39]

24.9 Industry associations

24.10 See also

- Adverse drug reaction
- Adverse effect (medicine)
- Boston Collaborative Drug Surveillance Program
- Consultant pharmacist
- COSTART – Coding Symbols for a Thesaurus of Adverse Reaction Terms
- DrugLogic
- European Medicines Agency
- EudraVigilance
- Food and Drug Administration (US FDA)
- International Society for Pharmacoepidemiology
- International Society of Pharmacovigilance
- Society of Pharmacovigilance, India (SoPI)
- MedDRA – Medical Dictionary for Regulatory Activities
- Medical device
- MedWatch
- National Drug & Safety League
- Public health
- Pharmacoepidemiology
- Pharmacotherapy
- Pharmacogenetics
- Pharmacogenomics
- Pharmaceuticals and personal care products in the environment

- Proportional reporting ratio
- Uppsala Monitoring Centre
- WHOART
- Yellow Card Scheme (UK)
- Black triangle (pharmacovigilance)

24.11 References

[1] Source: The Importance of Pharmacovigilance, WHO 2002

[2] "DIRECTIVE 2010/84/EU, Article 101" (PDF). ec.europa.eu. Retrieved 2015-10-26.

[3] "International Drug Monitoring : The Role of National Centers" (PDF). Who-umc.org. Retrieved 2015-02-27.

[4] "Preparing for a Safety Inspection" (PDF). Steve Jolley. 2011. Retrieved 2016-01-04.

[5] *Current Challenges in Pharmacovigilance: Pragmatic Approaches* (Report of CIOMS Working Group V), 2001 Geneva.

[6] "Common Terminology Criteria for Adverse Events (CTCAE) : Version 4.0" (PDF). Acrin.org. Retrieved 2015-02-27.

[7] Lindquist M. Vigibase, the WHO Global ICSR Database System: Basic Facts. *Drug Information Journal*, 2008, 42:409-419.

[8] Archived November 10, 2010, at the Wayback Machine.

[9] Archived August 23, 2010, at the Wayback Machine.

[10] "Uppsala Monitoring Centre". who-umc.org. Retrieved 2015-02-27.

[11] *Pharmacovigilance*. Mann RD, Andrews EB, eds. John Wiley & Sons Ltd, Chichester, 2002. spontaneous reports are very useful.

[12] "(REMS)Risk Evaluation Mitigation Strategies and ETASU programs". Ireminder.com. Retrieved 2015-02-27.

[13] Strom, Brian (2006). *Textbook of Pharmacoepidemiology*. West Sussex, England: John Wiley and Sons. p. 3. ISBN 978-0-470-02925-1.

[14] "Organisation". ICH. Retrieved 2015-02-27.

[15] "Steering Committee". ICH. Retrieved 2015-02-27.

[16] "ISoP : International Society of Pharmacovigilance - Home". Isoponline.org. Retrieved 2015-02-27.

[17] "The Importance of Pharmacovigilance - Safety Monitoring of Medicinal Products: Chapter 2 - A Short History of Involvement in Drug Safety Monitoring by WHO". Apps.who.int. 2015-02-10. Retrieved 2015-02-27.

[18] "Why Big Pharma Won't Get Its Piece Of The $1.2 Trillion Global Drug Market". Forbes. 2012-12-07. Retrieved 2015-02-27.

[19] "China's Pharmacovigilance System: The Hunger For Safety Insights". Clinicalleader.com. 2011-12-07. Retrieved 2015-02-27.

[20] "Navigating Emerging Markets — Latin America". Biopharminternational.com. 2013-04-01. Retrieved 2015-02-27.

[21] Hoffmann E, Fouretier A, Vergne C, et al. (2012). "Pharmacovigilance Regulatory Requirements in Latin America". *Pharm Med.* **26** (3): 153–164. doi:10.1007/bf03262389.

[22] "Pan American Network for Drug Regulatory Harmonization". Paho.org. Retrieved 2015-02-27.

[23] "Indian Pharmacopoeia Commission, Ministry of Health & Family Welfare, Government of India". Ipc.gov.in. Retrieved 2015-02-27.

[24] "Central Drugs Standard Control Organization". Cdsco.nic.in. Retrieved 2015-02-27.

[25] "Safety monitoring of medicines" European Medicines Agency

[26] "Heads of Medicines Agencies: Home". Heads.medagencies.org. Retrieved 2015-02-27.

[27] GUERRIAUD, Mathieu (2016-02-01). "Pharmacovigilance européenne, un système aux multiples visages". *BNDS* (in French). Retrieved 2016-07-05.

[28] "Generalidades". Se-fc.org. Retrieved 2015-02-27.

[29] "GCP and Pharmacovigilance Inspections - Swissmedic". Swissmedic.ch. Retrieved 2015-02-27.

[30] "Pharmacovigilance Electronic Reporting System - Kenya". Retrieved 2015-02-27.

[31] "Master of Pharmacy in Pharmacoepidemiology & Pharmacovigilance". Retrieved 28 Jun 2016.

[32] Holm, G; Snape, JR; Murray-Smith, R; Talbot, J; Taylor, D; Sörme, P (2013). "Implementing Ecopharmacovigilance in Practice: Challenges and Potential Opportunities". *Drug Saf.* **36**: 533–46. PMC 3691479 ə. PMID 23620169. doi:10.1007/s40264-013-0049-3.

[33] Rahman, SZ; Khan, RA (Dec 2006). "Environmental pharmacology: A new discipline". *Indian J Pharmacol.* **38** (4): 229–30. doi:10.4103/0253-7613.27017.

[34] Rahman, SZ; Khan RA; Gupta V; Misbah Uddin (2008). "Chapter 2: Pharmacoenvironmentology – Ahead of Pharmacovigilance". In Rahman SZ, Shahid M & Gupta A. *An Introduction to Environmental Pharmacology* (1st ed.). Aligarh: Ibn Sina Academy. pp. 35–52. ISBN 978-81-906070-4-9.

[35] Ruhoy, IS; Daughton, CG (2008). "Beyond the medicine cabinet: An analysis of where and why medications accumulate". *Environment International*. **34** (8): 1157–1169. doi:10.1016/j.envint.2008.05.002.

[36] Rahman, SZ; Khan, RA; Gupta, V; Uddin, Misbah (July 2007). "Pharmacoenvironmentology – A Component of Pharmacovigilance". *Environmental Health*. **6** (20). PMC 1947975 ☉. PMID 17650313. doi:10.1186/1476-069X-6-20.

[37] "Reporting By Health Professionals". Fda.gov. Retrieved 2015-02-27.

[38] WHO guidelines on safety monitoring of herbal medicines in pharmacovigilance systems, World Health Organization, Geneva, 2004

[39] S Z Rahman & K C Singhal, Problems in pharmacovigilance of medicinal products of herbal origin and means to minimize them, Uppsala Reports, WHO Collaborating Center for ADR monitoring, Uppsala Monitoring Centre, Sweden, Issue 17 January 2002: 1-4 (Supplement)

24.12 External links

- Current Problems in Pharmacovigilance
- "Guidance for Industry. Good Pharmacovigilance Practices and Pharmacoepidemiologic Assessment" (PDF). Food and Drug Administration. March 2005. Retrieved 2009-07-20.
- Public-access adverse event reference
- Volume 9 of "The rules governing medicinal products in the European Union" contains Pharmacovigilance guidelines for medicinal products for both human and veterinary use.

Chapter 25

Structural genomics

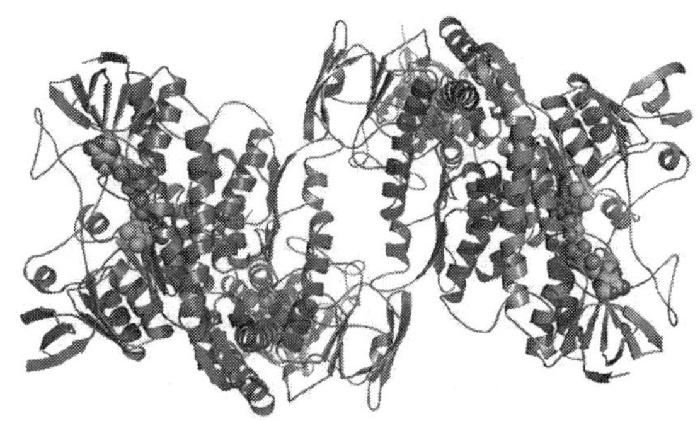

An example of a protein structure from Protein Data Bank.

Structural genomics has two meanings, one of which seeks to describe the 3-dimensional structure of every protein encoded by a given genome. This genome-based approach allows for a high-throughput method of structure determination by a combination of experimental and modeling approaches. The principal difference between structural genomics and traditional structural prediction is that structural genomics attempts to determine the structure of every protein encoded by the genome, rather than focusing on one particular protein. With full-genome sequences available, structure prediction can be done more quickly through a combination of experimental and modeling approaches, especially because the availability of large number of sequenced genomes and previously solved protein structures allows scientists to model protein structure on the structures of previously solved homologs.

Because protein structure is closely linked with protein function, the structural genomics has the potential to inform knowledge of protein function. In addition to elucidating protein functions, structural genomics can be used to identify novel protein folds and potential targets for drug discovery. Structural genomics involves taking a large number of approaches to structure determination, including experimental methods using genomic sequences or modeling-based approaches based on sequence or structural homology to a protein of known structure or based on chemical and physical principles for a protein with no homology to any known structure.

As opposed to traditional structural biology, the determination of a protein structure through a structural genomics effort often (but not always) comes before anything is known regarding the protein function. This raises new challenges in structural bioinformatics, i.e. determining protein function from its 3D structure.

Structural genomics emphasizes high throughput determination of protein structures. This is performed in dedicated centers of structural genomics.

While most structural biologists pursue structures of individual proteins or protein groups, specialists in structural genomics pursue structures of proteins on a genome wide scale. This implies large-scale cloning, expression and purification. One main advantage of this approach is economy of scale. On the other hand, the scientific value of some resultant structures is at times questioned. A *Science* article from January 2006 analyzes the structural genomics field.[1]

One advantage of structural genomics, such as the Protein Structure Initiative, is that the scientific community gets immediate access to new structures, as well as to reagents such as clones and protein. A disadvantage is that many of these structures are of proteins of unknown function and do not have corresponding publications. This requires new ways of communicating this structural information to the broader research community. The Bioinformatics core of the Joint center for structural genomics (JCSG) has recently developed a wiki-based approach namely Open protein structure annotation network (TOPSAN) for annotating protein structures emerging from high-throughput structural genomics centers.

25.1 Goals

One goal of structural genomics is to identify novel protein folds. Experimental methods of protein structure determination require proteins that express and/or crystallize well, which may inherently bias the kinds of proteins folds that this experimental data elucidate. A genomic, modeling-based approach such as *ab initio* modeling may be better able to identify novel protein folds than the experimental approaches because they are not limited by experimental constraints.

Protein function depends on 3-D structure and these 3-D structures are more highly conserved than sequences. Thus, the high-throughput structure determination methods of structural genomics have the potential to inform our understanding of protein functions. This also has potential implications for drug discovery and protein engineering.[2] Furthermore, every protein that is added to the structural database increases the likelihood that the database will include homologous sequences of other unknown proteins. The Protein Structure Initiative (PSI) is a multifaceted effort funded by the National Institutes of Health with various academic and industrial partners that aims to increase knowledge of protein structure using a structural genomics approach and to improve structure-determination methodology.

25.2 Methods

Structural genomics takes advantage of completed genome sequences in several ways in order to determine protein structures. The gene sequence of the target protein can also be compared to a known sequence and structural information can then be inferred from the known protein's structure. Structural genomics can be used to predict novel protein folds based on other structural data. Structural genomics can also take modeling-based approach that relies on homology between the unknown protein and a solved protein structure.

25.2.1 *de novo* methods

Completed genome sequences allow every open reading frame (ORF), the part of a gene that is likely to contain the sequence for the messenger RNA and protein, to be cloned and expressed as protein. These proteins are then purified and crystallized, and then subjected to one of two types of structure determination: X-ray crystallography and nuclear magnetic resonance (NMR). The whole genome sequence allows for the design of every primer required in order to amplify all of the ORFs, clone them into bacteria, and then express them. By using a whole-genome approach to this traditional method of protein structure determination, all of the proteins encoded by the genome can be expressed at once. This approach allows for the structural determination of every protein that is encoded by the genome.

25.2.2 Modelling-based methods

ab initio modeling

This approach uses protein sequence data and the chemical and physical interactions of the encoded amino acids to predict the 3-D structures of proteins with no homology to solved protein structures. One highly successful method for *ab initio* modeling is the Rosetta program, which divides the protein into short segments and arranges short polypeptide chain into a low-energy local conformation. Rosetta is available for commercial use and for non-commercial use through its public program, Robetta.

Sequence-based modeling

This modeling technique compares the gene sequence of an unknown protein with sequences of proteins with known structures. Depending on the degree of similarity between the sequences, the structure of the known protein can be used as a model for solving the structure of the unknown protein. Highly accurate modeling is considered to require at least 50% amino acid sequence identity between the unknown protein and the solved structure. 30-50% sequence identity gives a model of intermediate-accuracy, and sequence identity below 30% gives low-accuracy models. It has been predicted that at least 16,000 protein structures will need to be determined in order for all structural motifs to be represented at least once and thus allowing the structure of any unknown protein to be solved accurately through modeling.[3] One disadvantage of this method, however, is that structure is more conserved than sequence and thus sequence-based modeling may not be the most accurate way to predict protein structures.

Threading

Threading bases structural modeling on fold similarities rather than sequence identity. This method may help identify distantly related proteins and can be used to infer molecular functions.

25.3 Examples of structural genomics

There are currently a number of on-going efforts to solve the structures for every protein in a given proteome.

25.3.1 *Thermotogo maritima* proteome

One current goal of the Joint Center for Structural Genomics (JCSG), a part of the Protein Structure Initiative (PSI) is to solve the structures for all the proteins in *Thermotogo maritima*, a thermophillic bacterium. *T. maritima* was selected as a structural genomics target based on its relatively small genome consisting of 1,877 genes and the hypothesis that the proteins expressed by a thermophilic bacterium would be easier to crystallize.

Lesley *et al* used *Escherichia coli* to express all the open-reading frames (ORFs) of *T. martima*. These proteins were then crystallized and structures were determined for successfully crystallized proteins using X-ray crystallography. Among other structures, this structural genomics approach allowed for the determination of the structure of the TM0449 protein, which was found to exhibit a novel fold as it did not share structural homology with any known protein.[4]

25.3.2 *Mycobacterium tuberculosis* proteome

The goal of the TB Structural Genomics Consortium is to determine the structures of potential drug targets in *Mycobacterium tuberculosis*, the bacterium that causes tuberculosis. The development of novel drug therapies against tuberculosis are particularly important given the growing problem of multi-drug-resistant tuberculosis.

The fully sequenced genome of *M. tuberculosis* has allowed scientists to clone many of these protein targets into expression vectors for purification and structure determination by X-ray crystallography. Studies have identified a number of target proteins for structure determination, including extracellular proteins that may be involved in pathogenesis, iron-regulatory proteins, current drug targets, and proteins predicted to have novel folds. So far, structures have been determined for 708 of the proteins encoded by *M. tuberculosis*.

25.4 Protein structure databases and classifications

- Protein Data Bank (PDB): repository for protein sequence and structural information
- UniProt: provides sequence and functional information
- Structural Classification of Proteins (SCOP Classifications): hierarchical-based approach
- Class, Architecture, Topology and Homologous superfamily (CATH): hierarchical-based approach

25.5 See also

- Genomics
- Omics
- Structural proteomics
- Protein Structure Initiative

25.6 References

[1] Chandonia JM, Brenner SE (January 2006). "The impact of structural genomics: expectations and outcomes". *Science*. **311** (5759): 347–51. PMID 16424331. doi:10.1126/science.1121018.

[2] Kuhn P, Wilson K, Patch MG, Stevens RC (October 2002). "The genesis of high-throughput structure-based drug discovery using protein crystallography". *Curr Opin Chem Biol*. **6** (5): 704–10. PMID 12413557. doi:10.1016/S1367-5931(02)00361-7.

[3] Baker D, Sali A (October 2001). "Protein structure prediction and structural genomics". *Science*. **294** (5540): 93–6. PMID 11588250. doi:10.1126/science.1065659.

[4] Lesley SA, Kuhn P, Godzik A, et al. (September 2002). "Structural genomics of the Thermotoga maritima proteome implemented in a high-throughput structure determination pipeline". *Proc. Natl. Acad. Sci. U.S.A.* **99** (18): 11664–9. PMC 129326 ට. PMID 12193646. doi:10.1073/pnas.142413399.

25.7 Further reading

- Hooft RW, Vriend G, Sander C, Abola EE (May 1996). "Errors in protein structures". *Nature*. **381** (6580): 272. PMID 8692262. doi:10.1038/381272a0.
- Marsden RL, Lewis TA, Orengo CA (2007). "Towards a comprehensive structural coverage of completed genomes: a structural genomics viewpoint". *BMC Bioinformatics*. **8**: 86. PMC 1829165 ට. PMID 17349043. doi:10.1186/1471-2105-8-86.
- Baker EN, Arcus VL, Lott JS (2003). "Protein structure prediction and analysis as a tool for functional genomics". *Appl. Bioinform*. **2** (3 Suppl): S3–10. PMID 15130810.

- Goulding CW, Perry LJ, Anderson D, et al. (September 2003). "Structural genomics of Mycobacterium tuberculosis: a preliminary report of progress at UCLA". *Biophys. Chem.* **105** (2-3): 361–70. PMID 14499904. doi:10.1016/S0301-4622(03)00101-7.
- Skolnick J, Fetrow JS, Kolinski A (March 2000). "Structural genomics and its importance for gene function analysis". *Nat. Biotechnol.* **18** (3): 283–7. PMID 10700142. doi:10.1038/73723.

25.8 External links

- Protein Structure Initiative (PSI)
- PSI Structural Biology Knowledgebase: A Nature Gateway

Chapter 26

Toxicogenomics

Toxicogenomics is a field of science that deals with the collection, interpretation, and storage of information about gene and protein activity within particular cell or tissue of an organism in response to toxic substances. Toxicogenomics combines toxicology with genomics or other high throughput molecular profiling technologies such as transcriptomics, proteomics and metabolomics.[1][2] Toxicogenomics endeavors to elucidate molecular mechanisms evolved in the expression of toxicity, and to derive molecular expression patterns (i.e., molecular biomarkers) that predict toxicity or the genetic susceptibility to it.

In pharmaceutical research toxicogenomics is defined as the study of the structure and function of the genome as it responds to adverse xenobiotic exposure. It is the toxicological subdiscipline of pharmacogenomics, which is broadly defined as the study of inter-individual variations in whole-genome or candidate gene single-nucleotide polymorphism maps, haplotype markers, and alterations in gene expression that might correlate with drug responses.[3][4] Though the term toxicogenomics first appeared in the literature in 1999[5] it was already in common use within the pharmaceutical industry as its origin was driven by marketing strategies from vendor companies. The term is still not universal accepted, and others have offered alternative terms such as chemogenomics to describe essentially the same area.[6]

The nature and complexity of the data (in volume and variability) demands highly developed processes of automated handling and storage. The analysis usually involves a wide array of bioinformatics and statistics,[7] regularly involving classification approaches.[8]

In pharmaceutical drug discovery and development toxicogenomics is used to study adverse, i.e. toxic, effects, of pharmaceutical drugs in defined model systems in order to draw conclusions on the toxic risk to patients or the environment. Both the EPA and the U.S. Food and Drug Administration currently preclude basing regulatory decision making on genomics data alone. However, they do encourage the voluntary submission of well-documented, quality genomics data. Both agencies are considering the use of submitted data on a case-by-case basis for assessment purposes (e.g., to help elucidate mechanism of action or contribute to a weight-of-evidence approach) or for populating relevant comparative databases by encouraging parallel submissions of genomics data and traditional toxicologic test results.[9]

26.1 Public projects

- Chemical Effects in Biological Systems – Project hosted by the National Institute of Environmental Health Sciences building a knowledgebase of toxicology studies including study design, clinical pathology, and histopathology and toxicogenomics data.[10]
- InnoMed PredTox assessing the value of combining results from omics technologies together with the results from more conventional toxicology methods in more informed decision making in preclinical safety evaluation.[11]
- Open TG-GATEs (Toxicogenomics Project-Genomics Assisted Toxicity Evaluation System) is a Japanese public-private effort. They published gene expression and pathology information for more than 170 compounds (mostly drugs).[12]
- Predictive Safety Testing Consortium, aiming to identify and clinically qualify safety biomarkers for regulatory use as part of the FDA's "Critical Path Initiative"[11]
- ToxCast, program for Predicting Hazard, Characterizing Toxicity Pathways, and Prioritizing the Toxicity Testing of Environmental Chemicals at the United States Environmental Protection Agency[13]
- Tox21, a federal collaboration involving the NIH, Environmental Protection Agency (EPA), and Food and Drug Administration (FDA), is aimed at developing

better toxicity assessment methods.[14] Within this project the toxic effects of chemical compounds on cell lines derived from the 1000 Genomes Project individuals was assessed and associations with genetic markers were determined.[15] Parts of this data were used in the NIEHS-NCATS-UNC DREAM Toxicogenetics Challenge in order to determine methods for cytotoxicity predictions for individuals.[16][17]

26.2 See also

- Comparative Toxicogenomics Database
- Genomics
    - Chemogenomics
    - Structural genomics
    - Pharmacogenetics
    - Pharmacogenomics
- Toxicology

26.3 References

[1] The National Academies Press: Communicating Toxicogenomics Information to Nonexperts: A Workshop Summary (2005)

[2] Hamadeh HK, Afshari CA, eds. (2004). *Toxicogenomics: Principles and Applications*. Hoboken, NJ: Wiley-Liss. ISBN 0-471-43417-5.
Omenn GS (November 2004). "Book Review: Toxicogenomics: Principles and Applications". *Environ Health Perspect*. **112** (16): A962. PMC 1247673 ©.

[3] Lesko LJ, Woodcock J (2004). "Translation of pharmacogenomics and pharmacogenetics: a regulatory perspective". *Nature Reviews. Drug Discovery*. **3** (9): 763–9. PMID 15340386. doi:10.1038/nrd1499.

[4] Lesko LJ, Salerno RA, Spear BB, Anderson DC, Anderson T, Brazell C, Collins J, Dorner A, Essayan D, Gomez-Mancilla B, Hackett J, Huang SM, Ide S, Killinger J, Leighton J, Mansfield E, Meyer R, Ryan SG, Schmith V, Shaw P, Sistare F, Watson M, Worobec A (2003). "Pharmacogenetics and pharmacogenomics in drug development and regulatory decision making: report of the first FDA-PWG-PhRMA-DruSafe Workshop". *Journal of Clinical Pharmacology*. **43** (4): 342–58. PMID 12723455. doi:10.1177/0091270003252244.

[5] Nuwaysir EF, Bittner M, Trent J, Barrett JC, Afshari CA (1999). "Microarrays and toxicology: the advent of toxicogenomics". *Molecular Carcinogenesis*. **24** (3): 153–9. PMID 10204799. doi:10.1002/(SICI)1098-2744(199903)24:3<153::AID-MC1>3.0.CO;2-P.

[6] Fielden MR, Pearson C, Brennan R, Kolaja KL (2005). "Preclinical drug safety analysis by chemogenomic profiling in the liver". *American Journal of Pharmacogenomics : Genomics-related Research in Drug Development and Clinical Practice*. **5** (3): 161–71. PMID 15952870. doi:10.2165/00129785-200505030-00003.

[7] Mattes WB, Pettit SD, Sansone SA, Bushel PR, Waters MD (March 2004). "Database development in toxicogenomics: issues and efforts". *Environmental Health Perspectives*. **112** (4): 495–505. PMC 1241904 ©. PMID 15033600. doi:10.1289/ehp.6697.

[8] Ellinger-Ziegelbauer H, Gmuender H, Bandenburg A, Ahr HJ (January 2008). "Prediction of a carcinogenic potential of rat hepatocarcinogens using toxicogenomics analysis of short-term in vivo studies". *Mutation Research*. **637** (1-2): 23–39. PMID 17689568. doi:10.1016/j.mrfmmm.2007.06.010.

[9] Corvi R, Ahr HJ, Albertini S, Blakey DH, Clerici L, Coecke S, Douglas GR, Gribaldo L, Groten JP, Haase B, Hamernik K, Hartung T, Inoue T, Indans I, Maurici D, Orphanides G, Rembges D, Sansone SA, Snape JR, Toda E, Tong W, van Delft JH, Weis B, Schechtman LM (March 2006). "Meeting report: Validation of toxicogenomics-based test systems: ECVAM-ICCVAM/NICEATM considerations for regulatory use". *Environmental Health Perspectives*. **114** (3): 420–9. PMC 1392237 ©. PMID 16507466. doi:10.1289/ehp.8247.

[10] Collins BC, Clarke A, Kitteringham NR, Gallagher WM, Pennington SR (October 2007). "Use of proteomics for the discovery of early markers of drug toxicity". *Expert Opinion on Drug Metabolism & Toxicology*. **3** (5): 689–704. PMID 17916055. doi:10.1517/17425255.3.5.689.

[11] Mattes WB (2008). "Public Consortium Efforts in Toxicogenomics". In Mendrick DL, Mattes WB. *Essential Concepts in Toxicogenomics*. Methods in Molecular Biology. **460**. pp. 221–238. ISBN 978-1-58829-638-2. PMID 18449490. doi:10.1007/978-1-60327-048-9_11.

[12] Igarashi Y, Nakatsu N, Yamashita T, Ono A, Ohno Y, Urushidani T, Yamada H (January 2015). "Open TG-GATEs: a large-scale toxicogenomics database". *Nucleic Acids Research*. **43** (Database issue): D921–7. PMC 4384023 ©. PMID 25313160. doi:10.1093/nar/gku955.

[13] Dix DJ, Houck KA, Martin MT, Richard AM, Setzer RW, Kavlock RJ (January 2007). "The ToxCast program for prioritizing toxicity testing of environmental chemicals". *Toxicological Sciences*. **95** (1): 5–12. PMID 16963515. doi:10.1093/toxsci/kfl103.

[14] "Toxicology in the 21st century project" http://www.ncats.nih.gov/research/reengineering/tox21/tox21.html

[15] Abdo N, Xia M, Brown CC, Kosyk O, Huang R, Sakamuru S, Zhou YH, Jack JR, Gallins P, Xia K, Li Y, Chiu WA, Motsinger-Reif AA, Austin CP, Tice RR, Rusyn I, Wright

FA (May 2015). "Population-based in vitro hazard and concentration-response assessment of chemicals: the 1000 genomes high-throughput screening study". *Environmental Health Perspectives*. **123** (5): 458–66. PMC 4421772 . PMID 25622337. doi:10.1289/ehp.1408775.

[16] "NIEHS-NCATS-UNC-DREAM Toxicogenetics Challenge". Sage Bionetworks.

[17] "DeepTox: Deep Learning for Toxicity Prediction". Institute of Bioinformatics, Johannes Kepler University Linz.

26.4 External links

- Comparative Toxicogenomics Database – a public database that integrates toxicogenomic data for chemicals, genes, and diseases from the scientific literature.
- Center for Research on Occupational and Environmental Toxicology definition by the CROET Research Centers: (Neuro)toxicogenomics and Child Health Research Center.
- InnoMed PredTox – official project website
- Netherlands Toxicogenomics Centre – official project website
- ToxCast – official project website
- ToxExpress® Program – Gene Logic's ToxExpress® Program

Chapter 27

deCODE genetics

deCODE genetics, Inc. (Icelandic: ***Íslensk erfðagreining***) is a biopharmaceutical company based in Reykjavík, Iceland. The company was founded in 1996 by Kári Stefánsson[1] to identify human genes associated with common diseases using population studies, and apply the knowledge gained to guide the development of candidate drugs. The company isolated genes believed to be involved in cardiovascular disease, cancer and schizophrenia, among other diseases (the company's research concerning the latter[2] is said to represent the first time a gene has been identified by two independent studies to be associated with schizophrenia[3]).

deCODE's approach to identifying genes, and in particular its proposal to set up an Icelandic Health Sector Database (HSD) containing the medical records of all Icelanders, was controversial, and prompted national and international criticism for its approach to the concepts of privacy and consent.[4]

The company was removed from the NASDAQ Biotechnology Index in November 2008[5] In November 2009 the company filed for chapter 11 bankruptcy in a US court, listing total assets of $69.9 million and debt of $313.9 million. According to the American law deCODE was nevertheless allowed to continue its operations.[6] In January 2010 most of deCODE genetics Inc.'s assets were purchased by Saga Investments LLC – an investment company whose owners include Polaris Venture Partners and ARCH Venture Partners - who said they intended to continue most services including deCODE diagnostics and deCODEme™ personal genome scans[7] and management team.[8]

In December 2012, deCODE genetics was purchased by Amgen for $415 million[9] which in October 2013 spun off deCODE genetics' systems and database to a new company called NextCODE Health[10] which in turn was acquired in January 2015 by the Chinese company WuXi PharmaTech for $65 million.[11]

27.1 History

DeCODE was founded in 1996[12] by Ernir Kristján Snorrason, Kári Stefánsson, and Kristleifur Kristjánsson.[13]

In the late 1990s deCODE proposed to create the world's first population-wide genomic biobank by collecting data from the entire population of Iceland, which numbered 270,000 at the time.[14] The plan had these three major components: creating a genealogical database, collecting biobank specimens by means of which genotyping could be done, and creating a national electronic health record system to connect genetic data to each individual's phenotype.[14]

In December 1998 with lobbying from deCODE, the Icelandic Parliament passed the Act on Health Sector Database which permitted public bidding for the right of a company to create this health database and use it for various purposes.[14] The parliament shortly thereafter granted deCODE the right to create this database after the company made a successful bid to do so.[15]

27.2 deCODEme

As a step toward the personal genome, the company has announced that its deCODEme service is available for $985 to anyone who wishes to send a cheek swab to learn details about disease risk and ancestry. This service was launched in November 2007 and thereby became the first web-based service to offer a comprehensive genome scan and an online analysis of an individual's DNA. More than one million single nucleotide polymorphisms are included in the scan.[16] deCODEme claims that the DNA profile it provides can supply its customers with a basis from which they are able calculate the relative risk of developing these diseases and thereby enable them to make better informed decisions about medical prevention and treatment. The deCODEme service currently includes information on the genetic susceptibility to close to 45 common diseases such as myocardial infarction, atrial fibrillation, several types of cancers and type-2 diabetes as well as providing insights into distant ancestry and geographical origins.

The deCODEme service was discontinued in January 2013 and deCODE genetics stopped selling personal genetic tests altogether.[17]

27.3 Appearances in popular culture

The work of deCODE is criticised by Arnaldur Indriðason's novel *Jar City* from 2000, which was adapted as a film in 2006.[18]

deCODE and Kári Stefánsson are satirised as VikingDNA and Professor Lárus Jóhannsson in *Dauðans óvissi tími* by Þráinn Bertelsson (Reykjavík: JPV Útgáfu, 2004).

deCODE and specifically Kári Stefánsson is presented as the creator of monstrous genetic hybrids in Óttar M. Norðfjörð's satirical 2007 work *Jón Ásgeir & afmælisveislan* ([Reykjavík]: Sögur, 2007), and the history of De-CODE appears both directly and in allegorised form (under the fictional name OriGenes) in the same author's *Lygarinn: Sönn saga* (Reykjavík: Sögur, 2011).

27.4 See also

- Genomics
- Kári Stefánsson (co-founder and CEO of deCODE)
- 23andMe, a company offering a service similar to de-CODEme
- Navigenics, a company offering a service similar to deCODEme
- Genomic counseling

27.5 References

[1] Herper, Matthew (6 March 2001) DeCode-ing Schizophrenia Forbes, Retrieved 28 January 2015.

[2] Stefansson H, Sarginson J, Kong A, et al. (January 2003). "Association of neuregulin 1 with schizophrenia confirmed in a Scottish population". *Am. J. Hum. Genet.* **72** (1): 83–7. PMC 420015. PMID 12478479. doi:10.1086/345442.

[3] Sands, Tim (2008). "Gene surveys identify schizophrenia triggers". *Nature.* doi:10.1038/news.2008.994.

[4] David E. Winickoff (2006). "Genome and Nation: Iceland's Health Sector Database and its Legacy".

[5] Lee, Wayne (2008-11-14). "Semi-Annual Changes to the NASDAQ Biotechnology Index". *Press Release.* NASDAQ Newsroom. Retrieved 19 November 2008.

[6] "deCODE Genetics declare bankruptcy, will sell core business to US investors". scienceblogs.com.

[7] "Announcing the new deCODE". Archived from the original on 2011-11-17.

[8] "deCODE Management team". Archived from the original on 2012-10-01.

[9] Newswire, PR. "Amgen to Acquire deCODE Genetics, a Global Leader in Human Genetics". *amgen.com.* Amgen. Retrieved 3 May 2013.

[10] Proffit, Allison (24 October 2013) NextCODE Health Launches deCODE's Clinical Genomics Platform Bio IT World, Retrieved 28 January 2015.

[11] (9 January 2015) WuXi PharmaTech Acquires NextCODE Health to Create Global Leader in Genomic Medicine PR Newswire, Retrieved 28 January 2015.

[12] http://timarit.is/view_page_init.jsp?issId=129032&pageId=1866979&lang=is&q=erf%F0agreining

[13] 'Andlát: Dr. Ernir K. Snorrason', *Morgunblaðið* (30.4.2012), http://www.mbl.is/frettir/innlent/2012/04/30/andlat_dr_ernir_k_snorrason/

[14] Greely, H. T. (2007). "The Uneasy Ethical and Legal Underpinnings of Large-Scale Genomic Biobanks". *Annual Review of Genomics and Human Genetics.* **8**: 343–364. PMID 17550341. doi:10.1146/annurev.genom.7.080505.115721.

[15] Chadwick, R. (1999). "The Icelandic database—do modern times need modern sagas?". *BMJ.* **319** (7207): 441–444. PMC 1127047. PMID 10445931. doi:10.1136/bmj.319.7207.441.

[16] "deCODEme".

[17] decodeme.com

[18] Burke, Lucy, 'Genetics and the Scene of the Crime: De-CODING *Tainted Blood*', *Journal of Literary & Cultural Disability Studies*, 6 (2012), 193–208. doi:10.3828/jlcds.2012.16.

27.6 External links

- deCODE Genetics Inc. website
- deCODE Genetics Icelandic website
- 'The World's Most Successful Failure', *Newsweek*, 12 February 2010

Coordinates: 64°08′08″N 21°56′45″W / 64.13556°N 21.94583°W

Chapter 28

Navigenics

Navigenics, Inc. was a privately held personal genomics company, based in Foster City, California, that used genetic testing to help people determine their individual risk for dozens of health conditions.[1]

28.1 History

Navigenics was co-founded in 2006 by David Agus, M.D., a prostate cancer specialist who is a Professor of Medicine at the University of Southern California and Director of the USC Center for Applied Molecular Medicine and the USC Westside Prostate Cancer Center in Los Angeles, and Dietrich Stephan, Ph.D., member of the Board of Directors of the Personalized Medicine Coalition, current CEO of Silicon Valley Biosystems, former Chairman of Neurogenomics and Deputy Director for Discovery Research at the Translational Genomics Research Institute.[2]

In July 2012, Navigenics was acquired by Life Technologies,[3] which was acquired by Thermo Fisher Scientific in February, 2014.

28.2 Controversy in California

In June 2008, California health regulators sent cease-and-desist letters to Navigenics and 12 other genetic testing firms, including 23andMe.[4] The state regulators asked the companies to prove a physician was involved in the ordering of each test and that state clinical laboratory licensing requirements were being fulfilled. The controversy sparked a flurry of interest in the relatively new field, as well as a number of media articles, including an opinion piece on Wired.com entitled, "Attention, California Health Dept.: My DNA Is My Data."[5] In August 2008, Navigenics and 23andMe received state licenses allowing the companies to continue to do business in California.[6]

28.3 References

[1] Navigenics, Inc. "Navigenics launches Health Compass service." Retrieved 2008-10-15.

[2] Navigenics, Inc. "Navigenics launches with pre-eminent team of advisors, collaborators and investors." Press release. (2007-11-06.) Retrieved on 2008-10-24.

[3] https://www.bloomberg.com/news/2012-07-16/life-technologies-buys-navigenics-for-genetic-diagnostics.html

[4] Langreth, Robert. "California Orders Stop To Gene Testing." Forbes. (2008-06-14). Retrieved on 2008-10-15.

[5] Goetz, Thomas. "Attention, California Health Dept.: My DNA Is My Data." Wired Blog Network. (2008-06-17). Retrieved on 2008-08-27.

[6] Pollack, Andrew. "California Licenses 2 Companies to Offer Gene Services." New York Times. (2008-08-19). Retrieved 2008-10-15.

28.4 External links

- Official Navigenics site
- Official Navigenics blog
- NOVA scienceNOW profiles personal genetic services
- Charlie Rose discusses personal genetics

Chapter 29

23andMe

23andMe is a privately held personal genomics and biotechnology company based in Mountain View, California. The company is named for the 23 pairs of chromosomes in a normal human cell.[1] Its saliva-based direct-to-consumer genetic testing business was named Invention of the Year by *Time* magazine in 2008.[2]

In 2013 the US Food and Drug Administration (FDA) ordered 23andMe to discontinue marketing its personal genome service (PGS), as the company had not obtained the legally required regulatory approval. That resulted in concerns about the potential consequences of customers receiving inaccurate health results.[3] The company continued to sell a personal genome test without health-related results in the United States until October 21, 2015, when it announced that it would be including a revised health component with FDA approval.[4][5] 23andMe has been selling a product with both ancestry and health-related components in Canada since October 2014,[6][7][8] and in the United Kingdom since December 2014.[9]

29.1 Company history

The company was founded by Linda Avey, Paul Cusenza and Anne Wojcicki, ex-wife of Google founder Sergey Brin, in 2006 to provide genetic testing and interpretation to individual consumers.[10][11] In 2007, Google invested $3,900,000 in the company, along with Genentech, New Enterprise Associates, and Mohr Davidow Ventures.[12]

Cusenza left the company in 2007 and was appointed CEO of Nodal Exchange in 2008.[13] Avey left the company in 2009 and co-founded Curious, Inc. in 2011.[14]

In 2012, 23andMe raised $50 million in a Series D venture round, almost doubling its existing capital of $52.6 million.[15][16][17] In 2015, 23andMe raised $115 million in a Series E offering, increasing its total capital to $241 million.[5][18][19]

The company sponsored the PBS TV series "9 Months That Made You".[20]

The company had not turned a profit as of October 2015.[5]

29.2 Products and services

29.2.1 Direct to consumer genetic testing

23andMe began offering direct to consumer genetic testing in November 2007. Customers provide a saliva testing sample that is partially SNP genotyped and results are posted online.[10][21][22] In 2008, when the company was offering estimates of "predisposition for more than 90 traits and conditions ranging from baldness to blindness", *Time* magazine named the product Invention of the Year.[2]

Uninterpreted raw genetic data is posted online and may be downloaded by customers.[23] Customers who bought tests with an ancestry-related component have online access to genealogical DNA test results and tools including a relative-matching database. US customers who bought tests with a health-related component and received health-related results before November 22, 2013 have online access to an assessment of inherited traits and genetic disorder risks.[4][24][25] Health-related results for US customers who purchased the test from November 22, 2013 were suspended until late 2015 while undergoing an FDA regulatory review.[5][26][27] Customers who bought tests from 23andMe's Canadian and UK locations have access to some health-related results.[6][9]

As of June 2015, 23andMe has genotyped over 1,000,000 individuals.[28][29] FDA marketing restrictions reduced customer growth rates[30] but by 2017 they had tested 2,000,000 individuals.[31]

As of April 2017, FDA allowed marketing of 23andMe Personal Genome Service Genetic Health Risk (GHR) tests for 10 diseases or conditions. These are the first direct-to-consumer (DTC) tests authorized by the FDA that provide information on an individual's genetic predisposition to certain medical diseases or conditions, which may help to make decisions about lifestyle choices or to inform discussions with a health care professional.[32]

Product changes

In late 2009, 23andMe split its genotyping service into three products with different prices, an Ancestry Edition, a Health edition, and a Complete Edition.[33] This decision was reversed a year later when the different products were recombined.[34] In late 2010 the company introduced a monthly subscription fee for updates based on new medical research findings.[34][35] The subscription model proved unpopular with customers and was eliminated in mid-2012.[36]

23andMe sold only raw genetic data and ancestry-related results in the United States due to FDA restrictions from November 22, 2013 until October 21, 2015,[4][26][27] when it announced that it would resume providing health information in the form of carrier status and wellness reports with FDA approval.[37] Wojcicki said they still plan to report on disease risk, subject to future FDA approval.[5]

The price of the full direct-to-consumer testing service in the United States reduced from $999 in 2007 to $99 in 2012,[15] and was effectively being sold as a loss leader in order to build a valuable customer database.[23][38][39] In October 2015, the US price was raised to $199.[37] In September 2016, an ancestry-only version was once again offered at a lower price of $99 with an option to upgrade to include the health component for an additional $125 later.[40] The price for international customers was lowered from $199 to $149. To date this kit offers only ancestry information.[41]

The initial price of the product sold in Canada from October 2014, which includes health-related results, was C$199.[6][7] The initial price of the product sold in the UK from December 2014, which includes health-related results, was £125.[42]

Instrument and chip versions Up until 2010 Illumina sold only instruments that were labeled "for research use only"; in early 2010 Illumina obtained FDA approval for its BeadXpress system to be used in clinical tests.[43][44]

29.2.2 Medical research

Aggregated customer data is studied by scientific researchers employed by 23andMe for research on inherited disorders. The large pool of data in its customer database has also attracted the interest of academics and other partners,[30] including pharmaceutical and biotechnology companies.[23][46] In July 2012, 23andMe acquired the startup CureTogether, a crowdsourced treatment ratings website with data on over 600 medical conditions.[47]

23andMe provides services related to some specific medical research initiatives,[48] providing confidential customer datasets to and partnering with researchers to establish genetic associations with specific illnesses and disorders.[10] One analysis comparing 23andMe's Parkinson's disease research with a National Institutes of Health initiative suggested that the company's use of large amounts of computational power and datasets might offer comparable results, in much less time.[49] 23andMe has launched research initiatives enrolling patients into study populations for inflammatory bowel disease, myeloproliferative neoplasms, and lupus.[50][51] Papers on various genetic traits by 23andMe scientists were presented at the 2014 American Society of Human Genetics.[52]

In 2015, 23andMe made a business decision to pursue drug development themselves, under the direction of former Genentech executive Richard Scheller, as opposed to supplying pharmaceutical companies with raw data.[5][53]

29.3 Relationship with government regulators

The new genetic testing service and ability to map significant portions of the genome has raised controversial questions, including whether the results can be interpreted meaningfully and whether they will lead to genetic discrimination.[2][23] The regulatory environment for testing companies has been uncertain, and anticipated risk-based regulation catering for different types of genetic tests has not yet materialized.[34][54][55]

29.3.1 State regulators

In 2008 the states of New York and California each provided notice to 23andMe and similar companies, that they needed to obtain a CLIA license in order to sell tests in those states.[2][56][57] By August 2008, 23andMe had received licenses that allow them to continue to do business in California.[58]

29.3.2 FDA

According to Anne Wojcicki, 23andMe had been in dialogue with the FDA since 2008.[55] In 2010 the FDA notified several genetic testing companies, including 23andMe, that their genetic tests are considered medical devices and federal approval is required to market them; a similar letter was sent to Illumina, which makes the instruments and chips used by 23andMe in providing its service.[34][59][60]

23andMe first submitted applications for FDA clearance in July and September 2012.[3]

In November 2013, the FDA published a guidance on how it classified genetic analysis and testing services offered by companies using instruments and chips labelled for "research use only" and instruments and chips that had been approved for clinical use.[61]

At around the same time, after not hearing from 23andMe for six months, the FDA ordered 23andMe to stop marketing its Saliva Collection Kit and Personal Genome Service (PGS), as 23andMe had not demonstrated that they have "analytically or clinically validated the PGS for its intended uses" and the "FDA is concerned about the public health consequences of inaccurate results from the PGS device".[3][62][63] As of December 2, 2013, 23andMe had stopped all advertisements for its PGS test but is still selling the product.[64][65] As of December 5, 2013, 23andMe was selling only raw genetic data and ancestry-related results.[4][26][27]

23andMe publicly responded to media reports on November 25, 2013, stating, "We recognize that we have not met the FDA's expectations regarding timeline and communication regarding our submission. Our relationship with the FDA is extremely important to us and we are committed to fully engaging with them to address their concerns."[66][67][68] Anne Wojcicki subsequently posted an update on the 23andMe website, stating: "This is new territory for both 23andMe and the FDA. This makes the regulatory process with the FDA important because the work we are doing with the agency will help lay the groundwork for what other companies in this new industry do in the future. It will also provide important reassurance to the public that the process and science behind the service meet the rigorous standards required by those entrusted with the public's safety."[55]

On December 5, 2013, 23andMe announced that it had suspended health-related genetic tests for customers who purchased the test from November 22, 2013 in order to comply with the FDA warning letter while undergoing regulatory review.[4][26][27]

In May 2014 it was reported that 23andMe was exploring alternative locations abroad including Canada, Australia and the United Kingdom in which to offer its full genetic testing service.[69] 23andMe had been selling a product with both ancestry and health-related components in Canada since October 2014,[6][7][8] and in the United Kingdom since December 2014.[9]

In 2014 23andMe submitted a 510(k) application to the FDA to market a carrier test for Bloom syndrome, which included data showing that 23andme's results were consistent and reliable and that the saliva collection kit and instructions were easy enough for people to use without making mistakes that would affect the tests, and included citations to the scientific literature showing that the associations between the specific tests that 23andMe were relevant to Blooms.[70][71] The FDA cleared the test in February 2015; in the clearance notice the FDA said that it would not require similar applications for other carrier tests from 23andMe.[70][72] The FDA sent further clarification about regulation of the test to 23andMe on October 1, 2015.[73]

On October 21, 2015, 23andMe announced that it would begin marketing carrier tests in the US again.[5] CEO Anne Wojcicki said, "There was part of us that didn't understand how the regulatory environment works" in regards to the distributed laboratory regulatory functions of FDA and CMS.[74]

23andMe submitted de novo 510(k) applications to the FDA to market tests that provide people with information about whether they have gene mutations or alleles that put them at risk for getting or having certain diseases; the applications included data showing that 23andme's results were consistent and reliable, and that the saliva collection kit and instructions were easy enough for people to use without making mistakes that would affect the tests, and included with citations to the scientific literature showing that the associations between the specific tests that 23andMe were relevant to the diseases. In April 2017, the FDA approved the applications for ten tests: Late-Onset Alzheimer's Disease, Parkinson's Disease, Celiac disease, Hereditary Thrombophilia, Alpha-1 Antitrypsin Deficiency, Glucose-6-Phosphate Dehydrogenase deficiency, Early onset of Dystonia, Factor XI deficiency and Gaucher's Disease.[75][76] The FDA also said that it intended to exempt further 23andMe genetic risk tests from the needing 501(k) applications, and it clarified that it was only approving genetic risk tests, not diagnostic tests.[75]

29.4 References

[1] "Fact Sheet". 23andMe. Retrieved November 27, 2013.

[2] Hamilton, Anita (October 29, 2008). "Best Inventions of 2008". *Time*. Retrieved April 5, 2012.

[3] "Inspections, Compliance, Enforcement, and Criminal Investigations – 23andMe, Inc. 11/22/13". FDA. November 22, 2013. Retrieved November 25, 2013.

[4] Herper, Matthew (December 5, 2013). "23andMe Stops Offering Genetic Tests Related to Health". *Forbes*. Archived from the original on February 9, 2014. Retrieved December 6, 2013.

[5] Pollack, Andrew (2015-10-21). "23andMe Will Resume Giving Users Health Data". *The New York Times*. ISSN

0362-4331. Archived from the original on October 25, 2015. Retrieved 2015-10-21.

[6] Ubelacker, Sheryl (October 1, 2014). "U.S. company launches genetic health and ancestry info service in Canada". *Winnipeg Free Press*. The Canadian Press. Retrieved October 7, 2014.

[7] Hansen, Darah (October 2, 2014). "5Q: Anne Wojcicki, CEO 23andMe on knowing your DNA data (and being married to the boss of Google)". *Yahoo Finance Canada*. Retrieved October 7, 2014.

[8] "23andme genetic testing service raises ethical questions". *CBC News*. October 2, 2014. Retrieved October 7, 2014.

[9] Roberts, Michelle; Rincon, Paul (2 December 2014). "Controversial DNA test comes to UK". *BBC News*. Retrieved 2 December 2014.

[10] Goetz, Thomas (November 17, 2007). "23AndMe Will Decode Your DNA for $1,000. Welcome to the Age of Genomics". *Wired*. Archived from the original on March 12, 2014. Retrieved April 5, 2012.

[11] "Corporate Info". 23andMe. Retrieved November 27, 2013.

[12] "Google invests in genetics firm". BBC News. May 22, 2007. Retrieved June 28, 2007.

[13] "Board Of Directors". Nodal Exchange. Retrieved November 27, 2013.

[14] Curious: We've got questions

[15] Tsotsis, Alexia (December 11, 2012). "Another $50M Richer, 23andMe Drops Its Price To $99 Permanently. But Will The Average Dude Buy In?". *TechCrunch*. AOL. Retrieved December 12, 2012.

[16] "Press Release: 23andMe Raises More Than $50 Million in New Financing". 23andMe. December 11, 2012. Retrieved November 27, 2013.

[17] "23andMe". *CrunchBase*. AOL. Retrieved December 12, 2012.

[18] Chen, Caroline (October 14, 2015). "23andMe Funding Values Genetics Startup at $1.1 Billion". *Bloomberg Business*. Retrieved October 25, 2015.

[19] "Notice of Exempt Offering of Securities". U.S. Securities and Exchange Commission. Retrieved 11 July 2015.

[20] "9 Months That Made You". PBS. Retrieved July 14, 2016.

[21] "Our Service: Genotyping Technology". 23andMe. Retrieved November 27, 2013.

[22] Hadly, Scott (November 18, 2013). "23andMe's New Custom Chip". 23andMe. Retrieved November 27, 2013.

[23] Jeffries, Adrianne (December 12, 2012). "Genes, patents, and big business: at 23andMe, are you the customer or the product?". *The Verge*. Archived from the original on January 2, 2014. Retrieved July 17, 2014.

[24] Baertlein, Lisa (November 20, 2007). "Google-backed 23andMe offers $999 DNA test". *USA Today*. Archived from the original on July 17, 2014. Retrieved April 5, 2012.

[25] Swarns, Rachel L (23 January 2012). "With DNA Testing, Suddenly They Are Family". *The New York Times*. Archived from the original on July 17, 2014. Retrieved July 17, 2014.

[26] "23andMe, Inc. provides update on FDA regulatory review" (Press release). 23andMe. December 5, 2013. Retrieved December 6, 2013.

[27] Fung, Brian (December 6, 2013). "Bowing again to the FDA, 23andMe stops issuing health-related genetic reports". *The Washington Post*. Retrieved December 6, 2013.

[28] Wojcicki, Anne (June 18, 2015). "Power of One Million". Retrieved June 19, 2015.

[29] Ramsey, Lydia (July 7, 2015). "23andMe CEO defends practice of sharing genetic info with pharma companies". *Business Insider*. Archived from the original on July 8, 2015. Retrieved July 8, 2015.

[30] Kiss, Jemima (March 9, 2014). "23andMe admits FDA order 'significantly slowed up' new customers". *The Guardian*. Archived from the original on March 16, 2014. Retrieved March 10, 2014.

[31] "23andMe Breaks Two Million!". *The DNA Geek*. 2017-04-19. Retrieved 2017-04-24.

[32] "FDA allows marketing of first direct-to-consumer tests that provide genetic risk information for certain conditions". FDA.

[33] Wu, Shirley (November 13, 2009). "Get Just the Information You Want: 23andMe To Offer Separate Health and Ancestry Editions". 23andMe. Retrieved November 29, 2013.

[34] Vorhaus, Dan (November 23, 2010). "A Thanksgiving Tradition: 23andMe Repackages Product, Raises Prices". *Genomics Law Report*. Robinson Bradshaw & Hinson. Archived from the original on December 3, 2013. Retrieved November 29, 2013.

[35] MacArthur, Daniel (November 24, 2010). "News from 23andMe: a bigger chip, a new subscription model and another discount drive". *Wired*. Archived from the original on June 29, 2013. Retrieved November 27, 2013.

[36] "23andMe Eliminates Subscription Model". *GenomeWeb Daily News*. May 10, 2012. Retrieved November 27, 2013.

[37] "23andMe reboots genetic health testing, now with FDA approval". *Ars Technica*. Retrieved 2015-10-21.

[38] Hamilton, David (September 10, 2008). "23andMe's Price Cut: The End of Personal Genomics?". *CBSNews.com*. Archived from the original on July 17, 2014. Retrieved July 17, 2014.

[39] Krol, Aaron (March 24, 2014). "23andMe Pursues Health Research in the Shadow of the FDA". *Bio-IT World*. Archived from the original on July 17, 2014. Retrieved July 17, 2014.

[40] Ramsey, Lydia (September 22, 2016). "23andMe is now offering a $99 genetics test again – but it's very different from the original". *Business Insider*. Retrieved September 26, 2016.

[41] https://www.23andme.com/en-int/

[42] Gibbs, Samuel (December 2, 2014). "DNA-screening test 23andMe launches in UK after US ban". *The Guardian*. Retrieved October 26, 2015.

[43] Petrone, Justin (May 4, 2010). "FDA Clears Illumina's BeadXpress System for Clinical Use". *GenomeWeb*.

[44] "510(k) Premarket Notification K093128". FDA. Retrieved 7 April 2017.

[45] "23andMe". *isogg.org*. Retrieved 9 September 2015.

[46] McBride, Ryan (November 29, 2012). "23andMe sets stage for stronger ties with pharma". *FierceBiotech*. Archived from the original on August 8, 2013. Retrieved July 17, 2014.

[47] "23andMe Makes First Acquisition, Nabs CureTogether To Double Down On Crowdsourced Genetic Research = Jul 11, 2012". *TechCrunch*. Retrieved Feb 18, 2015.

[48] "23andWe Research". 23andMe. Retrieved April 5, 2012.

[49] Goetz, Thomas (June 22, 2010). "Sergey Brin's Search for a Parkinson's Cure". *Wired*. Archived from the original on July 17, 2014. Retrieved April 5, 2012.

[50] "People Powered IBD Research". *23andMe blog*. 23andMe. October 28, 2014. Retrieved 2014-12-21.

[51] "23andMe Launches Myeloproliferative Neoplasms Research Initiative". August 3, 2011. Retrieved Dec 21, 2014.

[52] "Science On the Beach". *23andMe Blog*. September 26, 2014. Retrieved Dec 21, 2014.

[53] Herper, Matthew (March 12, 2015). "In Big Shift, 23andMe Will Invent Drugs Using Customer Data". *Forbes*. Retrieved 28 October 2015.

[54] Greely, Hank (November 25, 2013). "The FDA drops an anvil on 23andMe – now what?". Stanford University. Retrieved November 29, 2013. FDA had promised a risk-based regulatory scheme, but we don't know what it is.

[55] Wojcicki, Anne (November 26, 2013). "An Update Regarding The FDA's Letter to 23andMe". 23andMe. Retrieved November 27, 2013.

[56] Robert Langreth; Matthew Herper (April 18, 2008). "States Crack Down On Online Gene Tests". *Forbes*.

[57] Jason Kincaid (June 18, 2008). "Cease And Desist: California Tries to Unravel 23andMe's Genetic Testing". *The Washington Post*. TechCrunch.com.

[58] Pollack, Andrew (August 20, 2008). "California Licenses 2 Companies to Offer Gene Services". *The New York Times*.

[59] "FDA cracking down on genetic tests". *NBC*. June 11, 2010. Retrieved November 27, 2013.

[60] Pollack, Andrew (11 June 2010). "F.D.A. Faults 5 Companies on Genetic Tests". *The New York Times*.

[61] Malone, Bill (February 1, 2014). "A New Chapter in FDA Regulation - AACC.org". *Clinical Laboratory News*.

[62] Perrone, Matthew (November 25, 2013). "FDA Tells 23andMe to Halt Sales of Genetic Test". ABC News. Retrieved November 25, 2013.

[63] Gray, Tyler (November 25, 2013). "FDA To 23andMe Founder Anne Wojcicki: Stop Marketing $99 DNA Test Or Face Penalties". *Fast Company (magazine)*. Retrieved November 25, 2013.

[64] Garde, Damian (December 3, 2013). "23andMe pulls ads after FDA warning, but sales roll on". *FierceMedicalDevices*. FierceMarkets. Retrieved December 4, 2013.

[65] del Castillo, Michael (December 3, 2013). "Calm down about 23andMe, the media is getting it wrong". *Upstart Business Journal*. Retrieved December 5, 2013.

[66] Khan, Razib (November 25, 2013). "The FDA's Battle With 23andMe Won't Mean Anything in the Long Run". *Slate Magazine*. Retrieved November 25, 2013.

[67] Etherington, Darrell (November 25, 2013). "DNA Testing Startup 23andMe Hits A Snag As FDA Shuts Down Sales Of Home Testing Kit". *TechCrunch*. Retrieved November 25, 2013.

[68] Young, Susan (November 25, 2013). "Updated: FDA Orders 23andMe to Stop Genetic Tests". *Technology Review*. Retrieved November 25, 2013.

[69] Farr, Christina (May 6, 2014). "Gene startup 23andme casts eyes abroad after U.S. regulatory hurdle". *Reuters*. Archived from the original on May 27, 2014. Retrieved July 17, 2014.

[70] "FDA permits marketing of first direct-to-consumer genetic carrier test for Bloom syndrome". *FDA News Release*. February 19, 2015.

[71] "Device Classification under Section 513(f)(2)(de novo)". *www.accessdata.fda.gov*. FDA. Retrieved 7 April 2017.. 23andMe's Autosomal Recessive Carrier Screening Gene Mutation Detection System in FDA database

[72] "23andMe Gets FDA Clearance to Market Bloom Syndrome Carrier Test Directly to Consumers". *GenomeWeb*. February 19, 2015.

[73] "Letter re DEN140044" (PDF). FDA. October 1, 2015.. Decision Summary: Evaluation of DEN140044 revising February 2015 evaluation.

[74] Bensinger, Greg (2016-10-26). "Disconnect Between Silicon Valley and Regulators Over Health Technologies, 23andMe CEO Says". *Wall Street Journal*. ISSN 0099-9660. Retrieved 2016-11-23.

[75] "Press Announcements - FDA allows marketing of first direct-to-consumer tests that provide genetic risk information for certain conditions". FDA. April 6, 2017.

[76] Kolata, Gina (6 April 2017). "F.D.A. Will Allow 23andMe to Sell Genetic Tests for Disease Risk to Consumers". *The New York Times*.

29.5 Further reading

- 23andMe's New Formula: Patient Consent = $. Antonio Regalado, *MIT Technology Review*

29.6 External links

- Official website

29.7 Text and image sources, contributors, and licenses

29.7.1 Text

- **Pharmacogenomics** *Source:* https://en.wikipedia.org/wiki/Pharmacogenomics?oldid=778385812 *Contributors:* Ahoerstemeier, Robbot, Techelf, MSGJ, Jfdwolff, Beland, Onco p53, PDH, TheObtuseAngleOfDoom, Rhobite, Bender235, Macowell, Velella, Ceyockey, Acer-peri, Joerg Kurt Wegner, BD2412, Rjwilmsi, Koavf, Bgwhite, NSR, YurikBot, Mushin, Bhny, Derek.cashman, CarolineFThorn, Banus, KnightRider~enwiki, SmackBot, CommodiCast, Cessator, Onebravemonkey, Edgar181, Apers0n, Jcuticchia, AmiDaniel, Gogo Dodo, Thijs!bot, Opabinia regalis, Andyjsmith, Headbomb, Nick Number, Scientific American, Jj137, Ph.eyes, Flowanda, MartinBot, Sjjupadhyay~enwiki, MoritzE~enwiki, Montie06, J.delanoy, Boghog, Xris0, Rod57, Oceanflynn, Ffrueh, Synthebot, !dea4u, AlleborgoBot, Pharmtao, ShelleyAdams, MenoBot, Ozarkridgerunner, Mild Bill Hiccup, CrazyChemGuy, SchreiberBike, Johnuniq, Jytdog, Addbot, SpBot, Quercus solaris, Shakiestone, Yunxiang987, Luckas-bot, Yobot, In7sky, Oleginger, AnomieBOT, LilHelpa, Bio-ITWorld, DSisyphBot, Bufftybuffty, Actonbiotech, GenOrl, FrescoBot, Jonesey95, Tom.Reding, Vivek BMS, TobeBot, Grow60, Brambleclawx, Aircorn, Klbrain, Dcirovic, Jahub, H3llBot, AManWithNo-Plan, Saxenasandeep, DrKC MD, ClueBot NG, Jon7245, Jonsynergy, Mjones1V1, BG19bot, Rick823, Assurerx, Estevezj, Rob Hurt, Batty-Bot, ChrisGualtieri, Me, Myself, and I are Here, Joeinwiki, Randykitty, Biology editor, Garzfoth, Tom (LT), Seppi333, MarcelBeauchamp94, Alanasappleby, Elisemary, Mfareedk, A.leatherd, Lauraannw, Éffièdaligrh, Phleg1, Monkbot, Dolleyj, Socialscienceinmedicine, Socialscience-andmedicine, KM1977, Medgirl131, Erinb520, Jarslan, Universitystudentone, L0st H0r!z0ns, CV9933, Dks5413, Prosto aneg, Xcodescience, Rap17 and Anonymous: 58

- **Metabolic pathway** *Source:* https://en.wikipedia.org/wiki/Metabolic_pathway?oldid=775315681 *Contributors:* Lexor, Dan Koehl, Kku, Ahoerstemeier, Ronz, TUF-KAT, Zoicon5, Marshman, Robbot, Unfree, Centrx, Bensaccount, Jfdwolff, Delta G, Stevietheman, Pgan002, HorsePunchKid, Snobscure, Ukexpat, Rich Farmbrough, Bobo192, Whosyourjudas, Arcadian, Jag123, Giraffedata, La goutte de pluie, Alansohn, Thebeginning, Ceyockey, RyanGerbil10, Fenteany, Al E., Chobot, CambridgeBayWeather, Sentausa, Chakazul, Kkmurray, Leptictidium, Modify, GraemeL, Albert.so, Dposse, Sameetmehta, SmackBot, Jamiestroud69, Delldot, Edgar181, Zephyris, Dreg743, Chlewbot, Drphilharmonic, Clicketyclack, Sir marek, Robofish, Muadd, Monkeyflower, Parakkum, Brendanliamboyle, Shrimp wong, CmdrObot, Raz1el, Was a bee, Tkynerd, Barticus88, David D., TimVickers, SoumenRoy, MER-C, Robin S, Klunz, Daylite, VolkovBot, AlnoktaBOT, Broadbot, Gregogil, ThinkerThoughts, SieBot, Beth Rogers, Correogsk, ClueBot, The Thing That Should Not Be, Zomno, Qinatan, MystBot, Addbot, Shahriyar alavi, TutterMouse, Fieldday-sunday, Laurinavicius, Swarm, Luckas-bot, Vini 17bot5, İnfoCan, Castingpagina~enwiki, Materialscientist, Xqbot, 4twenty42o, N419BH, A.amitkumar, D'ohBot, Girlwithgreeneyes, I dream of horses, Calmer Waters, SpunkyLepton, Anneli2, Benedict905, EmausBot, WikitanvirBot, Rcaspi, Tommy2010, ZéroBot, ClueBot NG, Bjc86, Stjernerlever~enwiki, NotWith, BattyBot, Cyberbot II, Padenton, FoCuSandLeArN, Blythwood, Tylercrumpton, Evolution and evolvability, Seppi333, SageRad, Ian (Wiki Ed), K scheik, Stavroskg, Preciouschioma, GreenC bot, Sazhnyev, Bear-rings and Anonymous: 90

- **Pharmacokinetics** *Source:* https://en.wikipedia.org/wiki/Pharmacokinetics?oldid=787487909 *Contributors:* Edward, Securiger, Graeme Bartlett, Techelf, Nick04, Erich gasboy, Bender235, Rcsheets, Arcadian, SpeedyGonsales, Hooperbloob, Pen1234567, Psiphim6, GregorB, Rjwilmsi, RussBot, Kkmurray, Itub, Fuzzyrandom, SmackBot, AndreasJS, Edgar181, Dmng, Chris the speller, RDBrown, Snowmanradio, Radagast83, DMacks, Vina-iwbot~enwiki, Alexander Iwaschkin, Nick Y., Rifleman 82, Corpx, Thijs!bot, Wikid77, Headbomb, Nick Number, MER-C, WolfmanSF, DerHexer, WLU, DGG, ChemNerd, Social tamarisk, Boghog, Mikael Häggström, Oceanflynn, Chiswick Chap, STBotD, Marksale, Technopat, Shanata, Ohiostandard, Crunkmonkey, SieBot, Dhusereau, Purplesplat, ClueBot, LAX, SummitPK, Excirial, En.Dev, KLassetter, Rdphair, Jonathan Laventhol, Addbot, DOI bot, AkhtaBot, LinkFA-Bot, Quercus solaris, Tide rolls, Alfie66, Luckas-bot, Anypodetos, Uofal, AnomieBOT, Citation bot, Frederic Y Bois, Smilingsandy, Omnipaedista, A45b22chp, Dsfarrier, FrescoBot, D'ohBot, Citation bot 1, Tom.Reding, RedBot, WikiUser853, TheSciolist, English Fig, Jhargrov, Peteronium, EmausBot, Klbrain, Dcirovic, Donner60, Bobthefish2, Xto 999, ClueBot NG, Mirisaamali, Baaphi, JimmyBWiki, Bibcode Bot, Dczock, Johnsonthomas, Aisteco, CarlWesolowski, Tutelary, DarafshBot, Khazar2, Dexbot, FoCuSandLeArN, Pharmacomagician, Mark viking, Mattkosloski, Ppilotte, Amr94, Pktastic, Ttocserp, Monkbot, BethNaught, KasparBot, EnriqueWikiEdit, JJMC89, Chelb, KTsaioun, PrimeBOT, CubeSat4U and Anonymous: 109

- **Germline mutation** *Source:* https://en.wikipedia.org/wiki/Germline_mutation?oldid=756820121 *Contributors:* The Anome, AdamRetchless, Thue, Larsie, Stemonitis, Stevenfruitsmaak, RussBot, Malcolma, SmackBot, Commander Keane bot, Bazonka, Onceler, Adj08, Erados, Cydebot, Alaibot, WhatamIdoing, R'n'B, Katharineamy, Mikael Häggström, Jasonasosa, Anniepema, Ltyeruham, Chmyr, LAX, Arjayay, Addbot, Julia W, Haetmachine, FrescoBot, ZéroBot, Tolly4bolly, Fixuture and Anonymous: 11

- **Mutation** *Source:* https://en.wikipedia.org/wiki/Mutation?oldid=787763023 *Contributors:* Magnus Manske, Kpjas, Marj Tiefert, Derek Ross, Mav, Bryan Derksen, Taw, Malcolm Farmer, Ed Poor, Rgamble, PierreAbbat, Mjb, Gog, Axel Driken, Michael Hardy, Palnatoke, Lexor, Gabbe, Ixfd64, Cyde, Skysmith, Ahoerstemeier, Stevenj, TUF-KAT, Plop, JWSchmidt, Nikai, Andres, Mxn, Nikola Smolenski, Quizkajer, Timwi, Steinsky, Markhurd, Furrykef, Omegatron, Samsara, Thue, Bevo, Pakaran, Robbot, Sander123, Romanm, Chopchopwhitey, Sverdrup, Academic Challenger, Hadal, Wikibot, Lupo, JerryFriedman, Pengo, Giftlite, Kim Bruning, Mintleaf~enwiki, Nunh-huh, Brian Kendig, Bensaccount, Cantus, Duncharris, Mboverload, Bobblewik, Wmahan, Adenosine, Andycjp, Quadell, Antandrus, Beland, PDH, PhDP, Oneiros, Joyous!, Ukexpat, Rahul sig, Stephenpratt, Mike Rosoft, D6, ClockworkTroll, Ocon, DanielCD, Jiy, Discospinster, Oliver Lineham, LindsayH, Ivan Bajlo, Bender235, ESkog, Mashford, Violetriga, Brian0918, Livajo, Gilgamesh he, Jpgordon, Guettarda, Bobo192, Smalljim, Arcadian, ParticleMan, Tgr, Nhandler, 4v4l0n42, Orangemarlin, Abstraktn, Alansohn, Etxrge, Plumbago, Wouterstomp, ClockworkSoul, Shogun~enwiki, Cal 1234, Amorymeltzer, Jon Cates, RainbowOfLight, LFaraone, Ndteegarden, MIT Trekkie, Embryomystic, Yurivict, Ceyockey, Andygainey, Camw, TomTheHand, Urod, WadeSimMiser, MONGO, Zzyzx11, Turnstep, GSlicer, RichardWeiss, BorisTM, Ketiltrout, Rjwilmsi, Koavf, Rogerd, Wikibofh, Stardust8212, Crazynas, Cww, Ian Dunster, Sango123, Yamamoto Ichiro, Dinosaurdarrell, FlaBot, Naraht, Nihiltres, GünniX, Dantecubed, RexNL, Gurch, Zsingaya, Nabarry, Terrace4, Daycd, Chobot, DVdm, Bgwhite, Gwernol, Whosasking, YurikBot, Wavelength, Sceptre, Phantomsteve, Hede2000, Mark Ironie, Kodemage, Ansell, Shell Kinney, Eleassar, Wimt, Shanel, NawlinWiki, EWS23, Dysmorodrepanis~enwiki, Wiki alf, Dureo, Irishguy, Ragesoss, Dhollm, Misza13, Semperf, DeadEyeArrow, Bota47, Phenz, Juicy fisheye, Wknight94, WAS 4.250, Zargulon, Johndburger, Deville, Ninly, Bayerischermann, Closedmouth, JoanneB, Alasdair, Guillom, RenamedUser jaskldjslak904, ArielGold, Johnpseudo, Allens, Sancassania, Kungfuadam, Roke, Jungenbergs, DVD R W, Brentt, SmackBot, Reedy, TestPilot, Hydrogen Iodide, Bomac, ScaldingHotSoup, Jrockley, Delldot, Hanmi74, Zephyris, JosephCCampana, Yamaguchi??, Gilliam, Rmosler2100, Sveika,

Kazkaskazkasako, Chris the speller, Keegan, RDBrown, JohnRobertMartin, Jprg1966, Miquonranger03, MalafayaBot, Wedian, Zven, Gracenotes, John Reaves, Can't sleep, clown will eat me, Nixeagle, Voyajer, Rrburke, Tommyjb, Xyzzyplugh, GVnayR, Radagast83, Makemi, Jiddisch~enwiki, Dreadstar, Smokefoot, DMacks, LeoNomis, Alan G. Archer, Madeleine Price Ball, Dncarley, LestatdeLioncourt, Robert Stevens, Mgiganteus1, Seb951, Dingopup, Waggers, Ryulong, Hogyn Lleol, Novangelis, MTSbot~enwiki, Ryanjunk, Hu12, White Ash, Roland Deschain, Poechalkdust, JoeBot, Jason7825, Toastybread, Courcelles, Frank Lofaro Jr., Tawkerbot2, Dlohcierekim, Geezerbill, JForget, CmdrObot, Tanthalas39, SimonSayz, Fedir, Lavateraguy, Agathman, ShelfSkewed, Metzenberg, Standonbible, Seven of Nine, Equendil, Icek~enwiki, Kupirijo, Clappingsimon, Ppgardne, Thewall2, Carifio24, Gogo Dodo, Was a bee, Khatru2, Anthonyhcole, Flowerpotman, Corpx, Tawkerbot4, Clovis Sangrail, Chrislk02, Narayanese, Sharonlees, InfoCan, Michael Johnson, Oleksii0, Thijs!bot, Epbr123, Peachiezworld, Mojo Hand, Marek69, James086, Nezzadar, Tellyaddict, Davidhorman, SineCurve, Sithu.Win, Capeo, Escarbot, Mentifisto, David D., AntiVandalBot, Guy Macon, Seaphoto, Gusgould, Quintote, TimVickers, Scepia, Danger, Credema, Myanw, Joycierules, ClassicSC, Ioeth, Erxnmedia, Solenoozerec, Gcm, MER-C, Plantsurfer, Sheitan, Instinct, Arch dude, Alex tudor, Agreene175, Dcooper, Frankie816, PhilKnight, Savant13, .anacondabot, Somnathroy, Magioladitis, Bongwarrior, VoABot II, DWIII, Tupeliano~enwiki, Culverin, WhatamIdoing, C.-ting Wu, Faustnh, Cgingold, CodeCat, Ciar, Daarznieks, Allstarecho, Emw, DerHexer, Saganaki-, Peter coxhead, 0612, MartinBot, STBot, Arjun01, Lid6, Tuganax~enwiki, Middlenamefrank, Mike6271, R'n'B, Thirdright, Anastacie, J.delanoy, Pharaoh of the Wizards, Filll, Sikatriz, Synapomorphy, Boghog, Uncle Dick, Jmjanzen, Netnuevo, Gr8niladri, Katalaveno, Dr d12, Mikael Häggström, AntiSpamBot, Chiswick Chap, Hut 6.5, Plindenbaum, Kdawson66, Gcrossan, Bioephemera, Aled 12345, VolkovBot, Jeff G., Pparazorback, Philip Trueman, DoorsAjar, TXiKiBoT, Oshwah, Zurrr, Tameeria, PizzaBox, Ajoust, Rei-bot, Sk741~enwiki, Naohiro19 revertvandal, Vanished user ikijeirw34iuaeolaseriffic, Lradrama, Sintaku, Abdullais4u, Cremepuff222, Watchdogb, Meepmoop, Daveheathr, Meters, ITxT, Enviroboy, Ceranthor, HiDrNick, AlleborgoBot, Karajade, SMC89, SieBot, VK35, Chaosof99, Dawn Bard, Eagleal, Yintan, Ncfuzzy101, LeadSongDog, Thebestlaidplans, Happysailor, Flyer22 Reborn, Radon210, Oda Mari, Hzh, Biskot~enwiki, Oxymoron83, Genenome, Faradayplank, Peternewell, Techman224, Hobartimus, OKBot, Chain funds, Alikingofkings, Gamall Wednesday Ida, Anchor Link Bot, Alchemes, Rabo3, Denisarona, Forluvoft, SallyForth123, Rustygin, Ratemonth, ClueBot, Snigbrook, The Thing That Should Not Be, Matdrodes, Lewa.27, YassineMrabet~enwiki, Drmies, Boing! said Zebedee, CounterVandalismBot, Harland1, Excirial, Jusdafax, Leonard^Bloom, Lartoven, Rhododendrites, ParisianBlade, NuclearWarfare, Arjayay, Micha, Fungusnunchuck, ChrisHodgesUK, Thingg, Aitias, Christomaniac76, Johnuniq, Egmontaz, DumZiBoT, Graevemoore, Ktfreese, MeSoDark, Crazy Boris with a red beard, EdChem, Rror, Mouse555~enwiki, Hylobius, TFOWR, Firebat08, WikiDao, Aunt Entropy, KnightofZion, Klgroom, Elvin fimel, Addbot, DOI bot, Jojhutton, Wickey-nl, Atethnekos, Jason.Rafe.Miller, Shirtwaist, CanadianLinuxUser, Imarockstarfoo, Jakeisbaked, Thedude212, Cst17, Bernstein0275, Quercus solaris, Lordlosss2, Ehrenkater, Tide rolls, Botanist3, Krano, Luckas Blade, Ngsmart, Gail, Zorrobot, AdamXtreme18, Ettrig, LuK3, Planettelex23, Ben Ben, Legobot, Luckas-bot, Yobot, Tohd8BohaithuGh1, Julia W, Mauler90, Anypodetos, Ayrton Prost, Azcolvin429, Eric-Wester, AnomieBOT, Haetmachine, Phlyght, Jim1138, Piano non troppo, AdjustShift, Flewis, Materialscientist, The High Fin Sperm Whale, Citation bot, Brightgalrs, Swastikde, ArthurBot, WaffleMaster44, LovesMacs, LilHelpa, Gsmgm, Sionus, Capricorn42, Kahkooi, Millahnna, Oxwil, Pb0823, Gatorgirl7563, Shirik, SassoBot, Methcub, Joejoeftw, SchnitzelMannGreek, Dante mars, FrescoBot, DoctorDNA, Riventree, Lothar von Richthofen, Ifiber, HJ Mitchell, Letatcestmoi94, Drew R. Smith, Matthew Ackerman, Citation bot 1, Trichodinamitra, Pinethicket, I dream of horses, Naif1989, Rushbugled13, Serols, Meaghan, Curehd, Jauhienij, Trappist the monk, MiRroar, Ooikahkooi0507, Specs112, Jcorry10, Bhawani Gautam, GoodScienceForYou, Bradshej, Hajatvrc, Wintonian, EmausBot, Liseranius, Orphan Wiki, WikitanvirBot, 478jjjz, Brendanology, Super48paul, RA0808, Teerickson, Christinebenson58, Winner 42, Jakirfirozkamal, Wikipelli, Dcirovic, ChrisTheBrown, WhatTheBlack, Tuxedo junction, Josve05a, Martboy722, John Mackenzie Burke, Wayne Slam, Aidarzver, Abergabe, Standinguptoit, Brandmeister, Donner60, Orange Suede Sofa, Iltoffier, TYelliot, DASHBotAV, Rocketrod1960, PTaschner, Woodsrock, Will Beback Auto, ClueBot NG, Gareth Griffith-Jones, Jack Greenmaven, MelbourneStar, Ur moma123456789, Frietjes, Mesoderm, Go Phightins!, Widr, Scaple10, Helpful Pixie Bot, Excerpted31, Titodutta, Plantdrew, Jaba96, Stacyroy, RobLandau, Scratlikesacorns, MusikAnimal, AwamerT, Mark Arsten, P'tit Pierre, MrBill3, Glacialfox, Klilidiplomus, Achowat, Biosthmors, Icodemachine, David.moreno72, MyNameIsLukeMonet, ChrisGualtieri, Professorrus, Fiona126, BrightStarSky, Dexbot, Sminthopsis84, Mogism, Lugia2453, Sfgiants1995, [illegible], The Anonymouse, Passengerpigeon, CsDix, Shahram.shirazi, Iztwoz, LHSaavedra, Xinyukiwi, Realguy12, DavidLeighEllis, Hi my names bobt42, CensoredScribe, Nightlight77, Evolution and evolvability, GirlWithWings, Zenibus, Spggoggles, Finnusertop, Jianhui67, Acalycine, Elizabeth sunny, OccultZone, Fundon1, StefanieStrohl, Jmmy2013, Csutric, Drericklagos, Chaya5260, Yahadzija, Monkbot, JRElings, AKS.9955, NewEnglandDr, Dsprc, AdamoulasA, Squad1234, Amortias, Tetucarlos, HMSLavender, Eakhiro, Hello1234566, Mike Swagger, Crystallizedcarbon, !0CaterPillar0!, Cavefish777, Giraffesyrup, YeOldeGentleman, LarryBoy79, ScienceGuru123, Somecoolman7888, K scheik, GeneralizationsAreBad, Krmeyer17, CREASHUNIST, Proyecto genómica, D.g.lab., HeirOfSumer, Diptanshu mandal, CAPTAIN RAJU, Lemondoge, FallenAngelXXXX, Duchamp 7, Ryan778, Historynerd1738, Austin2003 53, Hullo97, MichaelJCarroll, PrimeBOT, Dante123dmc, Bgendreau, Jdw13b, It'sAllinthePhrasing, Solt.kristen, Daniela.barreto, Dishabhavsar, Derianc12, PrateekKumarDas, Hh14b, Brinrodrgz and Anonymous: 862

- **Pharmacology** *Source:* https://en.wikipedia.org/wiki/Pharmacology?oldid=787487900 *Contributors:* Paul Drye, TwoOneTwo, Kpjas, The Cunctator, WojPob, Mav, Bryan Derksen, The Anome, Alex.tan, Jaknouse, Patrick, Kwertii, Kku, Gabbe, Qaz, Kosebamse, 168..., Looxix~enwiki, Ahoerstemeier, Mac, Snoyes, Glenn, Nikai, Tristanb, Evercat, Samw, [212], Heidimo, Tarka, Nohat, Bemoeial, Head, Kev, Robbot, Fredrik, Sharkey, ZimZalaBim, Academic Challenger, Rasmus Faber, Mendalus~enwiki, Hadal, Jmieres~enwiki, DocWatson42, Christopher Parham, Techelf, Ksheka, Tarek, Wolfkeeper, HangingCurve, Karn, Everyking, Bkonrad, Bensaccount, Jfdwolff, St3vo, Softssa, Erich gasboy, Antandrus, Alteripse, Mako098765, Karol Langner, APH, Oneiros, Frunge, PFHLai, Icairns, Zfr, Sam Hocevar, Anodyne, Joyous!, Mschlindwein, TheObtuseAngleOfDoom, Freakofnurture, MattKingston, Noisy, Discospinster, Ivan Bajlo, Bender235, Mashford, Aranel, Remember, ~K, CDN99, Truthflux, Davidruben, Maurreen, Arcadian, KBi, Obradovic Goran, Helix84, Haham hanuka, Geschichte, FMephit, Alan Isherwood, Jumbuck, Alansohn, Gary, Benjah-bmm27, MarkGallagher, Kurieeto, Hu, GJeffery, Gdavidp, Versageek, Johntex, Chrysaor, Red dwarf, Aeroflot, Nakos2208~enwiki, Eleassar777, Miroku Sanna, Graham87, BD2412, BorgHunter, Sjakkalle, Rjwilmsi, Mayumashu, Arisa, Stevenscollege, Miserlou, DoubleBlue, AED, Pitamakan, Dogbertd, SouthernNights, RexNL, Stevenfruitsmaak, Dayed, Roboto de Ajvol, Wavelength, Jlittlet, RussBot, Backburner001, Spaully, Ibpassociation, Ziddy, Badagnani, Pixiequix, Stevenwmccrary58, Tony1, Tsalman, DeadEyeArrow, Aprabhala, Scott Adler, Zzuuzz, Jolt76, MrTroy, GraemeL, TBadger, Carlwfbird, Mais oui!, Spliffy, Citylover, RichG, Joshbuddy, SmackBot, Jamott, Bobet, Unyoyega, Jagged 85, CommodiCast, Vilerage, Edgar181, Chef Ketone, Gilliam, Bluebot, Bartimaeus, MK8, Jprg1966, MalafayaBot, SchfiftyThree, Deli nk, TorW, Epastore, Zsinj, Rrburke, MrPMonday, RandomP, BullRangifer, Weregerbil, Fuzzypeg, DMacks, Acdx, Pilotguy, Lapaz, Mr. Lefty, Stoa, Slakr, Waggers, RBJ, Peter Horn, Kvng, OnBeyondZebrax, Pixi, IvanLanin, Tawkerbot2, JForget, Mig11, Pomping, Argon233, Cydebot, Rifleman 82, Master son, Doug Weller, RXPhd, Lindsay658, Chairman Natto, Missvain, AntiVandalBot, Drakonicon, TimVickers, KParks7, Waterthedog, I'll bring the food, JAnDbot, Kuzad, WmRowan, Dave Nelson, Magioladitis,

WolfmanSF, Bubba hotep, WhatamIdoing, Lethaniol, Bigbuck, Ksvaughan2, DGG, Pvosta, Anuragi, CWinslow, R'n'B, Nono64, Smokizzy, Mausy5043, Manticore, Vardar, Mikael Häggström, Oceanflynn, Alphapeta, Bob, Treisijs, Diego, MantleX, VolkovBot, Iosef, Jeff G., Fences and windows, Sirmelle~enwiki, Philip Trueman, Anonymous Dissident, Kburns81, Shanata, JameeBD, Yk Yk Yk, Falcon8765, Doc James, Fowerman, Steveking 89, Flyer22 Reborn, Antzervos, Lmc169, Fistencounter, UCSF CMP, Iain99, Lymeca, Szrahman, KingCuongL, Aliverius, Nancy, Pinkadelica, ClueBot, Abhinav, Mild Bill Hiccup, Eipnatas, Niceguyedc, ChandlerMapBot, Lilypink, Excirial, Jusdafax, CrazyChemGuy, Zeke8472, SpikeToronto, Llecount, Rhododendrites, KLassetter, SchreiberBike, Ottawa4ever, Aleksd, Taranet, Snake66, SoxBot III, Jytdog, EastTN, Ps256, Addbot, Willking1979, Ka Faraq Gatri, Favonian, O2demand, یامح, Frehley, Legobot, Cote d'Azur, Fraggle81, Nursereviewdotorg, Anypodetos, AnakngAraw, Proche01, Uofal, Raimundo Pastor, AnomieBOT, LlywelynII, Materialscientist, Citation bot, WillHHudson, Andrew-gus, Nasa-verve, GrouchoBot, Omnipaedista, Tobby72, Citation bot 1, Åkebråke, Pinethicket, Bernarddb, Pharmcog, Weeeeahmed, Dxar, Trappist the monk, Lotje, Vrenator, Radio89, Clarkcj12, Seahorseruler, Diannaa, Ivanvector, Mean as custard, Belomorkanal, Techhead7890, Thekady, EmausBot, Ajraddatz, ThF, DimaBorger, Klbrain, Dcirovic, Hashemi1971, Ryan Kaldari (WMF), Pedramgf, Coasterlover1994, 2tuntony, ChuispastonBot, Ejono, Rocketrod1960, Will Beback Auto, ClueBot NG, Inam marwat333, Irvin calicut, O.Koslowski, Karoutsos, Widr, Oddbodz, Helpful Pixie Bot, BG19bot, Mm smile, Northamerica1000, PhnomPencil, JohnChrysostom, AvocatoBot, Mark Arsten, MrBill3, NotWith, Shisha-Tom, EricEnfermero, ~riley, None but shining hours, ASPETwiki, Dexbot, Peripattikos, Hintzensam, DMField, Epicgenius, Sizzleking, I am One of Many, AmericanLemming, Tentinator, Seppi333, Amr94, Vinny Lam, Mfareedk, Marc.andre.f, Monkbot, Farahmustafa76, The temps des cerises, Jos Bessems, Julietdeltalima, Paewiki, Shikang Liu, Lokolkok, Zakhodgson, Sizeofint, L0st H0r!z0ns, KasparBot, Silicon Based, Marianna251, PrimeBOT, Physicalmathematics, Μαριλιάνα Σερέτη, Dwillsion and Anonymous: 386

- **Pharmaceutical drug** *Source:* https://en.wikipedia.org/wiki/Pharmaceutical_drug?oldid=787487927 *Contributors:* Kpjas, General Wesc, Toby Bartels, Deb, Karen Johnson, Ewen, Michael Hardy, Vaughan, Kku, Ixfd64, Dcljr, Karada, Pagingmrherman, Ahoerstemeier, Mac, Ronz, Tristanb, Samw, MichaK, Heidimo, Bemoeial, Harris7, Marshman, Topbanana, Robbot, Chealer, Fredrik, Chocolateboy, Mayooranathan, Tea2min, Filemon, Alan Liefting, Christopher Parham, Bensaccount, Jfdwolff, Edcolins, Andycjp, Beland, Onco p53, OwenBlacker, Rich Farmbrough, Cacycle, YUL89YYZ, Bender235, ESkog, RJHall, Bobo192, Davidruben, PeteThePill, Arcadian, Nk, Ranveig, Agjchs, Rodw, Wouterstomp, Thoric, SlimVirgin, Gene Nygaard, Blaxthos, Ceyockey, Woohookitty, Guy M, Sonelle, Al E., Wikiklrsc, Laurel Bush, DESiegel, Ruziklan, Mitomac, Magister Mathematicae, Josh Parris, Rjwilmsi, Pound, Astronaut, Amire80, Ttwaring, Mishuletz, Gurch, M7bot, DVdm, Bgwhite, WriterHound, YurikBot, Wavelength, Sceptre, Bhny, Yllosubmarine, Rsrikanth05, Wimt, ENeville, Dialectric, Joel7687, Terfili, Ragesoss, THB, Oribeta, PhilipO, Dbfirs, Tachs, Kkmurray, Boivie, Zzuuzz, GraemeL, LeonardoRob0t, Fram, Willtron, Garion96, Allens, Lyrl, Alexandrov, DVD R W, Veinor, SmackBot, Lestrade, Hydrogen Iodide, Blaine21886, Pgk, Leridant, Mscuthbert, JD, CommodiCast, Cessator, RobotJcb, Edgar181, Müslimix, Hmains, Alias777, Chris the speller, Deli nk, TorW, Hallenrm, A. B., Jaaf, Милан Јелисавчић, Frap, Chlewbot, Yidisheryid, RedHillian, DR04, COMPFUNK2, Decltype, Aotake, Pgillman, Cache22, Kukini, Ohconfucius, Daydreambeliever, Nathanael Bar-Aur L., Rklawton, Kuru, IronGargoyle, Erwin, Dicklyon, Stainedglasscurtain, Meco, Peter Horn, Quaeler, Iridescent, Bwalters, Schizmatic, Courcelles, Fvasconcellos, JForget, Wolfdog, Kylu, MarsRover, Seven of Nine, Meodipt, Gogo Dodo, ST47, Skittleys, RXPhd, Richard416282, Lindsay658, Oleksii0, Thijs!bot, Edupedro, Headbomb, John254, Chris goulet, Drugshome network, Deipnosophista, AntiVandalBot, QuiteUnusual, Drakonicon, Mauron~enwiki, LibLord, Piribo, Gökhan, JAnDbot, Jimothytrotter, MER-C, Bozman007, Michig, PhilKnight, Acroterion, Bencherlite, Magioladitis, VoABot II, Vitaminman, Randomazn69, WhatamIdoing, 28421u2232nfenfcenc, Cisum.ili.dilm, User A1, Khalid Mahmood, WLU, Pax:Vobiscum, Calltech, Drcaldev, Pvosta, Yobol, ChemNerd, JulianHensey, R'n'B, Nono64, EverSince, Thirdright, J.delanoy, CFCF, Svetovid, Ineck, Philologia Sæculārēs, Boghog, Nigholith, Chemistbrian, McSly, Mikael Häggström, DadaNeem, Jamesontai, DorganBot, Treisijs, Mike V, JavierMC, Necromance, VolkovBot, Quentonamos, QuackGuru, Philip Trueman, JuneGloom07, Andrew Su, Sankalpdravid, Charlesdrakew, OlavN, Una Smith, AmberEditor, Karenbenn, LetTheSunshineIn, Doc James, AlleborgoBot, Drake144, SieBot, Pkgx, Flyer22 Reborn, Wikibruger, Yerpo, Hello71, Szrahman, Torchwoodwho, Hariva, Literaturegeek, Heinebravo, Touchstone42, Standardname, ClueBot, Speedymarathon, Binksternet, GorillaWarfare, The Thing That Should Not Be, ImperfectlyInformed, Unbuttered Parsnip, Mild Bill Hiccup, LizardJr8, Ottawahitech, Leonard^Bloom, Bilal 902, Squared Away, Mayur sarolkar, Санта Клаус, S19991002, SchreiberBike, Carriearchdale, Aitias, Versus22, SDY, Finalnight, XLinkBot, Jytdog, Dthomsen8, Dgmvee, Dr.Soft, SilvonenBot, JinJian, Zodon, Sgpsaros, Good Olfactory, Addbot, C6541, Manuel Trujillo Berges, Blechnic, Nohomers48, Metsavend, Diptanshu Das, LinkFA-Bot, Quercus solaris, H kinani, Mfhulskemper, یامح, Ralf Roletschek, Ayacop, Delta 51, Ettrig, Luckas-bot, Fraggle81, TaBOT-zerem, Alfonso Márquez, Dowdow~enwiki, Eric-Wester, Raimundo Pastor, AnomieBOT, Daniele Pugliesi, Jim1138, Gate2wiki, Shogatetus, Flewis, Materialscientist, Onewiseman90, Citation bot, Xqbot, Timir2, Gilo1969, Jü, GrouchoBot, Pwthomas99, Ace111, Luciandrei, Sahehco, MerlLinkBot, Bekus, Vvdvet, Jatlas, Tranletuhan, Orhanghazi, Dimeron63, Citation bot 1, MarB4, Diwas, Ssspam, HRoestBot, Pikiwyn, Jccburnett, Full-date unlinking bot, Miguel Escopeta, Jauhienij, C messier, FoxBot, TobeBot, Francis E Williams, Darigan, Triok, Monkeyassault, Mean as custard, The Utahraptor, RjwilmsiBot, Smd75jr, EmausBot, Madoreck, Dewritech, Syncategoremata, GoingBatty, Tucci78, Klbrain, Tommy2010, Winner 42, TuHan-Bot, Wikipelli, Dcirovic, K6ka, Cavallad, Mbeaulieu, Stmc~enwiki, Tolly4bolly, ShantanuSingh198, Jonjtripp, Nijusby, 28bot, Karthikrrd, ClueBot NG, CocuBot, BarrelProof, Delusion23, Cntras, Anandjrao, Widr, سجاد, سجاد امجد, سجاد, Helpful Pixie Bot, Wbm1058, BG19bot, Jacopo.luppino, Fildovdokaz, CitationCleanerBot, Takerx77, ZRRSZR, Min.neel, NotWith, Glacialfox, RGloucester, BattyBot, Rlattuad, Toploftical, Chris-Gualtieri, LHcheM, Theomegapoint2012, EuroCarGT, IsraphelMac, Dexbot, Mogism, Lugia2453, Frosty, Romie234, Utscecc, I am One of Many, Iztwoz, DavidLeighEllis, Amalmehad, Wipur, Garzfoth, Pinkyt2008, Gensourcerxusa, Glaisher, Seppi333, Kind Tennis Fan, Elizabeth sunny, RuleTheWiki, Dpd1999, صعب, Monkbot, Renamed user 51g7z61hz5af2azs6k6, Kacampbell13, Luckydad, NQ-Alt, Aethyta, Jorge Guerra Pires, Narky Blert, Exiled Encyclopedist, Loraof, Benrusholme, Ursula Tschorn, KcBessy, Antsiepantsie, Timothyjosephwood, KasparBot, CAPTAIN RAJU, Clearfile86, Boilingorangejuice, Jkbuer, Tstric6, InternetArchiveBot, Jaafar A S, Wicodric, Hdlin, Bender the Bot, Anchitagrawall, PrimeBOT, Morganbohannon, Fuxkmejerry, B-Alberts, Importexportdata, Dwillsion and Anonymous: 331

- **Drug metabolism** *Source:* https://en.wikipedia.org/wiki/Drug_metabolism?oldid=783147073 *Contributors:* Rsabbatini, Ahoerstemeier, Bearcat, Robbot, Lysy, Tarek, Pashute, Jfdwolff, DanielCD, Discospinster, Cacycle, Bender235, JackWasey, CDN99, Bobo192, Axl, Boyd Steere, Alai, Stuartyeates, TigerShark, Joerg Kurt Wegner, Dysepsion, Rjwilmsi, Bhadani, Stevenfruitsmaak, Bgwhite, Todfox, User27091, Leptictidium, Closedmouth, Haymaker, Marc Kupper, RDBrown, Samkung86, Jedgold, Jtm71, Dr. Sunglasses, Spikeman, Shackers192, Keitei, Biomedeng, Luke poa, SingingOphelia, TimVickers, Alphachimpbot, LinkinPark, WolfmanSF, Tripbeetle, Faizhaider, WhatamIdoing, Djr5353, R'n'B, Thirdright, J.delanoy, Boghog, Mikael Häggström, LeaveSleaves, Ilyushka88, Arjayay, Jytdog, Thatguyflint, Sami Lab, Addbot, Thompsontc, DOI bot, Diptanshu Das, Madamecp, Quercus solaris, Ettrig, Yobot, AnomieBOT, Materialscientist, Citation bot, Xqbot, Meewam, Mattashner, RibotBOT, Shadowjams, Citation bot 1, Nirmos, Pinethicket, Suffusion of Yellow, RjwilmsiBot, Klbrain, Dcirovic, StopCyberBullies, A930913, Hazard-Bot, Teaktl17, ClueBot NG, SentientSystem, Miguelferig, BG19bot, NotWith, ChrisGualtieri, Isarra (HG), Andyhowlett, Ginsuloft, Meteor sandwich yum, Arvindnegi2301, Monkbot, L0st H0r!z0ns, KasparBot, InternetArchiveBot, GreenC bot, ImproveEverything, PrimeBOT and Anonymous: 82

- **Xenobiotic** *Source:* https://en.wikipedia.org/wiki/Xenobiotic?oldid=787140652 *Contributors:* AxelBoldt, Jll, Nikai, Bearcat, Robbot, Fuelbottle, Quartertone, Cacycle, Lycurgus, Kwamikagami, Bobo192, Ronline, Stemonitis, Benbest, Rjwilmsi, Margosbot~enwiki, YurikBot, Wavelength, Yrithinnd, El Cazangero, VIGNERON, Malcolma, Tony1, Bota47, Os2man, SmackBot, Zserghei, Bomac, Bluebot, Kristenq, Clicketyclack, Condem, Tomwood0, Cydebot, TimVickers, LookingGlass, JaGa, Rod57, Aboosh, Squids and Chips, Vipinhari, Swaranjit, SieBot, Mike2vil, Ken123BOT, ClueBot, Pakaraki, Tnxman307, Mlaffs, Addbot, ERK, DOI bot, KIRTIMAAN.MICRO, Luckas-bot, Ptbotgourou, AnomieBOT, AdjustShift, ArthurBot, Amoraea, Pinethicket, MastiBot, Suffusion of Yellow, EmausBot, WikitanvirBot, Dcirovic, ZéroBot, Triazine degrader, Fagopyrum, Orange Suede Sofa, Widr, Ed7654, Andrux, Caspase9, Jwratner1 and Anonymous: 56

- **Biotransformation** *Source:* https://en.wikipedia.org/wiki/Biotransformation?oldid=784366901 *Contributors:* Edward, Alan Liefting, Graeme Bartlett, Rich Farmbrough, CanisRufus, Snowolf, Sophisticated penguin, Woohookitty, Jclemens, Rjwilmsi, Ground Zero, Bgwhite, Wavelength, Stormbay, SmackBot, Gilliam, NickPenguin, Drphilharmonic, Alaibot, TimVickers, JAnDbot, Fabrictramp, Adrian J. Hunter, Capitaljay, Mikael Häggström, Manelcampos, Mowtown philippe, Funandtrvl, VolkovBot, SieBot, Flyer22 Reborn, Seuraza, Jesusbvf, Touchstone42, Animeronin, Mild Bill Hiccup, George moorey, Addbot, Koppas, Luckas-bot, Yobot, Templatehater, Aishking, Mattashner, Buchters, Pinethicket, Bernarddb, Jonesey95, Mean as custard, EmausBot, Pbdragonwang, ClueBot NG, Snotbot, Estopedist1, BG19bot, BattyBot, KirkMalone, Btpc.deepanshusoni, SirBigglesHemmingburgThePompous, Monkbot, Pharmaraj, L0st H0r!z0ns, InternetArchiveBot and Anonymous: 34

- **Multiple drug resistance** *Source:* https://en.wikipedia.org/wiki/Multiple_drug_resistance?oldid=759641910 *Contributors:* The Anome, Skysmith, CesarB, Jptwo, Bloodshedder, Michael Snow, Aetheling, Jfdwolff, Beland, PDH, Sam Hocevar, Smyth, Davidruben, Arcadian, Pearle, Axl, Walker44, SCEhardt, Joerg Kurt Wegner, MD2004, Rjwilmsi, Mohawkjohn, YurikBot, MagneticFlux, SmackBot, Radagast83, Drphilharmonic, MegaHasher, CmdrObot, MaxEnt, Headbomb, TimVickers, Hoffmeier, R'n'B, CFCF, Boghog, MistyMorn, Rod57, Msweany, VanBuren, MCTales, Graham Beards, Mimihitam, Touchstone42, CounterVandalismBot, CaptainCarolus, Coinmanj, Addbot, C6541, DOI bot, Quercus solaris, Wikimono111, Yobot, Imtechchd, AnomieBOT, Bluerasberry, Citation bot, Carturo222, Shadowjams, FrescoBot, Jeffrd10, John of Reading, Dewritech, Ckocks71, Klbrain, Dcirovic, ProfessorAM, MrBill3, Yaara dildaara, Reatlas, Jodosma, Wuerzele, Tom (LT), Monkbot, Χρυσάνθη Λυκούση, LeBleuDeVenus, Microbiology-scientist, Liechtenstein96, Anamfija and Anonymous: 32

- **Drug interaction** *Source:* https://en.wikipedia.org/wiki/Drug_interaction?oldid=784225196 *Contributors:* Bryan Derksen, Paul A, Ronz, Robbot, Unfree, Xezbeth, Longhair, Mareino, Agjchs, Lectonar, Axl, ReyBrujo, Ceyockey, Jeffrey O. Gustafson, Stevey7788, BD2412, Rjwilmsi, Stevenfruitsmaak, Bgwhite, Wavelength, Anomalocaris, Tevildo, Veinor, SmackBot, AndreasJS, Edgar181, Gilliam, Ohnoitsjamie, RDBrown, Uthbrian, Smallbones, Kristenq, LeeNapier, Pjvpjv, Nick Number, AntiVandalBot, Arch dude, Cooldesk, Cgingold, Boghog, VolkovBot, JhsBot, GCNYC, Epocrates, Doc James, Calliopejen1, Tatterfly, PipepBot, Jhaochen, Wnt, Addbot, C6541, Low-frequency internal, LinkFA-Bot, OlEnglish, Zorrobot, Yobot, Raimundo Pastor, AnomieBOT, Benhen1997, Xqbot, FrescoBot, LucienBOT, Redrose64, Serols, Trappist the monk, Teamabby, English Fig, Johncartmell, Klbrain, Dcirovic, Checkingfax, Netha Hussain, Tajdink, Hazard-Bot, Gorilla ch, ClueBot NG, Calabe1992, Plarmuseau, Wikih101, Snow Blizzard, MrBill3, Vartan Balian, Sidrah2012, ChrisGualtieri, Dexbot, Andyhowlett, Corn cheese, Everymorning, Esteed6608, Oleg ayranov, Monkbot, عبد الحكيم الألباني, Biochemistry&Love, Eurodyne, KasparBot, InternetArchiveBot, GreenC bot, Hdlin, Bender the Bot, PrimeBOT, Magic links bot and Anonymous: 37

- **Bioremediation** *Source:* https://en.wikipedia.org/wiki/Bioremediation?oldid=787707650 *Contributors:* Bryan Derksen, Lexor, Kku, Minesweeper, StAkAr Karnak, Big Bob the Finder, Vespristiano, Stewartadcock, Alan Liefting, Graeme Bartlett, Christopher Parham, Guanaco, JRR Trollkien, Onco p53, Tomruen, Ariff, Sonett72, Discospinster, Vsmith, CanisRufus, Pearle, Jumbuck, Alansohn, Paleorthid, ASK~enwiki, Velella, Benbest, GregorB, Macaddct1984, MrSomeone, Rjwilmsi, Goingin, Crazycomputers, Parutakupiu, Penguin, Dj Capricorn, Gwernol, Wavelength, Borgx, Samuel Wiki, RussBot, Pburka, Dialectric, Johann Wolfgang, Mikeblas, Bota47, IceCreamAntisocial, Closedmouth, Red Jay, Alexandrov, Jagz, SmackBot, Skaijo, McGeddon, Delldot, Gilliam, Skizzik, Bluebot, Oatmeal batman, Can't sleep, clown will eat me, Sunny17152, Brandymae, Answerthis, Plattbridger, Nick125, Smokefoot, Zeamays, Vylen, Anlace, Gobonobo, Bendzh, Saxbryn, Christian Roess, Joseph Solis in Australia, JoeBot, Basicdesign, Shoeofdeath, Dgw, Epbr123, Headbomb, Marek69, Majorly, Luna Santin, Cpelkas, Dreaded Walrus, JAnDbot, Husond, Kaushal mehta, Andonic, VoABot II, BrianGV, Fabrictramp, WhatamIdoing, C.lettinga, Cgingold, Mcfar54, WLU, MartinBot, R'n'B, Nono64, Jsmith86, J.delanoy, Enajy, Chriswiki, Chiswick Chap, HaruyukiFujimaki, Dorftrottel, FeralDruid, VolkovBot, MaD70, Scareth, TXiKiBoT, Oil Treatment International, Randomcoolkid, Gueneverey, Templationist, Leafyplant, Random Hippopotamus, Monty845, Billybobjrsrjr, Joemcjoejoe, Kksinter, Tiddly Tom, Skiview, Fratrep, Singmarlasing, Touchstone42, ClueBot, Tmol42, The Thing That Should Not Be, Mild Bill Hiccup, Hello Control, ParisianBlade, Tnxman307, Horselover Frost, Wnt, Roadrunner7002, XLinkBot, Jytdog, MystBot, Hawks22, Addbot, Polinizador, DOI bot, Jncraton, Granitethighs, Numbo3-bot, Caps910, Tide rolls, Ben Ben, Luckas-bot, Yobot, N1RK4UDSK714, AnomieBOT, Rubinbot, Dinesh smita, Giants27, Citation bot, ArthurBot, Xqbot, Ched, Smartbballchick74, Twirligig, Recycling-composting, Farglesword, HJ Mitchell, Citation bot 1, ScienceSolutions, Kgrad, Diannaa, Myeditsatwork, WikitanvirBot, WikiloverMe, Look2See1, ScottyBerg, GoingBatty, Dcirovic, Nz101, ClueBot NG, Widr, Helpful Pixie Bot, BG19bot, Undeadfenix, CitationCleanerBot, Sweetaarajesh, JYBot, Keys5954, Sminthopsis84, Sidelight12, Me, Myself, and I are Here, Joeinwiki, Microbeteacher, CensoredScribe, Napy65, Mykophile, Jwratner1, NottNott, Hbeshara, Myself zeeshanized, Anrnusna, Vaselineeeeeeee, Sockratic Method, Newconceptum, Argo-e, Grral, KasparBot, BluewaterShores, Xelkman, The Voidwalker, Bender the Bot, BroderickGerano, Alexandra.N and Anonymous: 201

- **Cytochrome P450** *Source:* https://en.wikipedia.org/wiki/Cytochrome_P450?oldid=779597089 *Contributors:* AxelBoldt, Michael Hardy, GTBacchus, Ronz, Julesd, Maximus Rex, Giftlite, Techelf, Jfdwolff, Mboverload, Wmahan, Alteripse, Beland, Richardsur, DragonflySixtyseven, Neutrality, Anodyne, Rich Farmbrough, Cacycle, Szquirrel, John Vandenberg, Arcadian, Jag123, Giraffedata, Wdfarmer, Beakerboy, Woohookitty, Ekem, Benbest, SDC, BD2412, BorisTM, Rjwilmsi, MarnetteD, Margosbot~enwiki, Krzysiu, Takometer, Physchim62, GangofOne, WriterHound, Wavelength, Mushin, Deeptrivia, Metalloid, Joelr31, Daniel Mietchen, Rmky87, Zwobot, Heathhunnicutt, Fuzzyrandom, Veinor, SmackBot, Slashme, Edgar181, Bartimaeus, Uthbrian, Jedgold, Smokefoot, Drphilharmonic, DMacks, Lobster101, Ligulembot, Gobonobo, Capmo, Kyoko, Meco, Fvasconcellos, CmdrObot, Bonás, Kupirijo, Ntsimp, Thijs!bot, Headbomb, Nick Number, Ju66l3r, Pro crast in a tor, TimVickers, Dougher, Qwerty Binary, MER-C, Fetchcomms, Magioladitis, Bikadi, ZackTheJack, Rlonsdale, Nono64, Adifeldman,

Boghog, Hodja Nasreddin, Ibrmrn3000, Pwnsey, 1000Faces, Mikael Häggström, Chiswick Chap, Belovedfreak, Chango369w, Malljaja, Monkey Bounce, Ninjatacoshell, Kristine1c, Countincr, Doc James, G00nsf, Petergans, Bform, Fimbriata, KoshVorlon, Koene, Interestedperson, Midtempo, EhJJ, Arjayay, Stone geneva, Trefork, DumZiBoT, EdChem, Dthomsen8, GroverPennyshaft, Erikamit, Addbot, DOI bot, Idiotchild, DavidLon, Quercus solaris, Numbo3-bot, Luckas-bot, Yobot, AnomieBOT, Citation bot, Xqbot, Skaaii, J04n, Abce2, Rünno, LucienBOT, Nacho Insular, Ricardo Ferreira de Oliveira, Citation bot 1, Ntse, Jonesey95, Tom.Reding, My very best wishes, RoadTrain, Paiamshadi, Mipah, RjwilmsiBot, John of Reading, Edu32, Johncartmell, Dcirovic, Fluffylumpy, MajorVariola, SporkBot, Ocdncntx, JoeSperrazza, Abergabe, WeigelaPen, Jparcon, Kinkreet, Teaktl17, ClueBot NG, JPBoyd, AaronLarsen, Rezabot, Swmmr1928, Bibcode Bot, Miguelferig, BG19bot, Mukeshsamani, NotWith, Jimw338, Rob Scarrow, Dexbot, Webclient101, The Anonymouse, Project Osprey, Randykitty, Sumraw, Evolution and evolvability, Stamptrader, Pitpeelorchard, YickChongLam, Monkbot, Hardiman12, Joflaher, Medgirl131, MicroMad, N8bt11, CV9933, Iamthepoopman69, Anarchyte, Krusec, Phylometab, At least I try, Rslateriii and Anonymous: 102

- **Transferase** *Source:* https://en.wikipedia.org/wiki/Transferase?oldid=785578542 *Contributors:* Nurg, Timrollpickering, Jfdwolff, Delta G, PFHLai, Remember, Jon the Geek, Arcadian, Jag123, Rjwilmsi, Margosbot~enwiki, Korg, YurikBot, Chaos, Zwobot, Edgar181, Lovecz, Drphilharmonic, Twas Now, Was a bee, Thijs!bot, Headbomb, Rhadamante, Drewmutt, R'n'B, PatríciaR, Boghog, 97198, STBotD, Idiomabot, AlnoktaBOT, Kyle the bot, TXiKiBoT, Broadbot, PixelBot, Addbot, Download, LaaknorBot, Quercus solaris, PRL42, Luckas-bot, Yobot, Citation bot, LilHelpa, Xqbot, FrescoBot, Logiphile, Trappist the monk, EmausBot, John of Reading, Timtempleton, Dcirovic, ZéroBot, Senator2029, The Banner Turbo, CitationCleanerBot, BattyBot, ChrisGualtieri, Mogism, Kenneth.jh.han, Lesadmick, WillPugarth, Galemu2, Adimart1, Monkbot, 65HCA7, Derek Johnson, Bender the Bot and Anonymous: 14
- **Glutathione S-transferase** *Source:* https://en.wikipedia.org/wiki/Glutathione_S-transferase?oldid=779145305 *Contributors:* Charles Matthews, Grendelkhan, Jotomicron, DragonflySixtyseven, Wisdom89, Arcadian, Jag123, Giraffedata, FMephit, Seans Potato Business, Rjwilmsi, Merv, Tony1, SmackBot, Chris the speller, Miguel Andrade, A. B., UniPharm, Smokefoot, Drphilharmonic, DMacks, Acdx, J. Finkelstein, Shackers192, JHunterJ, Martious, Fvasconcellos, CmdrObot, Phenylfairy, Headbomb, WolfmanSF, MiPe, Jack007, Nono64, Boghog, Hodja Nasreddin, Rod57, Pdcook, Malljaja, Kosigrim, Xiuyechen, Virtualt333, Cyberix, Toxicologyman, Biologe77, Alboyle, Addbot, Kinesin-8, Yobot, Ptbotgourou, AnomieBOT, Omnipaedista, Ganimede~enwiki, Danfa1971, FrescoBot, NSH002, Happiness4ever, Citation bot 1, HRoestBot, SF Gyros, Shawthorn, RjwilmsiBot, EmausBot, Pipino55, Allocati, Dcirovic, ZéroBot, SporkBot, Abergabe, WeigelaPen, Lv131, Frietjes, Helpful Pixie Bot, Siligang, Langing, Hjimker, Carliitaeliza, BattyBot, Biosthmors, Biolprof, Maximus155, Flemingrjf, Jnims, Gpruett2, MChapman5, Martchas, Monkbot, BethNaught, CV9933, Bpatte16 and Anonymous: 36
- **Efflux (microbiology)** *Source:* https://en.wikipedia.org/wiki/Efflux_(microbiology)?oldid=780150678 *Contributors:* Axl, Rjwilmsi, Erebus555, R.O.C, Chris Capoccia, PTSE, Carabinieri, SmackBot, Samkung86, Wedian, Smokefoot, Drphilharmonic, Thijs!bot, Headbomb, TimVickers, Nono64, Boghog, Hodja Nasreddin, Cspan64, Jeepday, NathanT17, Number774, XLinkBot, Schlenk, Addbot, Neodop, Ssschhh, AnomieBOT, Citation bot, Nanopharmacy, Quebec99, LilHelpa, Mrgionfriddo, Fdardel, 6amiesh, Amkilpatrick, Dcirovic, ZéroBot, Hazard-SJ, Michael Bailes, LordGud, Curb Chain, Paul Kemp~enwiki, Seppi333, Monkbot, Gaspanico, L0st H0r!z0ns, CV9933, Guomichelle18, Jhk7793 and Anonymous: 16
- **Antioxidant** *Source:* https://en.wikipedia.org/wiki/Antioxidant?oldid=782432753 *Contributors:* Carey Evans, Marj Tiefert, Mav, Andre Engels, SimonP, DennisDaniels, Michael Hardy, Dculberson, Axlrosen, Baylink, Bluelion, Julesd, Kimiko, TonyClarke, Tpbradbury, Phoebe, Samsara, Thue, J D, Owen, Jni, Chuunen Baka, Paranoid, Ee00224, Merovingian, Rasmus Faber, Sunray, Wikibot, Jleedev, Carnildo, Adam78, Fargoth~enwiki, Matthew Stannard, Centrx, Giftlite, DocWatson42, Haeleth, Wolfkeeper, Netoholic, No Guru, Mark T, Digital infinity, Jfdwolff, Ravn, Neilc, Utcursch, Andycjp, Geni, IdahoEv, Beland, Onco p53, OverlordQ, Chiu frederick, Wikimol, Elroch, Jessesamuel, Freakofnurture, Discospinster, Rich Farmbrough, Guanabot, Cacycle, Qutezuce, Vsmith, Night Gyr, Bender235, Sten, RJHall, El C, Lorem Ipsum~enwiki, CDN99, Spalding, Circeus, Wood Thrush, Brim, Scott Ritchie, Vanished user 19794758563875, Orangemarlin, Jumbuck, Abstraktn, MatthewEHarbowy, Dbeardsl, Keenan Pepper, Riana, Lightdarkness, Sligocki, Hu, Avenue, Suruena, Proski, Tony Sidaway, Frankg, Kazvorpal, Japanese Searobin, Tariqabjotu, Woohookitty, Kzollman, Benbest, Tmrobertson, FreplySpang, GrundyCamellia, DePiep, Icey, Edison, Graniterock, Jorunn, Rjwilmsi, Kinu, Miserlou, SeanMack, Brighterorange, Jbamb, Yamamoto Ichiro, Crizz, RobertG, AED, CalJW, Musical Linguist, Nihiltres, Intgr, Consumed Crustacean, Srleffler, BradBeattie, Physchim62, Chobot, Bgwhite, YurikBot, Wavelength, Borgx, Mikalra, Brandmeister (old), J. M., Malevious, Chris Capoccia, GLaDOS, Chrispounds, Gaius Cornelius, Shaddack, Wimt, DarkPhoenix, Pproctor, Welsh, Clam0p, BirgitteSB, Brandon, PeepP, JoshuaTree, Dbfirs, BOT-Superzerocool, Derek.cashman, Bluecricket, Personalvoice, BorgQueen, GraemeL, Rlove, Allens, Katieh5584, Alexandrov, Patiwat, CIreland, N3362, Mhardcastle, Yvwv, Veinor, Crystallina, SmackBot, RhaneC, M dorothy, Tarret, Prodego, K-UNIT, FlipOne, Edgar181, TheTweaker, Magwich77, Gilliam, Markzero, Hmains, David.Throop, Tyciol, Chris the speller, Bluebot, Audacity, RDBrown, Deli nk, Uthbrian, Jerome Charles Potts, Jstdadd, Egads, Rlevse, Zymatik, Langbein Rise, Onceler, Can't sleep, clown will eat me, Vanished User 0001, Nixeagle, JonHarder, RespekT, Parent5446, Xyzzy n, Monotonehell, RandomP, ShahJahan, Smokefoot, Mwtoews, SteveLower, Slotaa, Kukini, Ohconfucius, Harryboyles, Anlace, Apalapala, Kereish~enwiki, John, StevenRobertson, Ampersand777, Peterlewis, Ekrub-ntyh, Slakr, Beetstra, Mr Stephen, SandyGeorgia, Spook`, Ryulong, Peyre, Dfred, BranStark, Iridescent, Joseph Solis in Australia, JoeBot, Bdusel, Blehfu, Velocipedia, Linkspamremover, Tawkerbot2, Chetvorno, Mjunaidbabar, Fvasconcellos, Prithason, CmdrObot, Leevanjackson, Banedon, Dgw, Green caterpillar, Lentower, TheTito, DShantz, Johner, Guitarmankev1, Cydebot, Kupirijo, WillowW, Rhode Island Red, Rifleman 82, Red Director, CuTop, ST47, Christian75, FastLizard4, Jluciano~enwiki, Cakulbet, Brad101, Pustelnik, Casliber, Thijs!bot, Opabinia regalis, HappyInGeneral, CopperKettle, Chasingtulips, Oliver202, Headbomb, Gnurkel, Nick Number, Transhumanist, LigerThai, BCable, Mentifisto, Flosseveryday, Cyclonenim, AntiVandalBot, The Obento Musubi, Genmail2000, Majorly, Luna Santin, KP Botany, TimVickers, Paul144, MER-C, Sophie means wisdom, Sangak, Lenny Kaufman, Jaysweet, Bongwarrior, VoABot II, Meredyth, Myxoma, Wikidudeman, Saabisu, Nucleophilic, Nposs, Dirac66, A3nm, DerHexer, WLU, Eeatkinson, Gwern, Snoader, Dr. Morbius, Wiki wiki1, Hdt83, MartinBot, Mermaid from the Baltic Sea, Sui1989, ChemNerd, Bissinger, Chickenboner, Trixt, Gidip, Uriel8, Nono64, Fconaway, PStrait, J.delanoy, DrKay, Boghog, Aetkin, Maproom, Gzkn, BaseballDetective, Mikael Häggström, Wikiwopbop, Itecle, Bushcarrot, Alnokta, Williamfarrell, Juliancolton, STBotD, DorganBot, Treisijs, Kphillipps, Bonadea, Pdcook, VolkovBot, Regulations, Ttfan12311, Khochman, Soliloquial, Philip Trueman, TXiKiBoT, Oshwah, Kriak, Despres, Freedomforall227, Someguy1221, Littlealien182, Kroarick, PointSkull, Denaun, Geometry guy, MsMichele, ACEOREVIVED, Madhero88, Costacs, Lmuenchen, Doc James, AlleborgoBot, Resurgent insurgent, Treesaredying, Wikiapg, Britzingen, SieBot, StAnselm, Rukiddingme?, Antiox, WereSpielChequers, Sebastiano venturi, Lizjanda, Phe-bot, Kkrouni, Araignee, LeadSongDog, France3470, Cuvette, Qst, Arbor to SJ, PhilMacD, Hello71, Azghari, Smartsarang, OKBot, Brice one, Spartan-James, Maderibeyza, JmalcolmG, Iwouldknow, Modelun88, KyanFiFi, Cmshatto, Dust Filter, Treekids, Forluvoft, MenoBot, LikeFunYouAre, ClueBot, The Thing That Should Not Be, Drmies, SuperHamster, Doseiai2, Boing! said Zebedee, Niceguyedc, Phenylalanine, Somno, Sgv 6618, Pstrous,

Alexbot, Akane700, Panoramix303, NuclearWarfare, Arjayay, Medos2, Singhalawap, TimothyJacobson, BOTarate, Thingg, Pzoxicuvybtnrm, Winebloom, Blow of Light, Omer88f, SoxBot III, MasterOfHisOwnDomain, Press olive, win oil, Fiquei, Jytdog, Vanished 45kd09la13, Gggh, Addbot, Grayfell, DOI bot, StayHealthyTV, Download, LaaknorBot, LAAFan, Favonian, LemmeyBOT, LinkFA-Bot, Tide rolls, Ufocason, Teles, Legobot, आशीष भटनागर, Luckas-bot, Yobot, Ptbotgourou, TaBOT-zerem, Freikorp, AnomieBOT, Nutriveg, 1exec1, JackieBot, Kingpin13, Materialscientist, The High Fin Sperm Whale, Citation bot, Eumolpo, Maniadis, ArthurBot, Korin tiger, Jay L09, MauritsBot, Xqbot, Zad68, TinucherianBot II, Capricorn42, Elvim, M.Ebner, Skaaii, J04n, GrouchoBot, Zefr, SassoBot, Amaury, Wiki emma johnson, Sophus Bie, Chongkian, 33rogers, Brett Lally, WhatisFeelings?, Dunxian, FrescoBot, Hardtheslam, Citation bot 1, DrilBot, Pinethicket, HRoestBot, Sctechlaw, BeVeryGoodMan, Beverygoodman1, 11ken11, TobeBot, Trappist the monk, Venturisebastiano, Overagainst, Janescott145, Feelthegreens, Tbhotch, Rajakarthis14, RjwilmsiBot, Bento00, DASHBot, Vinnyzz, EmausBot, WikitanvirBot, Chelcal, RA0808, Rkb 205, Dcirovic, K6ka, Serketan, Golakers94, Hazard-SJ, H3llBot, AManWithNoPlan, Erianna, Bandn, Rcsprinter123, Donner60, Ego White Tray, ChuispastonBot, Llightex, Michambers2013, Rajakarthis, Michael Bailes, Petrb, ClueBot NG, Rainbowwrasse, E-hrana, Dictabeard, Tammyregis, Widr, Acanderson18, Helpful Pixie Bot, Novusuna, Wbm1058, Bibcode Bot, Angus2006, BG19bot, Nirupama.ARC, Iselilja, IraChesterfield, AdventurousSquirrel, Dustinlull, Johnt9000, Biosthmors, W.D., Illia Connell, Jarlaxll, Dexbot, FoCuSandLeArN, Ventur-sebastiano, Darryl from Mars, Wainwright1956, Everything Is Numbers, Wateresque, Cathry, Hokila, Qwh, Everymorning, Blythwood, RomanGrandpa, Clr324, AresLiam, Seppi333, R. U. Abhishek, Anrnusna, Navneetsoni451978, Monkbot, Renamed user 51g7z61hz5af2azs6k6, Venturi Seba, Negril123, Davidcarroll, Sensitive Corridor, A Great Catholic Person, Sandeepdhall, Wlemmo, 333Monu, MicroPaLeo, Skylord a52, Adityadipas, MAJakobs, Ggux, InternetArchiveBot, Entranced98, Dr hits p, Bpatte16, Great floors, MikeNbirmingham, OAbot, PolyphenolNet, Purificateurdair, Magic links bot, Garg himanshu and Anonymous: 598

- **Biodegradation** *Source:* https://en.wikipedia.org/wiki/Biodegradation?oldid=786968908 *Contributors:* Marj Tiefert, Mav, Bryan Derksen, Roadrunner, Europrobe, D, Kku, MartinHarper, TakuyaMurata, Skysmith, Ronz, Bogdangiusca, Daniel Quinlan, Timc, SEWilco, Fredrik, Alan Liefting, DocWatson42, Michael Devore, Guanaco, JRR Trollkien, Onco p53, MacGyverMagic, Adziura, Gscshoyru, Sonett72, Ouro, CALR, Discospinster, Vsmith, Pedant, Kaszeta, El C, Prainog, Vortexrealm, Gyrus~enwiki, Cohesion, Alphatwo~enwiki, Jumbuck, Danski14, Alansohn, AnnaP, Paleorthid, Kurieeto, Sciurinæ, Guthrie, Mindmatrix, LOL, Hdante, Bowman, V8rik, Enzo Aquarius, Rjwilmsi, Koavf, DeadlyAssassin, Rydia, HappyCamper, Brighterorange, RobertG, Karelj, Wavelength, Sceptre, Toquinha, Pseudomonas, Plhofmei, Bota47, Arthur Rubin, LeonardoRob0t, DoriSmith, NeilN, NickelShoe, SmackBot, Elizabeth ackley, Eakspeasy, Chris the speller, Bluebot, Dlohcierekim's sock, Can't sleep, clown will eat me, Chlewbot, Rrburke, Phaedriel, Smooth O, Smokefoot, DMacks, Terminator50, Gobonobo, Minna Sora no Shita, Geologyguy, Argento, BranStark, V111P, Tawkerbot2, The Letter J, DaveSumpner, Insanephantom, Casper2k3, Callsign, MC10, Jelwell, Gogo Dodo, Odie5533, Tawkerbot4, Thijs!bot, Epbr123, N5iln, JNighthawk, John254, Manthanfadia, Orfen, Omegakent, Gökhan, Dreaded Walrus, Mikenorton, Albertoeba, MER-C, Magioladitis, VoABot II, Engelbaet, Farquaadhnchmn, Jim Douglas, Adrian J. Hunter, DerHexer, Eugenwpg, MartinBot, Shentino, EdBever, J.delanoy, Svetovid, MrBell, Extransit, Afaber012, Cometstyles, VolkovBot, Jeff G., Indubitably, Bkengland, Philip Trueman, TXiKiBoT, Ann Stouter, Someguy1221, LeaveSleaves, ^demonBot2, Wiae, RiverStyx23, Madhero88, Eubulides, Lamro, Locke9k, Triggersite, Logan, SieBot, Zephyrus67, One more night, Pkgx, Yintan, Copycatken, Pinkadelica, Touchstone42, ClueBot, The Thing That Should Not Be, Cpachon, MikeVitale, Ndenison, CounterVandalismBot, Arunsingh16, Auntof6, Excirial, Jusdafax, Protozoon, Versus22, Methanus, Alexius08, Evelyne V, ZooFari, RyanCross, Addbot, Some jerk on the Internet, Zoommer, Glane23, Tide rolls, Bartledan, Carterjj33, Kaysons, Ben Ben, Legobot, Luckas-bot, Yobot, Carleas, AnomieBOT, Shootbamboo, Rubinbot, Junket76, Ulric1313, Limideen, Citation bot, Neurolysis, Wololo, Xqbot, Schmutz MDPI, Sionus, Anna Frodesiak, GrouchoBot, Twirligig, RibotBOT, Martin fed, Captain-n00dle, Hairbearhog, Citation bot 1, Pinethicket, ENSO Bottles, HRoestBot, Jhog1978, A8UDI, EdoDodo, Bpi communications, OxobioEPI, Yangsang20001, GossamerBliss, Viskocity, Fpintl67, StorkenSture, พุทธพร ส่องศรี, Oxobio PA, Kooljoe, Slon02, EmausBot, Ahmed adel1, Renewalwater, Bsqueen, RA0808, Wikipelli, Jasonanaggie, JoeSperrazza, ClueBot NG, Gareth Griffith-Jones, Satellizer, Esebi95, Tideflat, Dnpatton, Estopedist1, Wikiuser2k11, Widr, LaureenBuckhe, Helpful Pixie Bot, Curb Chain, Calabe1992, Northamerica1000, Hallows AG, Prof. Hein, Indah blestari, Rm1271, NotWith, TBrandley, Nitrobutane, Anigelseth, ENSOPlastics, Mogism, GGIwin, TheRealWallyrus, Piluke, Vaccinationist, Tentinator, Microbeteacher, Napy65, Amr94, AntiCompositeNumber, Meteor sandwich yum, Argus3, Randy Glad, Vaselineeeeeeee, KH-1, Tymon.r, Azealia911, Daniel Dela Pasion, CAPTAIN RAJU, Kurousagi, Joe2719, Tfgfhgg, Bear-rings, Concus Cretus, Timeflow X and Anonymous: 309

- **Microbial biodegradation** *Source:* https://en.wikipedia.org/wiki/Microbial_biodegradation?oldid=787791326 *Contributors:* Alan Liefting, Rich Farmbrough, CanisRufus, Mandarax, Rjwilmsi, Nihiltres, Scott Dial, Sadads, TimVickers, R'n'B, Touchstone42, Avenged Eightfold, Mild Bill Hiccup, Niceguyedc, Bioguz, Addbot, Koppas, AnomieBOT, Citation bot, Twirligig, Trappist the monk, Dcirovic, Sphingomonas, ChuispastonBot, BG19bot, CatPath, CitationCleanerBot, WikiHannibal, Teammm, SirBigglesHemmingburgThePompous, Microbeteacher, Monkbot, Vaselineeeeeeee, Pariah24, Timmer26 and Anonymous: 12

- **Genotyping** *Source:* https://en.wikipedia.org/wiki/Genotyping?oldid=783899944 *Contributors:* Zoe, Michael Hardy, Meisterflexer, Pgan002, Xezbeth, Dmanning, Dirac1933, Zenkat, Smoe, Ground Zero, Chobot, Paphrag, Pseudomonas, Rsriprac, SmackBot, ApersOn, Kaiwen1, Bluebot, Jethero, Persian Poet Gal, Miguel Andrade, JonHarder, Radagast83, TedE, Gosolowe, Dgw, N2e, Narayanese, Wmasterj, Alphachimpbot, Sauff, Squidonius, Grook Da Oger, James obejas, PCock, Nbauman, Hans Dunkelberg, Boghog, Artgen, GcSwRhIc, Ivajdo, Canoepan, Upsud, Lmsutton1, Flyer22 Reborn, Gustavocarra, Joncurtis7, Erikamit, Albambot, Addbot, PaleWhaleGail, Micromaster, Ronhjones, Fieldday-sunday, Yobot, Ptbotgourou, Dchai, FrescoBot, Ktpickard, ZéroBot, ClueBot NG, Maikroeder, Vigi.limi, Anthdoyle, Mwsal, Paulorapazote, Wordstorn, InternetArchiveBot, Rtrust, Albahrani Batool and Anonymous: 37

- **Chemogenomics** *Source:* https://en.wikipedia.org/wiki/Chemogenomics?oldid=783037145 *Contributors:* Steinbeck~enwiki, Lfh, Joerg Kurt Wegner, Rjwilmsi, Biochemza, Bhny, Derek.cashman, Kkmurray, Moez, Radcen, Marcuscalabresus, Fl, Z22, Boghog, Ospjuth, Muro Bot, Addbot, Dawynn, Omnipaedista, Tom.Reding, RjwilmsiBot, GoingBatty, Klbrain, Noot al-ghoubain, KLBot2, Dexbot, Éffièdaligrh, KasparBot, SciDragon and Anonymous: 10

- **Personalized medicine** *Source:* https://en.wikipedia.org/wiki/Personalized_medicine?oldid=786944783 *Contributors:* AxelBoldt, Ronz, Robbot, HaeB, Michael Devore, Jfdwolff, Chowbok, Rich Farmbrough, Rhobite, Bender235, Arcadian, Giraffedata, Sriram sh, Pearle, Alansohn, Ceyockey, Damicatz, BD2412, Rjwilmsi, Ground Zero, Idaltu, Shaggyjacobs, Mushin, Jlittlet, Kkmurray, Andrew73, Izayohi, SmackBot, KnowledgeOfSelf, Edgar181, Yamaguchi??, Boul22435, Deli nk, Colonies Chris, Kuru, John, Robofish, JHunterJ, Andrew Davidson, CmdrObot, Crimson Observer, Green caterpillar, Cydebot, Gogo Dodo, Skittleys, Sigchiguy, Headbomb, Jayron32, Spartaz, Arsenikk, MER-C, Ph.eyes, Albany NY, Coolhandscot, Freshacconci, Syaskin, JamesBWatson, Jllm06, Tedickey, WhatamIdoing, JaGa, Pvosta, R'n'B, Joshuaali,

MistyMorn, Maurice Carbonaro, Rod57, Katharineamy, SteveChervitzTrutane, Chiswick Chap, KylieTastic, Caddymob, Scewing, Anas.sal, Kuebi, Amaher, Karlengblom, Lamro, Jpmaytum, Fudzter, Sunrise, Pharmtao, ClueBot, Mild Bill Hiccup, Wsmith4474, Jeffemiller, XLinkBot, Jytdog, Clinicaljournal, Sgpsaros, Addbot, Nestorius, MrOllie, HerculeBot, Yobot, Kendallmorgan, Medical geneticist, AnomieBOT, Jim1138, Gallowolf, Bluerasberry, Citation bot, La comadreja, Gginsburg1, Neurolysis, LilHelpa, Twirligig, Alvin Seville, Jdegreef, Sahehco, Quikfastgoninja, Erik9bot, DoctorDNA, Steve Quinn, WAGAHAGA123, Citation bot 1, Tom.Reding, Kuak, Serols, Rosaryville, Justjennifer, Mean as custard, Aircorn, EmausBot, Snyde001, Daskalak, Ajs123, Dcirovic, Jahub, PwCHealth, H3llBot, Nkamireddi, PMADVO, Yossi Kimchi, Jesanj, Δ, Rangoon11, Quantumash, ClueBot NG, This lousy T-shirt, Ettuquoque, MerlIwBot, Helpful Pixie Bot, Wbm1058, KLBot2, BG19bot, Virtualerian, BattyBot, Ossip Groth, Fraulein451, Rp1989, TylerDurden8823, Dexbot, Dunnoitall, Hydra Rain, SFK2, Leemon2010, Me, Myself, and I are Here, EngEdit, 19876BV, So26, Trthths, Monkbot, Filedelinkerbot, Followww, Wuser6, Wiki CRUK John, Davidambareensamiepriyanka, KM1977, SCHAPPY, BrettofMoore, Brianbleakley, Boneywoulded, Mineben256, Goodmorningfrank, Dmichalski, SciDragon, InternetArchiveBot, Art379m, ISRbenh, PaulinaZQ, Franz101085, CancerSurvivor54, Sumana S, SafinaR, Rap17, Andrew McAdams, Personalizedmedicine and Anonymous: 134

- **Pharmacovigilance** *Source:* https://en.wikipedia.org/wiki/Pharmacovigilance?oldid=787892500 *Contributors:* Edward, Kku, Chuunen Baka, Rholton, Alan Liefting, Giftlite, Falcon Kirtaran, Remuel, Arcadian, Jhertel, Sabine's Sunbird, Joeggi, Ceyockey, GraemeLeggett, Marudubshinki, BD2412, Rjwilmsi, Jivecat, Mikalra, RussBot, Gaius Cornelius, Malcolma, Zwobot, Rathfelder, Andrew73, Lcarsdata, Edgar181, Chris the speller, TimBentley, Deli nk, Epastore, Khukri, G716, Derek R Bullamore, Jklin, Ohconfucius, G-Bot~enwiki, Ediefienduk, Voceditenore, Beetstra, DabMachine, Skapur, Fvasconcellos, Uptoeleven, CmdrObot, Cahk, Kanags, Jiankabe, Hopping, RXPhd, Inside x, Thomrossi, Porqin, Docx43, Barek, Ajordanl, Magioladitis, WallyFromColumbia, Pvosta, Flowanda, Gmurimi, Helen Politis-Norton, Montie06, Mojodaddy, Terrek, Spellcast, Fences and windows, TXiKiBoT, Doc James, Levinem, Flyer22 Reborn, Ldoan, Szrahman, Dmannsanco, Explicit, Nnsevencom, MenoBot, Usctrojan, Neil hariharan, Bhajish, Thuja, Skbkekas, Anotheruserhere, Qwfp, Mellebga, XLinkBot, Jytdog, Dthomsen8, Avoided, Cg2p0B0u8m, Wegely, Addbot, Some jerk on the Internet, Alfie66, Luckas-bot, Yobot, Anand.sequeira, Anypodetos, THEN WHO WAS PHONE?, EricP, Raimundo Pastor, AnomieBOT, Materialscientist, Nomoremissniceguy, LilHelpa, Xqbot, TheAMmollusc, Kereul, J04n, RibotBOT, Diwas, Hiresavi, 10metreh, Tom.Reding, Onel5969, Mean as custard, RjwilmsiBot, حامد صوفی, Klbrain, Tranphuongthuy, Potionism, Hashemi1971, Araghuram19, Ramansoz, ClueBot NG, Digger65, Snotbot, Widr, Ryan Vesey, BG19bot, Sapharm, Cyberbot II, ChrisGualtieri, Tow, Lugia2453, Ndave1404, Arjunsingar, Me, Myself, and I are Here, Leprof 7272, Passengerpigeon, Faizan, Michipedian, Sceaf, Health Canada, Naren001, Marketing Employee, Pharmanerd, Pharmerbill, Ginsuloft, Denis Lebel, Fixuture, Ulrich Hagemann, PhV-specialist, Mr. Smart LION, Monkbot, Lulaare, Banumakki, DIA-DC, The Quixotic Potato, Riccardo Lora, Danielfb156, GreenC bot, Natureium, Dept. of Pharmacology, B-Alberts, Satyamspot and Anonymous: 165
- **Structural genomics** *Source:* https://en.wikipedia.org/wiki/Structural_genomics?oldid=780374265 *Contributors:* Miguel~enwiki, Hephaestos, Lexor, Docu, Eugene~enwiki, Giftlite, Dmb000006, Mpw115, CALR, Viriditas, Mindmatrix, Jeff3000, Rjwilmsi, Biochemza, Jongbhak, Lustig~enwiki, Kerowyn, Carrionluggage, Bgwhite, Martin.jambon, Kkmurray, Occhanikov, SmackBot, Kjaergaard, Chris the speller, RDBrown, Martin Jambon, Ben Moore, Headbomb, Tholton, WhatamIdoing, Boghog, Hodja Nasreddin, Pjftheclimber, Alexbateman, SieBot, BobShair, Webridge, Ideal gas equation, Vriend, Bioinfoguy1, Addbot, Leszek Jańczuk, LaaknorBot, आशीष भटनागर, Yobot, Themfromspace, ??, Cycormier, FrescoBot, LucienBOT, Citation bot 1, Psisgkb, My very best wishes, Nat we, Polshgrl, Dcirovic, ClueBot NG, Widr, Mdomagalski, Sminthopsis84, Joeinwiki, Éffièdaligrh, Monkbot, Kevo Strevin, Hankwbass and Anonymous: 31
- **Toxicogenomics** *Source:* https://en.wikipedia.org/wiki/Toxicogenomics?oldid=772408595 *Contributors:* Nikai, PuzzletChung, Phil Boswell, Jfdwolff, Zhuuu, TheObtuseAngleOfDoom, Rich Farmbrough, Bender235, Ceyockey, Zenkat, Joerg Kurt Wegner, Marudubshinki, Rjwilmsi, Gmosaki, Elkman, Kkmurray, RDBrown, Kristenq, Alaibot, AM Smith, SNP, Scottalter, Boghog, Hodja Nasreddin, Anuacharya, Addbot, Mai-tai-guy, Luckas-bot, Yobot, Oleginger, Citation bot, Cygnenoir, Mamaberry11, Citation bot 1, Mcrosenstein, Cyberic71, Dcirovic, AManWithNoPlan, Yaser hassan 2006, Wingman417, Gklambauer, Moldeck, Dexbot, Me, Myself, and I are Here, Randykitty, Éffièdaligrh, Monkbot and Anonymous: 11
- **DeCODE genetics** *Source:* https://en.wikipedia.org/wiki/DeCODE_genetics?oldid=777505274 *Contributors:* AxelBoldt, Bjarki S, Nagelfar, Larus~enwiki, Orangemike, ConradPino, D6, Rich Farmbrough, Aecis, Jumbuck, PaulHanson, Ceyockey, Stemonitis, Benbest, Rjwilmsi, MZMcBride, FlaBot, PaulWicks, Gsch, Alarichall, Carabinieri, Mais oui!, SmackBot, Twynholm, RDBrown, George Church, Maltodextrin, Tryggvia, Wen D House, TPO-bot, CastorCanada, Robofish, Ben Moore, Chris55, JForget, Cydebot, Hebrides, After Midnight, ThevikasIN, TillJE, Karlpearson, Funandtrvl, SylviaStanley, S.Örvarr.S, Meightysix, Gimlifilmfestival, Jóhann Heiðar Árnason, Gigacephalus, DaveBurstein, Doprendek, MarmadukePercy, Addbot, Bragiw, Lightbot, Yobot, Mayflower3, Hypermorph, Bluerasberry, Citation bot, Sciencechick, Ahtremblay, Zikaron, Citation bot 1, RjwilmsiBot, Nick Moyes, GoingBatty, Dcirovic, Mccapra, ClueBot NG, BG19bot, Dexbot, Monkbot, Tigercompanion25, ArtZ72, InternetArchiveBot, Mrbergstrom, Thordurkristjans and Anonymous: 16
- **Navigenics** *Source:* https://en.wikipedia.org/wiki/Navigenics?oldid=769559875 *Contributors:* Zundark, Ihcoyc, Ukexpat, D6, Dols, Viriditas, Rjwilmsi, PaulWicks, Malcolma, Brandon, Nikkimaria, Nimbex, SmackBot, Hmains, Chris the speller, A. Parrot, Argotechnica, JForget, Patho~enwiki, Alaibot, Andyjsmith, John254, Adavidb, SteveChervitzTrutane, Mercurywoodrose, Jdcrutch, Meightysix, Gigacephalus, Addbot, Beeswaxcandle, Mission Fleg, AnomieBOT, Sarah.arora, Woo1000, Sciencechick, Ahtremblay, DoctorDNA, DarylJThomas, Laurverr, ZéroBot, Maureen1307, Y.golovko, Renamed user 51g7z61hz5af2azs6k6, Bender the Bot, Jeanjung212 and Anonymous: 14
- **23andMe** *Source:* https://en.wikipedia.org/wiki/23andMe?oldid=788264385 *Contributors:* AxelBoldt, Edward, Michael Hardy, Jerzy, Nagelfar, Drant, Simplicius, MementoVivere, AnandKumria, Pmsyyz, Xezbeth, Gronky, Vipul, Reidhoch, Kocio, Ceyockey, Richard Arthur Norton (1958-), Starblind, Josephf, Ma Baker, Tabletop, DESiegel, Behun, BD2412, Edison, Rjwilmsi, Vegaswikian, PaulWicks, Bgwhite, WriterHound, Ericorbit, Mohsen Basirat, Shanel, Schlafly, Malcolma, Abune, Urchin, Thespian, Sbassi, Sarah, SmackBot, Eperotao, InverseHypercube, Cutter, Zyxw, Septegram, Ohnoitsjamie, Hmains, Droll, George Church, Threeafterthree, Ambix, Ohconfucius, Ser Amantio di Nicolao, Gugel~enwiki, Tereufuk, Iridescent, Tmangray, Bobamnertiopsis, Chris55, JForget, Jiwhit01, Cognizance, EqualRights, Ebyabe, Tewapack, Thijs!bot, TPREX, FirefoxRocks, Soren121, Karlpearson, Ph.eyes, Magioladitis, WolfmanSF, Waacstats, Cardamon, WLU, Church of emacs, Nono64, Plorenzini, Dahliarose, Betswiki, Nwbeeson, 83d40m, Robert Kosara, Jevansen, Meadors, Oshwah, Immerpank~enwiki, UnitedStatesian, Ploxhoi, SylviaStanley, Ponyo, Promethius3, Meightysix, Flyer22 Reborn, Enti342, Ward20, Xnatedawgx, Forluvoft, Sfan00 IMG, Gigacephalus, Sitonera, DerekMorr, Dekisugi, Doprendek, Stepheng3, Bhickey, Bjdehut, DumZiBoT, XLinkBot, Auto469680, Jytdog, Addbot, GUI Hopper, Zhitelew, Yobot, Mayflower3, AnomieBOT, Sakdfjk43rsd, Bluerasberry, Sciencechick, CalSGWorker, Coretheapple, Ahtremblay, Alboran, MCBot, Cannolis, I dream of horses, Kops2222, Jmc200, Meiguess, Accarmichael, Catchnaveen, HelenOnline, Dethroned Buoy,

Carniolus, Ilovshuz, RjwilmsiBot, DASHBot, J36miles, Timtempleton, Travelinjoe, Dcirovic, Djembayz, Josve05a, WeijiBaikeBianji, Alpha Quadrant, EWikist, Timetraveler3.14, Abergabe, Geneager, ClueBot NG, Helpful Pixie Bot, Erik.Bjareholt, TelomeraseMe, Ginger Maine Coon, 220 of Borg, BattyBot, Mollskman, EuroCarGT, IjonTichyIjonTichy, Boseritwik, Vellvisher, Shymaster9, Clerkerina, Tentinator, RaphaelQS, RunningRedRhino, Garzfoth, NorthBySouthBaranof, NoApostropheInIts, Will-o-the-west, 666informer, SJ Defender, Stamptrader, Fixuture, X2Y2k6, Djosopolar, KH-1, Wallacewhite, CityKeys2015, InternetArchiveBot, GreenC bot, Guan yu, Bender the Bot, Metaphysicswar, Jean-jung212, Jui89, Kaylee and me and Anonymous: 97

29.7.2 Images

- **File:ATCase_reaction.jpg** *Source:* https://upload.wikimedia.org/wikipedia/commons/f/f1/ATCase_reaction.jpg *License:* Public domain *Contributors:* Own work *Original artist:* Sasata
- **File:Aegopodium_podagraria1_ies.jpg** *Source:* https://upload.wikimedia.org/wikipedia/commons/b/bf/Aegopodium_podagraria1_ies.jpg *License:* CC-BY-SA-3.0 *Contributors:* Own work *Original artist:* Frank Vincentz
- **File:Agonist_Antagonist.svg** *Source:* https://upload.wikimedia.org/wikipedia/commons/2/24/Agonist_Antagonist.svg *License:* CC-BY-SA-3.0 *Contributors:* File:Agonist_Antagonist.png *Original artist:* ES:Usuario:House
- **File:Alpha-Amanitin–RNA_polymerase_II_complex_1K83.png** *Source:* https://upload.wikimedia.org/wikipedia/commons/5/56/Alpha-Amanitin%E2%80%93RNA_polymerase_II_complex_1K83.png *License:* Public domain *Contributors:* From PDB entry 1K83. *Original artist:* Fvasconcellos 21:15, 14 November 2007 (UTC)
- **File:Ambox_important.svg** *Source:* https://upload.wikimedia.org/wikipedia/commons/b/b4/Ambox_important.svg *License:* Public domain *Contributors:* Own work based on: Ambox scales.svg *Original artist:* Dsmurat
- **File:Amphibolic_Properties_of_the_Citric_Acid_Cycle.gif** *Source:* https://upload.wikimedia.org/wikipedia/commons/c/cd/Amphibolic_Properties_of_the_Citric_Acid_Cycle.gif *License:* CC BY-SA 4.0 *Contributors:* Own work *Original artist:* Sazhnyev
- **File:Antioxidant.png** *Source:* https://upload.wikimedia.org/wikipedia/commons/4/49/Antioxidant.png *License:* Public domain *Contributors:* Own work *Original artist:* Smokefoot
- **File:Antioxidant_pathway.svg** *Source:* https://upload.wikimedia.org/wikipedia/commons/9/9a/Antioxidant_pathway.svg *License:* Public domain *Contributors:* w:Image:Antioxidant pathway.png *Original artist:* Tim Vickers, vectorized by Fvasconcellos
- **File:Argonne'{}s_Midwest_Center_for_Structural_Genomics_deposits_1,000th_protein_structure.jpg** *Source:* https://upload.wikimedia.org/wikipedia/commons/8/86/Argonne%27s_Midwest_Center_for_Structural_Genomics_deposits_1%2C000th_protein_structure.jpg *License:* CC BY-SA 2.0 *Contributors:* Argonne's Midwest Center for Structural Genomics deposits 1,000th protein structure *Original artist:* Matt Howard
- **File:Aspartate_aminotransferase_reaction.png** *Source:* https://upload.wikimedia.org/wikipedia/commons/a/a0/Aspartate_aminotransferase_reaction.png *License:* CC BY-SA 3.0 *Contributors:* Own work *Original artist:* Benzenamino
- **File:Benzopyrene_DNA_adduct_1JDG.png** *Source:* https://upload.wikimedia.org/wikipedia/commons/d/d8/Benzopyrene_DNA_adduct_1JDG.png *License:* CC-BY-SA-3.0 *Contributors:* ? *Original artist:* ?
- **File:Biodegradation_of_Pollutants.png** *Source:* https://upload.wikimedia.org/wikipedia/commons/3/3d/Biodegradation_of_Pollutants.png *License:* CC BY-SA 4.0 *Contributors:* Own work *Original artist:* Timmer26
- **File:Branched_ABS.png** *Source:* https://upload.wikimedia.org/wikipedia/commons/e/ed/Branched_ABS.png *License:* Public domain *Contributors:* Own work *Original artist:* Smokefoot
- **File:CYP2C9_1OG2.png** *Source:* https://upload.wikimedia.org/wikipedia/commons/2/29/CYP2C9_1OG2.png *License:* Public domain *Contributors:* From PDB 1OG2. More information: *Original artist:* Fvasconcellos 03:26, 7 October 2007 (UTC)
- **File:Chemical_Genomics_Robot.jpg** *Source:* https://upload.wikimedia.org/wikipedia/commons/2/27/Chemical_Genomics_Robot.jpg *License:* Public domain *Contributors:* http://www.genome.gov/dmd/img.cfm?node=Photos/Technology/Research%20laboratory&id=79299 *Original artist:* Maggie Bartlett, National Human Genome Research Institute
- **File:Chromosomes_mutations-en.svg** *Source:* https://upload.wikimedia.org/wikipedia/commons/2/26/Chromosomes_mutations-en.svg *License:* Public domain *Contributors:* Own work based on Chromosomenmutationen.png *Original artist:* This vector image was created with Inkscape.
- **File:Citrus_paradisi_(Grapefruit,_pink)_white_bg.jpg** *Source:* https://upload.wikimedia.org/wikipedia/commons/d/d0/Citrus_paradisi_%28Grapefruit%2C_pink%29_white_bg.jpg *License:* CC BY-SA 2.5 *Contributors:*
- Citrus_paradisi_(Grapefruit,_pink).jpg *Original artist:* Citrus_paradisi_(Grapefruit,_pink).jpg: ℵ(Aleph)
- **File:Commons-logo.svg** *Source:* https://upload.wikimedia.org/wikipedia/en/4/4a/Commons-logo.svg *License:* PD *Contributors:* ? *Original artist:* ?
- **File:DFE_in_VSV.png** *Source:* https://upload.wikimedia.org/wikipedia/commons/f/f9/DFE_in_VSV.png *License:* CC BY-SA 3.0 *Contributors:* Own work *Original artist:* Fiona126
- **File:Dna.png** *Source:* https://upload.wikimedia.org/wikipedia/commons/9/91/Dna.png *License:* CC BY 3.0 *Contributors:* [1] [2] *Original artist:* BanzaiTokyo
- **File:Dopamine_degradation.svg** *Source:* https://upload.wikimedia.org/wikipedia/commons/e/e7/Dopamine_degradation.svg *License:* CC-BY-SA-3.0 *Contributors:* Own work *Original artist:* NEUROtiker
- **File:Drug_ampoule_JPN.jpg** *Source:* https://upload.wikimedia.org/wikipedia/commons/a/a1/Drug_ampoule_JPN.jpg *License:* CC-BY-SA-3.0 *Contributors:* Own work *Original artist:* ignis

- **File:Earth_Day_Flag.png** *Source:* https://upload.wikimedia.org/wikipedia/commons/6/6a/Earth_Day_Flag.png *License:* Public domain *Contributors:* File:Earth flag PD.jpg, File:The Earth seen from Apollo 17 with transparent background.png *Original artist:* NASA (Earth photograph) SiBr4 (flag image)
- **File:Edit-clear.svg** *Source:* https://upload.wikimedia.org/wikipedia/en/f/f2/Edit-clear.svg *License:* Public domain *Contributors:* The *Tango! Desktop Project*. *Original artist:*

  The people from the Tango! project. And according to the meta-data in the file, specifically: "Andreas Nilsson, and Jakub Steiner (although minimally)."
- **File:Farmacocinética_lineal.svg** *Source:* https://upload.wikimedia.org/wikipedia/commons/7/70/Farmacocin%C3%A9tica_lineal.svg *License:* GFDL *Contributors:* Own work *Original artist:* User: Miguel A. Ortiz Arjona, vectorizada y subida por Racso
- **File:Farmacocinética_no_lineal.svg** *Source:* https://upload.wikimedia.org/wikipedia/commons/6/62/Farmacocin%C3%A9tica_no_lineal.svg *License:* GFDL *Contributors:* Own work *Original artist:* User: Miguel A. Ortiz Arjona, vectorizada y subida por Racso
- **File:FeIVO 2.tif** *Source:* https://upload.wikimedia.org/wikipedia/commons/9/91/FeIVO_2.tif *License:* CC BY-SA 4.0 *Contributors:* Own work *Original artist:* Pwnsey
- **File:Folder_Hexagonal_Icon.svg** *Source:* https://upload.wikimedia.org/wikipedia/en/4/48/Folder_Hexagonal_Icon.svg *License:* Cc-by-sa-3.0 *Contributors:* ? *Original artist:* ?
- **File:Galactose-1-phosphate_uridylyltransferase_1GUP.png** *Source:* https://upload.wikimedia.org/wikipedia/commons/c/ce/Galactose-1-phosphate_uridylyltransferase_1GUP.png *License:* Public domain *Contributors:* From PDB 1GUP. *Original artist:* Fvasconcellos (talk · contribs)
- **File:Gene_structure_eukaryote_2_annotated.svg** *Source:* https://upload.wikimedia.org/wikipedia/commons/5/54/Gene_structure_eukaryote_2_annotated.svg *License:* CC BY 4.0 *Contributors:* Shafee T, Lowe R (2017). "Eukaryotic and prokaryotic gene structure". *WikiJournal of Medicine* **4** (1). DOI:10.15347/wjm/2017.002. ISSN 20024436. *Original artist:* Thomas Shafee
- **File:Gluconeogenese_Schema_2.png** *Source:* https://upload.wikimedia.org/wikipedia/commons/d/da/Gluconeogenese_Schema_2.png *License:* CC-BY-SA-3.0 *Contributors:* de:Datei:Gluconeogenese_Schema_2.jpg *Original artist:* C. Mönchmeier (de:Benutzer:Luziferase)
- **File:Glutathione-3D-vdW.png** *Source:* https://upload.wikimedia.org/wikipedia/commons/3/3c/Glutathione-3D-vdW.png *License:* Public domain *Contributors:* Own work *Original artist:* Ben Mills
- **File:Gray1128.png** *Source:* https://upload.wikimedia.org/wikipedia/commons/0/02/Gray1128.png *License:* Public domain *Contributors:* Henry Gray (1918) *Anatomy of the Human Body* (See "Book" section below)

  Original artist: Henry Vandyke Carter
- **File:Gráfica_Km.png** *Source:* https://upload.wikimedia.org/wikipedia/commons/6/67/Gr%C3%A1fica_Km.png *License:* CC BY 2.5 *Contributors:* Own work *Original artist:* user:Gonn
- **File:Gtk-dialog-info.svg** *Source:* https://upload.wikimedia.org/wikipedia/commons/b/b4/Gtk-dialog-info.svg *License:* LGPL *Contributors:* http://ftp.gnome.org/pub/GNOME/sources/gnome-themes-extras/0.9/gnome-themes-extras-0.9.0.tar.gz *Original artist:* David Vignoni
- **File:Hybridogenesis_in_water_frogs_gametes.gif** *Source:* https://upload.wikimedia.org/wikipedia/commons/7/7f/Hybridogenesis_in_water_frogs_gametes.gif *License:* CC BY-SA 4.0 *Contributors:* Own work *Original artist:* Darekk2
- **File:Hypericum_moserianum0.jpg** *Source:* https://upload.wikimedia.org/wikipedia/commons/4/4d/Hypericum_moserianum0.jpg *License:* CC-BY-SA-3.0 *Contributors:* caliban.mpiz-koeln.mpg.de/mavica/index.html part of www.biolib.de *Original artist:* Kurt Stüber [1]
- **File:Implicaciones_terapéuticas_farmacocinética_lineal.svg** *Source:* https://upload.wikimedia.org/wikipedia/commons/2/2e/Implicaciones_terap%C3%A9uticas_farmacocin%C3%A9tica_lineal.svg *License:* GFDL *Contributors:* Own work *Original artist:* User: Miguel A. Ortiz Arjona, vectorizada y subida por Racso
- **File:Interactive_icon.svg** *Source:* https://upload.wikimedia.org/wikipedia/commons/e/e6/Interactive_icon.svg *License:* CC BY 4.0 *Contributors:* Own work *Original artist:* Thomas Shafee
- **File:Intermediate_Metabolizer_Genotype_to_Reporting_Process_Example.jpg** *Source:* https://upload.wikimedia.org/wikipedia/commons/2/2f/Intermediate_Metabolizer_Genotype_to_Reporting_Process_Example.jpg *License:* CC BY 3.0 *Contributors:* Adapted from Nature Publishing Group (Clinical Pharmacology & Therapeutics) *Original artist:* Crews KR, Hicks JK, Pui CH, Relling MV, Evans, WE
- **File:Issoria_lathonia.jpg** *Source:* https://upload.wikimedia.org/wikipedia/commons/2/2d/Issoria_lathonia.jpg *License:* CC-BY-SA-3.0 *Contributors:* ? *Original artist:* ?
- **File:Iv_time_conc_curve.svg** *Source:* https://upload.wikimedia.org/wikipedia/commons/e/e7/Iv_time_conc_curve.svg *License:* CC BY-SA 3.0 *Contributors:* Own work *Original artist:* Boghog
- **File:Karl_Johanssvamp,_Iduns_kokbok.png** *Source:* https://upload.wikimedia.org/wikipedia/commons/4/42/Karl_Johanssvamp%2C_Iduns_kokbok.png *License:* Public domain *Contributors:* Iduns kokbok, scanned by Project Runeberg; transparency added by Ilmari Karonen *Original artist:* Elisabeth Östman (1869–1933)
- **File:L-ascorbic-acid-3D-balls.png** *Source:* https://upload.wikimedia.org/wikipedia/commons/d/da/L-ascorbic-acid-3D-balls.png *License:* Public domain *Contributors:* ? *Original artist:* ?
- **File:Linear_PK_Example.png** *Source:* https://upload.wikimedia.org/wikipedia/commons/e/ee/Linear_PK_Example.png *License:* CC BY 3.0 *Contributors:* Own work *Original artist:* <a href='//commons.wikimedia.org/wiki/User:Alfie66' title='User:Alfie66'>Alfie</a><a href='//commons.wikimedia.org/wiki/User_talk:Alfie66' title='User talk:Alfie66'>↑↓</a>© (Helmut Schütz)
- **File:Lipid_peroxidation.svg** *Source:* https://upload.wikimedia.org/wikipedia/commons/9/9e/Lipid_peroxidation.svg *License:* Public domain *Contributors:* w:Image:Lipid peroxidation v2.png *Original artist:* Tim Vickers, after Young IS, McEneny J (2001). "Lipoprotein oxidation and atherosclerosis". *Biochem Soc Trans* **29** (Pt 2): 358–62. PMID 11356183.

- **File:Lock-green.svg** *Source:* https://upload.wikimedia.org/wikipedia/commons/6/65/Lock-green.svg *License:* CC0 *Contributors:* en:File: Free-to-read_lock_75.svg *Original artist:* User:Trappist the monk
- **File:MAPK-pathway-mammalian.png** *Source:* https://upload.wikimedia.org/wikipedia/commons/d/d6/MAPK-pathway-mammalian.png *License:* CC BY-SA 3.0 *Contributors:* Own work *Original artist:* Bubus12
- **File:Metabolic_Metro_Map_(no_legends).svg** *Source:* https://upload.wikimedia.org/wikipedia/commons/4/41/Metabolic_Metro_Map_%28no_legends%29.svg *License:* CC BY-SA 4.0 *Contributors:* Own work *Original artist:* Chakazul
- **File:Metabolic_metro_blue.svg** *Source:* https://upload.wikimedia.org/wikipedia/commons/1/1b/Metabolic_metro_blue.svg *License:* CC BY-SA 4.0 *Contributors:* Own work *Original artist:* Thomas Shafee
- **File:Metabolic_metro_brown.svg** *Source:* https://upload.wikimedia.org/wikipedia/commons/9/99/Metabolic_metro_brown.svg *License:* CC BY-SA 4.0 *Contributors:* Own work *Original artist:* Thomas Shafee
- **File:Metabolic_metro_green.svg** *Source:* https://upload.wikimedia.org/wikipedia/commons/9/97/Metabolic_metro_green.svg *License:* CC BY-SA 4.0 *Contributors:* Own work *Original artist:* Thomas Shafee
- **File:Metabolic_metro_grey.svg** *Source:* https://upload.wikimedia.org/wikipedia/commons/c/c8/Metabolic_metro_grey.svg *License:* CC BY-SA 4.0 *Contributors:* Own work *Original artist:* Thomas Shafee
- **File:Metabolic_metro_orange.svg** *Source:* https://upload.wikimedia.org/wikipedia/commons/a/a0/Metabolic_metro_orange.svg *License:* CC BY-SA 4.0 *Contributors:* Own work *Original artist:* Thomas Shafee
- **File:Metabolic_metro_pink.svg** *Source:* https://upload.wikimedia.org/wikipedia/commons/d/db/Metabolic_metro_pink.svg *License:* CC BY-SA 4.0 *Contributors:* Own work *Original artist:* Thomas Shafee
- **File:Metabolic_metro_purple.svg** *Source:* https://upload.wikimedia.org/wikipedia/commons/e/e7/Metabolic_metro_purple.svg *License:* CC BY-SA 4.0 *Contributors:* Own work *Original artist:* Thomas Shafee
- **File:Metabolic_metro_red.svg** *Source:* https://upload.wikimedia.org/wikipedia/commons/b/ba/Metabolic_metro_red.svg *License:* CC BY-SA 4.0 *Contributors:* Own work *Original artist:* Thomas Shafee
- **File:Metabolism_790px.png** *Source:* https://upload.wikimedia.org/wikipedia/commons/0/0d/Metabolism_790px.png *License:* CC-BY-SA-3.0 *Contributors:* Own work *Original artist:* Zephyris
- **File:Metabolism_pathways_(partly_labeled).svg** *Source:* https://upload.wikimedia.org/wikipedia/commons/5/5d/Metabolism_pathways_%28partly_labeled%29.svg *License:* CC BY-SA 4.0 *Contributors:* <a href='//commons.wikimedia.org/wiki/File:Metabolism_790px_partly_labeled.png' class='image'><img alt='Metabolism 790px partly labeled.png' src='https://upload.wikimedia.org/wikipedia/commons/thumb/5/50/Metabolism_790px_partly_labeled.png/100px-Metabolism_790px_partly_labeled.png' width='100' height='73' srcset='https://upload.wikimedia.org/wikipedia/commons/thumb/5/50/Metabolism_790px_partly_labeled.png/150px-Metabolism_790px_partly_labeled.png 1.5x, https://upload.wikimedia.org/wikipedia/commons/thumb/5/50/Metabolism_790px_partly_labeled.png/200px-Metabolism_790px_partly_labeled.png 2x' data-file-width='790' data-file-height='573' /></a> *Original artist:* Fred the Oyster
- **File:Myoglobin.png** *Source:* https://upload.wikimedia.org/wikipedia/commons/6/60/Myoglobin.png *License:* Public domain *Contributors:* self made based on PDB entry *Original artist:* →<a href='//commons.wikimedia.org/wiki/User:AzaToth' title='User:AzaToth'>Aza</a><a href='//commons.wikimedia.org/wiki/User_talk:AzaToth' title='User talk:AzaToth'>Toth</a>
- **File:Net_reactions_for_glycolysis_of_glucose,_oxidative_decarboxylation_of_pyruvate,_and_Krebs_cycle..png** *Source:* https://upload.wikimedia.org/wikipedia/commons/b/b5/Net_reactions_for_glycolysis_of_glucose%2C_oxidative_decarboxylation_of_pyruvate%2C_and_Krebs_cycle..png *License:* CC BY-SA 4.0 *Contributors:* Own work *Original artist:* Carolineneil
- **File:Notable_mutations.svg** *Source:* https://upload.wikimedia.org/wikipedia/commons/7/72/Notable_mutations.svg *License:* Public domain *Contributors:* Derivative of en:File:GeneticCode21-version-2.svg (Public Domain license) *Original artist:* Mikael Häggström
- **File:Open_book_nae_02.svg** *Source:* https://upload.wikimedia.org/wikipedia/commons/9/92/Open_book_nae_02.svg *License:* CC0 *Contributors:* OpenClipart *Original artist:* nae
- **File:P450cycle.svg** *Source:* https://upload.wikimedia.org/wikipedia/commons/3/39/P450cycle.svg *License:* Public domain *Contributors:* M.Sc. Thesis, David Richfield (User:Slashme) *Original artist:* Slashme at English Wikipedia
- **File:PDB_1aqy_EBI.jpg** *Source:* https://upload.wikimedia.org/wikipedia/commons/f/ff/PDB_1aqy_EBI.jpg *License:* Public domain *Contributors:* http://www.ebi.ac.uk/pdbe-srv/view/images/entry/1aqy600.png, displayed on http://www.ebi.ac.uk/pdbe-srv/view/entry/1aqy/summary *Original artist:* Jawahar Swaminathan and MSD staff at the European Bioinformatics Institute
- **File:PDB_1ek9_EBI.jpg** *Source:* https://upload.wikimedia.org/wikipedia/commons/d/d2/PDB_1ek9_EBI.jpg *License:* Public domain *Contributors:* http://www.ebi.ac.uk/pdbe-srv/view/images/entry/1ek9600.png, displayed on http://www.ebi.ac.uk/pdbe-srv/view/entry/1ek9/summary *Original artist:* Jawahar Swaminathan and MSD staff at the European Bioinformatics Institute
- **File:People_icon.svg** *Source:* https://upload.wikimedia.org/wikipedia/commons/3/37/People_icon.svg *License:* CC0 *Contributors:* OpenClipart *Original artist:* OpenClipart
- **File:Peroxiredoxin.png** *Source:* https://upload.wikimedia.org/wikipedia/commons/e/ee/Peroxiredoxin.png *License:* Public domain *Contributors:* Transferred from en.wikipedia to Commons. *Original artist:* TimVickers at English Wikipedia
- **File:Pharmacogenomics_challenges_from_research_to_practice.jpg** *Source:* https://upload.wikimedia.org/wikipedia/commons/7/74/Pharmacogenomics_challenges_from_research_to_practice.jpg *License:* CC BY 3.0 *Contributors:* Swen, Jesse J.; Huizinga, Tom W.; Gelderblom, Hans; de Vries, Elisabeth G. E.; Assendelft, Willem J. J.; Kirchheiner, Julia; Guchelaar, Henk-Jan (2007). "Translating Pharmacogenomics: Challenges on the Road to the Clinic". *PLoS Medicine* **4** (8): e209. DOI:10.1371/journal.pmed.0040209. *Original artist:* Swen JJ, Huizinga TW, Gelderblom H, de Vries EGE, Assendelft WJJ, Kirchheiner J, Guchelaar HK

- **File:Pharmacologyprism.jpg** *Source:* https://upload.wikimedia.org/wikipedia/commons/8/8f/Pharmacologyprism.jpg *License:* Public domain *Contributors:* ? *Original artist:* ?
- **File:Phytate.png** *Source:* https://upload.wikimedia.org/wikipedia/commons/7/71/Phytate.png *License:* Public domain *Contributors:* No machine-readable source provided. Own work assumed (based on copyright claims). *Original artist:* No machine-readable author provided. Shaddack assumed (based on copyright claims).
- **File:Pill_box_with_pills.JPG** *Source:* https://upload.wikimedia.org/wikipedia/commons/f/f9/Pill_box_with_pills.JPG *License:* CC BY-SA 3.0 *Contributors:* Own work *Original artist:* Dvortygirl
- **File:Polylactid_sceletal.svg** *Source:* https://upload.wikimedia.org/wikipedia/commons/f/fa/Polylactid_sceletal.svg *License:* Public domain *Contributors:* Own work *Original artist:* Polimerek
- **File:Portal-puzzle.svg** *Source:* https://upload.wikimedia.org/wikipedia/en/f/fd/Portal-puzzle.svg *License:* Public domain *Contributors:* ? *Original artist:* ?
- **File:Portulaca_grandiflora_mutant1.jpg** *Source:* https://upload.wikimedia.org/wikipedia/commons/e/e5/Portulaca_grandiflora_mutant1.jpg *License:* CC BY-SA 3.0 *Contributors:* Own work *Original artist:* JerryFriedman
- **File:Prodryas.png** *Source:* https://upload.wikimedia.org/wikipedia/commons/b/bc/Prodryas.png *License:* Public domain *Contributors:* Canadian Entomologist *Original artist:* Samuel Hubbard Scudder
- **File:PropDrugsMetabCYP.png** *Source:* https://upload.wikimedia.org/wikipedia/commons/9/98/PropDrugsMetabCYP.png *License:* CC0 *Contributors:* This file was derived from: Proportion of drugs metabolized by different CYPs.png
  Original artist: en:User:Ibrmrn3000
- **File:Question_book-new.svg** *Source:* https://upload.wikimedia.org/wikipedia/en/9/99/Question_book-new.svg *License:* Cc-by-sa-3.0 *Contributors:*
  Created from scratch in Adobe Illustrator. Based on Image:Question book.png created by User:Equazcion *Original artist:*
  Tkgd2007
- **File:Slime.mold.jpg** *Source:* https://upload.wikimedia.org/wikipedia/commons/f/f1/Slime.mold.jpg *License:* CC BY 3.0 *Contributors:* Own work *Original artist:* Red58bill
- **File:Sodium_dodecylbenzenesulfonate.png** *Source:* https://upload.wikimedia.org/wikipedia/commons/f/fa/Sodium_dodecylbenzenesulfonate.png *License:* Public domain *Contributors:* Own work *Original artist:* Smokefoot
- **File:Steroidogenesis.svg** *Source:* https://upload.wikimedia.org/wikipedia/commons/1/13/Steroidogenesis.svg *License:* CC-BY-SA-3.0 *Contributors:* Häggström M, Richfield D (2014). "Diagram of the pathways of human steroidogenesis". *WikiJournal of Medicine* **1** (1). DOI:10.15347/wjm/2014.005. ISSN 20024436. *Original artist:* David Richfield (User:Slashme) and Mikael Häggström. Derived from previous version by Hoffmeier and Settersr.
- **File:Sustainable_development.svg** *Source:* https://upload.wikimedia.org/wikipedia/commons/7/70/Sustainable_development.svg *License:* CC-BY-SA-3.0 *Contributors:*
- Inspired from Developpement durable.jpg *Original artist:*
- original: Johann Dréo (talk · contribs)
- **File:Symbol_book_class2.svg** *Source:* https://upload.wikimedia.org/wikipedia/commons/8/89/Symbol_book_class2.svg *License:* CC BY-SA 2.5 *Contributors:* Mad by Lokal_Profil by combining: *Original artist:* Lokal_Profil
- **File:Symbol_list_class.svg** *Source:* https://upload.wikimedia.org/wikipedia/en/d/db/Symbol_list_class.svg *License:* Public domain *Contributors:* ? *Original artist:* ?
- **File:Symbol_template_class.svg** *Source:* https://upload.wikimedia.org/wikipedia/en/5/5c/Symbol_template_class.svg *License:* Public domain *Contributors:* ? *Original artist:* ?
- **File:TPI1_structure.png** *Source:* https://upload.wikimedia.org/wikipedia/commons/1/1c/TPI1_structure.png *License:* Public domain *Contributors:* based on 1wyi (http://www.pdb.org/pdb/explore/explore.do?structureId=1WYI), made in pymol *Original artist:* →<a href='//commons.wikimedia.org/wiki/User:AzaToth' title='User:AzaToth'>Aza</a><a href='//commons.wikimedia.org/wiki/User_talk:AzaToth' title='User talk:AzaToth'>Toth</a>
- **File:TRNA-Phe_yeast_1ehz.png** *Source:* https://upload.wikimedia.org/wikipedia/commons/b/ba/TRNA-Phe_yeast_1ehz.png *License:* CC BY-SA 3.0 *Contributors:* Own work *Original artist:* Yikrazuul
- **File:Tabletten.JPG** *Source:* https://upload.wikimedia.org/wikipedia/commons/5/5f/Tabletten.JPG *License:* CC-BY-SA-3.0 *Contributors:* ? *Original artist:* ?
- **File:Text_document_with_page_number_icon.svg** *Source:* https://upload.wikimedia.org/wikipedia/commons/3/3b/Text_document_with_page_number_icon.svg *License:* Public domain *Contributors:* Created by bdesham with Inkscape; based upon Text-x-generic.svg from the Tango project. *Original artist:* Benjamin D. Esham (bdesham)
- **File:Transaldolase_reaction.svg** *Source:* https://upload.wikimedia.org/wikipedia/commons/2/2b/Transaldolase_reaction.svg *License:* CC BY-SA 3.0 *Contributors:* Own work *Original artist:* Yikrazuul
- **File:Tree_of_life.svg** *Source:* https://upload.wikimedia.org/wikipedia/commons/0/09/Tree_of_life.svg *License:* CC-BY-SA-3.0 *Contributors:* No machine-readable source provided. Own work assumed (based on copyright claims). *Original artist:* No machine-readable author provided. Vanished user fijtji34toksdcknqrjn54yoimascj assumed (based on copyright claims).
- **File:Vegetarian_diet.jpg** *Source:* https://upload.wikimedia.org/wikipedia/commons/4/4e/Vegetarian_diet.jpg *License:* Public domain *Contributors:* This image was released by the Agricultural Research Service, the research agency of the United States Department of Agriculture, with the ID K8234-2 <a class='external text' href='//commons.wikimedia.org/w/index.php?title=Category:Media_created_by_the_United_States_Agricultural_Research_Service_with_known_IDs,<span>,&,</span>,filefrom=K8234-2#mw-category-media'>(next)</a>. *Original artist:* Scott Bauer, USDA ARS

- **File:WHO_Rod.svg** *Source:* https://upload.wikimedia.org/wikipedia/commons/d/d6/WHO_Rod.svg *License:* Public domain *Contributors:* http://www.who.int/about/licensing/emblem/en/ *Original artist:* WHO
- **File:Wiki_letter_w_cropped.svg** *Source:* https://upload.wikimedia.org/wikipedia/commons/1/1c/Wiki_letter_w_cropped.svg *License:* CC-BY-SA-3.0 *Contributors:* This file was derived from Wiki letter w.svg: <a href='//commons.wikimedia.org/wiki/File:Wiki_letter_w.svg' class='image'><img alt='Wiki letter w.svg' src='https://upload.wikimedia.org/wikipedia/commons/thumb/6/6c/Wiki_letter_w.svg/50px-Wiki_letter_w.svg.png' width='50' height='50' srcset='https://upload.wikimedia.org/wikipedia/commons/thumb/6/6c/Wiki_letter_w.svg/75px-Wiki_letter_w.svg.png 1.5x, https://upload.wikimedia.org/wikipedia/commons/thumb/6/6c/Wiki_letter_w.svg/100px-Wiki_letter_w.svg.png 2x' data-file-width='44' data-file-height='44' /></a>
  Original artist: Derivative work by Thumperward
- **File:Wikivoyage-Logo-v3-icon.svg** *Source:* https://upload.wikimedia.org/wikipedia/commons/d/dd/Wikivoyage-Logo-v3-icon.svg *License:* CC BY-SA 3.0 *Contributors:* Own work *Original artist:* AleXXw
- **File:Xenobiotic_metabolism.png** *Source:* https://upload.wikimedia.org/wikipedia/commons/9/97/Xenobiotic_metabolism.png *License:* Public domain *Contributors:* Own work *Original artist:* TimVickers

29.7.3 Content license

- Creative Commons Attribution-Share Alike 3.0

54857390R00117

Made in the USA
Middletown, DE
07 December 2017